MEASURING THE SIZE OF THINGS IN THE UNIVERSE: HBT INTERFEROMETRY AND HEAVY ION PHYSICS

MEASURING THE SIZE OF THINGS IN THE UNIVERSE:

HBT INTERFEROMETRY AND HEAVY ION PHYSICS

Proceedings of

CRIS '98

2nd Catania Relativistic Ion Studies

Acicastello, Italy, June 8 –12, 1998

Editors

S. Costa, S. Albergo, A. Insolia, C. Tuvè

Università di Catania and INFN, Italy

World Scientific

Singapore • New Jersey • London • Hong Kong

Published by

World Scientific Publishing Co. Pte. Ltd.

P O Box 128, Farrer Road, Singapore 912805

USA office: Suite 1B, 1060 Main Street, River Edge, NJ 07661

UK office: 57 Shelton Street, Covent Garden, London WC2H 9HE

British Library Cataloguing-in-Publication Data
A catalogue record for this book is available from the British Library.

Measuring the size of things in the universe: HBT interferometry and heavy ion physics, proceedings of CRIS '98, 2.nd Catania Relativistic Ion Studies : Acicastello, Italy, June 8-12, 1998 / editor S. Costa ... [et al.]. - Singapore World scientific, 1999.
1. Fisica Nucleare. 2. Congressi - Aci Castello - 1998. I. Costa, Salvatore<1954>.
539.7 CDD-20 SBN Pa10158567

CIP - Biblioteca centrale della Regione siciliana

MEASURING THE SIZE OF THINGS IN THE UNIVERSE: HBT INTERFEROMETRY AND HEAVY ION PHYSICS

ISBN 981-02-4038-4

Printed in Singapore by Uto-Print

CRIS '98

International Advisory Committee:

P. Braun-Munzinger (GSI)
C.K. Gelbke (MSU)
O. Hansen (NBI)
J. Harris (Yale Univ.)
I. Iori (Milano Univ.)
E. Migneco (LNS)
G. Palumbo (Bologna Univ.)
R. Potenza (Catania Univ.)
H.G. Ritter (LBNL)
J. Schukraft (CERN)
P. Seyboth (MPI)
W. Zajc (Columbia Univ.)

Local Organizing Committee:

S. Albergo
S. Costa (Scientific Secretary)
A. Insolia (Chairman)
C. Tuvè

PREFACE

Two years ago, the First Catania Relativistic Ion Studies, CRIS '96, was devoted to the still unsettled question of the liquid-gas phase transition in finite nuclei, in connection with multifragmentation. According to many, CRIS '96 was quite successful.

Now, two years later, the Italian National Istitute of Nuclear Physics, or INFN, approved, funded, and supported this second CRIS. This way INFN has in practice recognized CRIS as a continuing series of Conferences, that we can now reasonably expect to have every other year in the Catania area.

The establishment of this regular series of topical Conferences in Catania testifies to the increasingly relevant role that Catania is playing in the Nuclear Physics community worldwide.

This second CRIS Conference is devoted to intensity interferometry applications in different fields of Physics, with the main emphasys on Nuclear Physics. There is no shortage of Conferences on any subject these days, and HBT is no exception. So we devised for CRIS '98 a program that in our intention should somewhat distinguish this Conference from the many other ones on the same subject. That's why the CRIS '98 program has a cross-disciplinary flavour, trying to draw an ideal bridge between the early astronomic measurements, the current or planned use of HBT techniques in Nuclear Physics and also some current and future use of interferometry in astrophysics.

We hope you, the reader, will find the interspersion of astrophysical subjects and of broad overviews in a program otherwise devoted to nuclear physics, intriguing and enticing rather than annoying. According to many participants, this cross-disciplinary thrust of the program was an appreciable feature of the Conference.

Certainly, it was a great honor for the CRIS '98 Organizers that Professor Emeritus Robert Hanbury Brown accepted to open the conference giving us a glimpse on how it all started.

We thank all the speakers for giving so interesting and excellent talks. And we thank the about 100 participants, whose interested attendance was high through all Sessions.

We are grateful to the members of the International Advisory Committee, whose opinions and suggestions were invaluable to setting up the program.

Coming now to the organizational matters, as anyone who has organized a Conference knows, this is the result of the cooperative work of many people. Thanks are due to the many agencies, both scientific institutions and administrative bodies which contributed either with their financial contribution or with their products, or with their work, or all of the above, to make CRIS '98 possible.

We have already aknowledged INFN's support. Let us at this point thank explicitly INFN's President, the whole INFN Management Board, represented at the Conference by Vice President Prof. Emilio Migneco, and the Director of the Catania Section of INFN, Prof. Maurizio Consoli.

Special mention deserve those people who really worked hard every day for this Conference: Angela Torrisi, Anna Linda Magrì and Tony Venasco, members of our local INFN Administrative Staff, who helped the Organizers for several months during the preparation of the Conference and all participants during the Conference week. And also Angela Di Silvestro and Elisabetta Di Silvestro who helped in the Office during the Conference week.

In addition, we had remarkable support from the University of Catania, the Sicilian Center for Nuclear Physics and Solid State Physics, the Sicilian regional Government, the Provincial Tourism Agency of Catania (which supplied the tourist material included in the Conference briefcase), the City of Catania, the City of Acicastello, and finally the Fratelli Averna Group, based in Caltanissetta, Sicily, which offered a small sample of sicilian gastronomic products.

Special thanks are due to Prof. Paolo Castorina, mayor of Acicastello and colleague physicist, for his kind hospitality and for originally encouraging us to hold the CRIS Conferences in Acicastello's Hotels.

Finally, we wish to thank Alessia Tricomi who designed the CRIS logo, the black 'Etna' volcano erupting nuclei and particles from a 'quark-gluon plasma' lava bed, seen from the surfs of the Ionian sea. This logo has become the distinctive emblem of the CRIS Conferences.

S.Albergo, S. Costa, A. Insolia, C. Tuvè
(CRIS '98 Local Organizing Committee)

Group picture of (some of) the participants to CRIS '98

The CRIS '98 Local Organizing Committee. Left to right: S. Albergo, C. Tuvè and S. Costa

Dr. S. Costa opens the Conference

Prof. R. Pucci, Dean of the Faculty of Science
University of Catania, welcomes the participants

Prof. E. Migneco, Vicepresident of INFN, welcomes the participants

Prof. R. Hanbury Brown gives his opening talk

Prof. G. Baym gives his overview talk

CONTENTS

PHOTONS, WAVES AND STARS

R.HANBURY BROWN

A brief account is given of the early development of a new technique, intensity interferometry, of measuring the angular size of radio sources. Following the chance discovery that it was unaffected by scintillation it was proposed to apply the same principle to measuring visible stars. This proposal met with vigorous opposition from physicists when it was realised that it implied that the time of arrival of photons in two mutually coherent beams of light must be correlated. Two laboratory experiments were done to demonstrate that this correlation does in fact take place. Then, after a pilot model had measured the angular size of Sirius, a full scale stellar intensity interferometer was built and installed at Narrabri in Australia. In a programme lasting 12 years it measured the angular diameters of 32 single stars in the spectral range O to F and established the first wholly empirical temperature scale for stars in that range.

1 Introduction

Most of the work which I am going to talk about was done rather a long time ago; indeed it was started in 1949 at the Jodrell Bank Experimental Station of the University of Manchester. At that time the most exciting problem in radio-astronomy was the nature of the so-called "radio stars" - the bright points of radio emission which had been discovered in the sky. What were they - galaxies, stars, or nebulae? We didn't know. Most of us thought they were some sort of invisible star and, as it turned out later, most of us were wrong.

I decided to measure their angular size - at least that would tell us if they were stars or galaxies. An obvious way of doing this was to make a radio analogue of Michelson's stellar interferometer using spaced radio antennas instead of spaced mirrors; but if it should turn out that these "radio stars" were really stars, as we suspected, then to measure angles of the order of 1/100th of a second of arc or perhaps 1/1000th of a second, the two antennas would have to be thousands of kilometres apart. In 1949 I could see no way of doing this; the principal technical difficulties were to provide independent local oscillators with sufficient stability and to equalise the very large and variable delays in the two channels. Nowadays all this can be done.

I worried over this problem for weeks until late one night I saw a way of making an interferometer which might perhaps work with a baseline of thousands of kilometres. In my mind I saw two people with identical radio receivers sitting in a field some distance apart and looking at cathode-ray tubes on which were displayed the noise from the same radio star; I wondered, whether both of them would see the same noise from the star? If so, maybe I could find the angular size of the star by comparing their two pictures?

To cut a long story short the answer was "yes - parts of the two pictures are the same when the two people are sufficiently close together". With the help of a friendly mathematician Richard Twiss, I developed the detailed theory [1] of a radically new type of instrument which would work with very long baselines - we called it an "intensity interferometer"

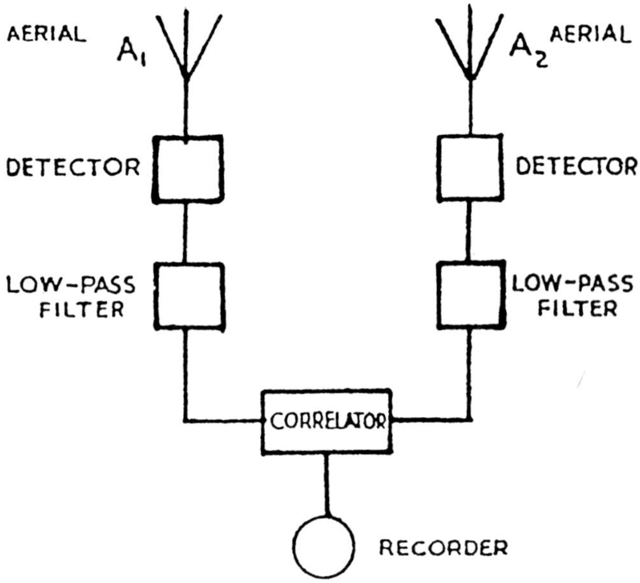

Figure 1: A simplified diagram of a radio Intensity Interferometer

Fig.1 shows a simplified diagram of this instrument. The radio frequency "noise" from the two separated antennas, which are both looking at the same radio star, are detected and then passed through low-pass filters to a correlator. We showed that the correlation, measured as a function of the separation between the two receivers, yields the angular diameter of the star.

In due course two of my students, R.C.Jennison and M.K.das Gupta [2], built an intensity interferometer on a wavelength of 2 metres and used it to measure the angular size of the two principal radio sources in the sky (Cygnus A and Cassiopeia A). To my great disappointment neither of these sources proved to be a star - one was a galaxy, the other a supernova remnant - there was no need for a very long baseline; it would have been much easier to make a radio version of Michelson's interferometer!

But all was not lost. One day while Richard Twiss and I were watching this radio interferometer working through a disturbed ionosphere we suddenly realised that the measurements were totally unaffected by twinkling in the ionosphere. A quick look at the theory showed that we had overlooked one of the most valuable advantages of an intensity interferometer - it can easily be made insensitive to twinkling. It dawned on us that the same principle might, perhaps, be used to crack the ancient and very difficult problem of measuring the angular size of the visible stars through a twinkling atmosphere.

Michelson's original stellar interferometer was limited by its size (20 ft), to measuring only six stars, all giants or supergiants. Efforts to enlarge it to 50 ft were made by Hale, but they failed and the work was discontinued in 1930. The major difficulties were, firstly, that the two paths of the light inside the instrument had to be held equal to a few microns and that proved to be too difficult mechanically; secondly, the effects of twinkling in the atmosphere were disastrous. What we now realised was that an intensity interferometer working with visible light should overcome both these difficulties; we decided to carry on where Michelson had left off.

2 Correlation between Photons

It was at this point that we ran into trouble. We had worked out the mathematics of the intensity interferometer for radio waves using Fourier analysis and now we had to do it for light and that meant that we must think about photons and quantum theory. So far the principle of our new instrument had been accepted by everyone without question, in those days radio engineers didn't worry about photons; but when we proposed to apply it to light, physicists became interested and they did worry about photons. Indeed they were very unhappy about what we proposed to do.

The trouble was that, when seen in terms of photons, our scheme looked like Fig.2. Individual photons from the star produce a pulse of current in the output of two separated phototubes and these pulses are fed to a coincidence counter. When the spacing between the two phototubes is large and the light on them is mutually incoherent then the coincidence rate is - obviously - that for two random streams of pulses. However, when the spacing is small and the light is mutually coherent, we showed theoretically that the times at which photons arrive at the two phototubes must be correlated; photons tend to arrive simultaneously at the two phototubes and the output of the coincidence counter is therefore increased. Furthermore we showed that this "excess" coincidence rate is a measure of the mutual coherence of the light at the two phototubes and so, by measuring it at different spacings, it is possible to find the angular diameter of the star.

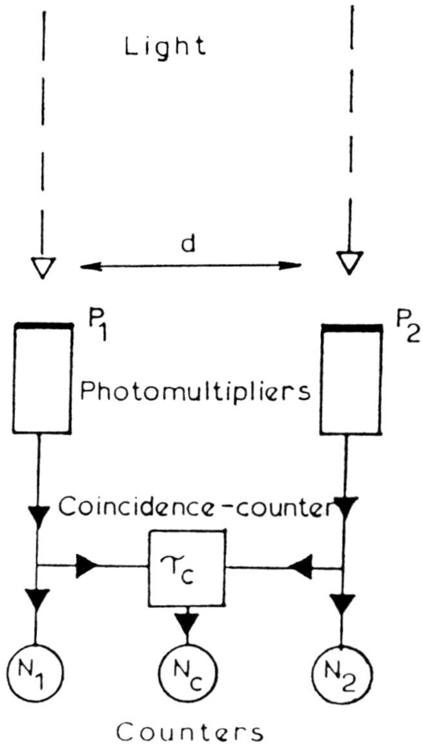

Figure 2: A coincidence-counting optical Intensity Interferometer

To most physicists our ideas were laughable. Their simplest argument, among many more highbrow ones, was that because photons are generated at random times in the source and all travel with the same velocity they must therefore arrive at random times. There is no conceivable way, we were told, in which they can arrive in pairs; to arrive hand in hand they would have to hang about waiting for each other! Obviously our proposal would not work. Furthermore on page 18 of one of the holy books of physics, "Quantum Mechanics", Dirac had written "interference between different photons can never occur" - and yet our theory involved the coordinated behaviour of different photons; indeed it involved their interference in time. Every time I stuck my nose inside the University, I was waylaid by an irate physicist brandishing some sacred text, Heitler or Dirac, which showed that the behaviour of photons could never be correlated. They told me this in person, in letters, in lectures and in print. Even after the publication of our first experiment, which I shall describe in a minute, elaborate laboratory experiments were carried out in Hungary and Canada which claimed to show that there was no correlation between photons. We analysed both these experiments in detail and published a paper pointing out that, with their experimental set-up, it would have taken the Canadians at least 1000 years to demonstrate the phenomenon and with their equipment the Hungarians would have had to observe for a time considerably longer than the age of the Earth! Neither of them had managed to work out the theoretical signal to noise ratio of their experiments - it's not easy - I couldn't do it, but my colleague Richard Twiss could!

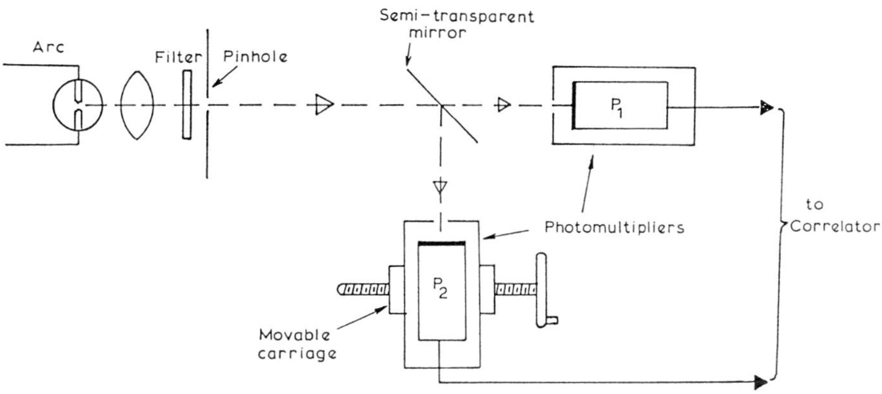

Figure 3: Optical system of first correlation experiment

The difficulty which many physicists had in accepting that the arrival of photons can be correlated was that most of them were particle physicists who thought of a photon as a real thing with its own properties, like a billiard ball, whereas it is better to think of a photon as an *event, not as a* thing, as something which happens when light is generated or detected. They hadn't grasped that "photon" and "wave" are both metaphors which describe the behaviour of light only in a limited context; outside that context those metaphors can be very misleading.

This opposition from physicists was a serious matter because to build an optical intensity interferometer to measure visible stars we needed lots of money, and somehow we had to convince the people who might give us that money that our ideas were sound. So we did some experiments.

For our first experiments we[3] built an optical analogue of our radio interferometer as shown in Fig.3. We made an artificial star by focussing the light from a high pressure mercury arc through a narrow band optical filter on to a pinhole. The light was then divided by a half-silvered mirror into two beams which were detected by identical phototubes. To vary the spacing between the detectors, as seen from the pinhole, one phototube could be translated sideways. In this experiment the number of photons was so great that the individual pulses in the output of a phototube were superimposed and formed a continuous waveform of noise which was fed to an electronic correlator with a bandwidth of about 5OMHz. The correlation between the two noise outputs was measured as a function of the spacing between the detectors.

We did two experiments (1955, 1957) both of which showed the expected correlation. Fig.4 shows the results of our experiment in 1957 together with the correlation expected theoretically. The theoretical curve was calculated on the simple assumption that the probability of emission of a photoelectron is proportional to the intensity of the light. The analysis shows that in the output current of a phototube there are two components, the classical "particle" or "shot" noise in a stream of electrons and the "wave" noise due to the fluctuations in the intensity of the light itself. Obviously the particle noises from the two phototubes are uncorrelated; but if the light on their photocathodes is mutually coherent then the two wave noises are correlated. I won't go through the rather cumbersome expressions for the correlation which we published, but simply point to the solid line in Fig.4 which shows the theoretical correlation for this experiment.

As you can see the experimental results agreed well with theory and most of the doubts which people had about our proposal for an optical interferometer were laid to rest. All the same, Twiss and I thought it would be interesting, and good for our public relations, to demonstrate more clearly that the arrival time of individual photons is actually correlated; we therefore set up the experiment shown by Fig.5.

The optical layout was similar to the first experiment but to achieve a workable signal noise ratio with individual photons the light source must have a very high specific intensity; we used an electrodeless mercury isotope lamp excited by radio-frequency. The output of each phototube was taken through wide-band amplif iers to a coincidence counter with a resolving time of about $3.5 \cdot 10^{-9} s$.

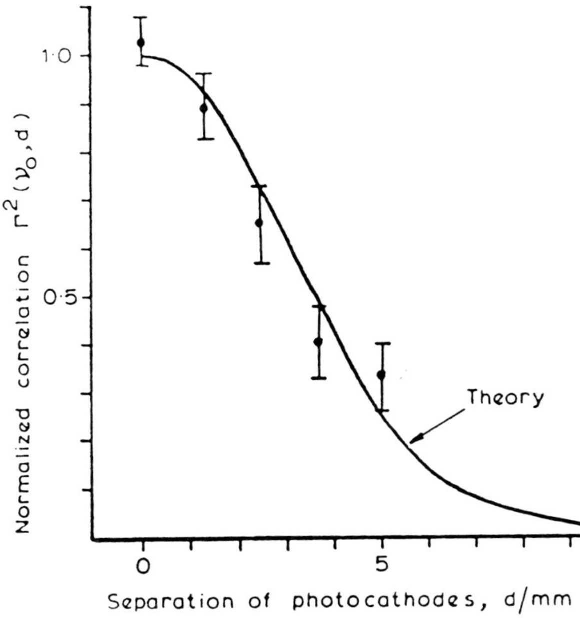

Figure 4: Results of first correlation experiment

Figure 5: Optical system of second correlation experiment

The rate of arrival of photoelectrons in each channel (N_1, N_2) and the number of coincident pulses (N_c) was measured for 2 minutes, first with the two photocathodes superimposed, as seen from the pinhole, and then with them sufficiently far apart to resolve the pinhole.

In a total exposure time of 8 hours we measured the ratio of the "excess" of correlated coincidences $(N_c - N_r)$ to the random coincidences (N_r) to be

$$(N_c - N_r)/N_r = 0.0193 \pm 0.0016(p.e.). \tag{1}$$

Theory showed us that the coincidence rate should be

$$N_c = N_1 \cdot N_2 \cdot 2T_c[1 + 1/2| \, Y_{12} \, |^2 T_o/2T_c] \tag{2}$$

where N_1, N_2 are the counting rates in the two channels, T_c is the resolving time of the coincidence counter, T_o is the coherence time of the light, $| \, Y_{12} \, |$ is the modulus of the mutual coherence of the light on the two photocathodes and it is assumed that in practice $T_c \gg T_o$.

The first term in equation (2) corresponds to the inevitable random coincidences between two streams of photoelectrons; the second term corresponds to the "bunching" of particles which obey Bose statistics and it increases the coincidence rate when the light is mutually coherent on the two photocathodes.

The expected value of $(N_c - N_r)/N_r$, calculated from equation (2) after allowing for the partial resolution of the pinhole by the photocathodes and a small loss of correlation due to polarisation in the half-silvered mirror, was

$$(N_c - N_r)/N_r = 0.020 \pm 0.002 \tag{3}$$

Within the rather wide limits of error, the observed correlation agreed with theory. Thus, although this second experiment was not as precise as the first, it was a clear demonstration that the time of arrival of individual photons in mutually coherent beams of light is actually correlated.

3 A Stellar Intensity Inteferometer

So far so good, but to raise money for a new interferometer we still had to show that we could actually measure a star. To do this I borrowed two very large searchlights from the Army, removed their arc lamps and put phototubes at the focus of their mirrors. With this equipment I measured the angular diameter of Sirius in 1956, the first time the angular size of a main sequence star had ever been measured. The whole thing worked perfectly; there was no need for high mechanical precision; furthermore Sirius twinkled violently most of the time but the twinkling had no effect on the measurements.

The stage was now set to build a full-scale instrument which we did in partnership with the University of Sydney. In 1962 it was installed in the Australian bush at Narrabri some 350 miles northwest of Sydney. It was a very striking instrument - the starlight was received by two 260 inch mosaic glass mirrors which ran on a circular railway track 200 metres in diameter. At the focus of each mirror there was a photomultiplier whose output current was carried by a catenary cable to a laboratory in the centre of the track. The correlation between the electrical "noise" in the two currents was measured by an electronic correlator with an electrical bandwidth of 10OMHz. The angular size of the star was measured by recording this correlation as a function of the spacing between the two mirrors.

It worked perfectly for 12 years during which time we measured the angular sizes of 32 of the brightest single stars in the southern sky in the spectral range O to F. Those measurements were the first ever to be made of main sequence stars. In fact they established the first wholly empirical temperature scale for stars; all previous scales had been based on theory. We did lots of other experiments, on double stars, spinning and emission- line stars and the Cerenkov radiation from gamma rays, but I won't bother you with all that.

The Stellar Intensity Interferometer at Narrabri was designed to measure stars brighter than magnitude +2.5 and when that was done I shut it down and sold it for scrap, leaving the foundations of its great circular track for future archaeologists to interpret.

The work we started at Narrabri goes on. After shutting down the original instrument we decided to measure fainter stars by going back to square one and finding out if the original difficulties of enlarging Michelson's stellar interferometer could be overcome by using all the latest gadgets which weren't around in his day - lasers, narrow-band filters, phototubes, computers, active optics and all that sort of thing. We believed that this would yield a cheaper and more sensitive instrument than a larger intensity interferometer.

The new instrument at Culgoora near Narrabri is called SUSI (Sydney University Stellar Interferometer) and is run by Professor John Davis. SUSI has 12 small mirrors - coelostats - in a line 640 metres long; they reflect the light from a star into an evacuated pipe which carries it to a central laboratory where there is a very impressive optical system. By selecting different pairs of mirrors it is possible to work with baselines in the range 5 to 640 metres. This instrument has already told us a lot about how to combat atmospheric scintillation; it is still too early to say how well it will eventually perform.

References

1. R.Hanbury Brown and R.Q. Twiss, *A new type of interferometer for use in radio-astronomy*, Phil. Mag. **45**, 663-682 (1954).
2. R.C. Jennison and M.K. Das Gupta, *The measurement of the angular diameter of two intense radio sources*, Phil. Mag. Ser. **8**, 1, 55 (1956).
3. R.Hanbury Brown *The Intensitv Interferometer*, (Taylor and Francis, London, 1974).

THE PHYSICS OF HANBURY BROWN–TWISS INTERFEROMETRY: FROM STARS TO NUCLEAR COLLISIONS TO ATOMIC AND CONDENSED MATTER SYSTEMS

GORDON BAYM
Department of Physics, University of Illinois
Urbana, IL 61801, USA

Since the fertile development by Hanbury Brown and Twiss in the 1950's of using correlations of signal intensities in independent detectors to determine the angular sizes of astronomical radio sources and stars, intensity interferometry has become a major technique in nuclear and particle collisions as well as atomic and condensed matter physics. In this talk we review the basic quantum physics of intensity interferometry, and its current applications.

1 Introduction

Hanbury Brown–Twiss (HBT) interferometry, measured in two identical particle correlations, is a fundamental process in quantum mechanics, carrying information about incoherent sources, e.g., their geometry. First developed, as Professor Hanbury Brown described in his introductory talk, to measure the angular diameter of radio sources [1,2] and then stars;[3] the technique has become widespread in particle and nuclear physics, and is of growing importance in atomic and condensed matter physics.

A remarkable outgrowth of the Hanbury Brown and Twiss development of optical intensity interferometry for stars was the birth of the field of quantum optics. While Hanbury Brown and Twiss had thoroughly demonstrated, both theoretically and experimentally, that intensity interferometry worked for radio waves, which were commonly understood as classical fields, it was not obvious in the mid 1950's that the effect should also work for light. Light being made of photons was more mysterious than radio signals made of classical electrical waves, connections now clear that were then obscure. The seminal step was Hanbury Brown and Twiss' simple tabletop experiment in which they used a beam from a mercury vapor lamp – a thermal source – and a half-silvered mirror to split the beam in two.[4] Measuring the intensity correlations between the two separated beams, they essentially compared the intensities at two different points in the unseparated beam, and by varying the relative path lengths between the mirror and the detectors they could vary the time separation, τ, of the points. While at large τ there were no intensity correlations, the correlations increased with decreasing τ. The characteristic timescale, the *co-*

herence time of the beam is here essentially $\hbar/T$, where T is the temperature of the source. This experiment demonstrated for the first time "photon bunching," i.e., that photons in a seemingly uncorrelated thermal beam tend to be detected in close-by pairs.

The detection of interferometry in particle collisions dates from 1960 when G. Goldhaber, S. Goldhaber, W. Y. Lee, and A. Pais [5] studied angular correlations of pions produced in $p\bar{p}$ collisions at the Bevatron. According to Pais,[6] the idea of exploring interferometry in particle physics, although so similar to that in astronomical observations, was conceived independently. The method is now a standard technique in high energy collisions, from heavy ions [see Refs. 7-11, as well as accompanying talks in this volume] to meson-nucleon interactions,[12] to electron-positron annihilation,[13] to anti-correlations of fermions in nuclear collisions [see Refs. 8 and 14, and accompanying talks in this volume].

The basic issue I will discuss in this talk is how and why HBT interferometry works. As we stress, the effect is in a unique class of experiments involving multiparticle correlations that are sensitive to the actual wave mechanics of particles as they stream out to the detectors. Normally, one imagines quantum mechanics as being important in high energy experiments only until the particles leave the interaction region; from then on one usually pictures them as little bullets on classical trajectories. (Quantum phenomena such as kaon regeneration and neutrino oscillations involve the internal degrees of freedom of the particles, rather than spatial amplitudes.) By considering the wave mechanics of the emitted particles in space and time one can understand questions such as why and when *independent* particle detectors give a greater signal when they are close together than far apart.[a]

2 Basic model of HBT intensity interferometry

In normal quantum mechanical interference, e.g., the canonical two-slit diffraction experiment, components of the amplitude of an *individual* particle interfere among themselves. The process is the quantum analogue of interference of classical light waves. The crucial significance of the discovery of Hanbury Brown and Twiss is that two *different*, but identical, particles can also produce interference. HBT interferometry differs from ordinary amplitude interferometry in that it does not compare amplitudes (as in a Michaelson interferometer) but rather correlated intensities or probabilities at different points. The simplest picture of HBT interferometry, from which one can see the fundamental

[a] A fuller account of the material in this talk can be found in Ref. 15. Much of the material reported here has been developed in Urbana in collaboration with A. Ayala and J. Popp, and reported in Refs. 16 and 17.

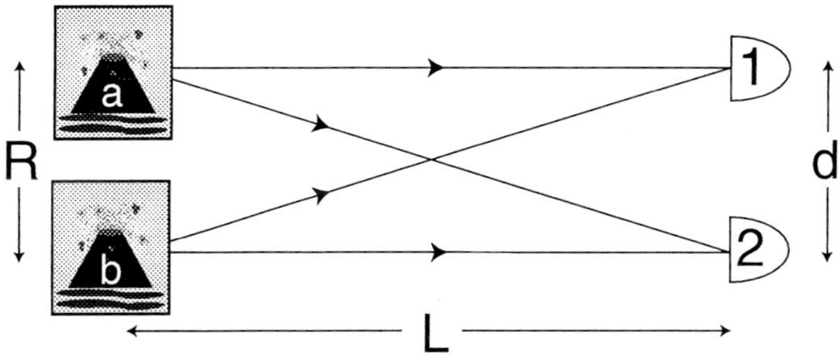

Fig. 1. Measurement of the separation of two sources, a and b, by correlation of intensities in detectors 1 and 2.

idea, is to consider the classical problem of measuring the light falling in two independent telescopes, 1 and 2 (not connected in any way), from two distant random point sources of light, a and b, or more realistically for a star, a distribution of point sources[18]: see Fig. 1. Assume that the sources are separated in space by $\vec{R}$, the two detectors by $\vec{d}$, and that the distance from the sources to the detectors is much larger than R. To be specific we choose the origin so that the sources are at $\pm\vec{R}/2$ and the detectors at $\vec{L}_\pm = \vec{L} \pm \vec{d}/2$.

Imagine that source a produces a spherical electromagnetic wave of amplitude $\alpha e^{ik|\vec{r}-\vec{r}_a|+i\phi_a}/|\vec{r}-\vec{r}_a|$, and source b a spherical wave of amplitude $\beta e^{ik'|\vec{r}-\vec{r}_b|+i\phi_b}/|\vec{r}-\vec{r}_b|$, where ϕ_a and ϕ_b are random phases (we ignore polarizations here). Let us calculate the correlation of the electromagnetic intensities in 1 and 2 as a function of the separation of the two telescopes. The total amplitude at detector 1 is

$$A_1 = \frac{\alpha e^{ikr_{1a}+i\phi_a}}{r_{1a}} + \frac{\beta e^{ik'r_{1b}+i\phi_b}}{r_{1b}}, \tag{1}$$

where r_{1a} is the distance from source a to detector 1, etc., and the total intensity in 1 is

$$I_1 = \frac{|\alpha|^2}{r_{1a}^2} + \frac{|\beta|^2}{r_{1b}^2} + \frac{\alpha^*\beta}{r_{1a}r_{1b}} e^{i(k'r_{1b}-kr_{1a}+\phi_b-\phi_a)} + \frac{\alpha\beta^*}{r_{1a}r_{1b}} e^{-i(k'r_{1b}-kr_{1a}+\phi_b-\phi_a)}, \tag{2}$$

with a similar result for I_2. On averaging over the random phases the latter exponential terms average to zero, and we find the average intensities in the two detectors (i=1,2),

$$\langle I_i \rangle = \frac{|\alpha|^2}{r_{ia}^2} + \frac{|\beta|^2}{r_{ib}^2}. \tag{3}$$

The product of the averaged intensities $\langle I_1 \rangle \langle I_2 \rangle$ is independent of the separation of the detectors.

On the other hand, multiplication of the intensities $I_1 I_2$ before averaging gives an extra non-vanishing term $\sim (\alpha^* \beta)(\alpha \beta^*)$, and we find after averaging over the phases that

$$\langle I_1 I_2 \rangle = \langle I_1 \rangle \langle I_2 \rangle + 2 \frac{|\alpha|^2 |\beta|^2}{r_{1a} r_{2a} r_{1b} r_{2b}} \cos\left(k(r_{1a} - r_{2a}) - k'(r_{1b} - r_{2b})\right). \tag{4}$$

Then the correlation function, defined by

$$C(\vec{d}\,) = \frac{\langle I_1 I_2 \rangle}{\langle I_1 \rangle \langle I_2 \rangle}, \tag{5}$$

is given, with $d \ll L$, by

$$C(\vec{d}\,) = 1 + 2 \frac{\langle |\alpha|^2 \rangle \langle |\beta|^2 \rangle}{(\langle |\alpha|^2 \rangle + \langle |\beta|^2 \rangle)^2} \cos\left(k(r_{1a} - r_{2a}) - k'(r_{1b} - r_{2b})\right). \tag{6}$$

For source separation, $\vec{d}$, much smaller than the distance to the detectors, as is always the case in both astronomical measurements and collision experiments, we have

$$\begin{aligned} &k(r_{1a} - r_{2a}) - k'(r_{1b} - r_{2b}) \\ &\qquad \simeq (k' - k)(|\vec{L} + \vec{d}/2| - |\vec{L} - \vec{d}/2|) + K\vec{R} \cdot (\hat{L}_- - \hat{L}_+), \end{aligned} \tag{7}$$

where $K = (k + k')/2$. The first term is a phase offset proportional to the difference in arrival times in the two detectors of a signal from the origin. The second is of the form $\vec{R} \cdot (\vec{k}_2 - \vec{k}_1)$, where $\vec{k}_{1,2} = K\hat{L}_\pm$ are the (mean) wavevectors of the light seen in individual detectors. [For detector separation, d, small compared with L, the argument of the cosine reduces to $\simeq (k' - k)\vec{d} \cdot \hat{L} - K(\vec{R} \cdot \vec{d} - \vec{R} \cdot \hat{L}\vec{d} \cdot \hat{L})$, displaying explicit symmetry under interchange of the sources and detectors.] The correlated signal in Eq. (6) varies as a function of the detector separation d on a characteristic length scale

$$d = \lambda/\theta, \tag{8}$$

where $\lambda = 2\pi/K$ is the (mean) wavelength of the light, and $\theta = R/L$ is the angular size of the sources as seen from the detectors. Thus by varying the separation of the detectors one learns the apparent angle between the two sources, and with a knowledge of the individual wavevectors, the physical size of the source.

If instead of two discrete sources, one has a distribution of sources, $\rho(\vec{r})$, then averaging over the distribution, one finds that the correlation function measures the Fourier transform of the source distribution:

$$C(\vec{d}) - 1 \sim \left| \int d^3 r \rho(\vec{r}) e^{i(\vec{k}_1 - \vec{k}_2) \cdot \vec{r}} \right|^2 . \tag{9}$$

In particle experiments, one measures the correlation function in the form

$$C(p, p') \equiv \frac{\{\langle n_{\vec{p}_1} n_{\vec{p}_2} \rangle\}}{\{\langle n_{\vec{p}_1} \rangle \langle n_{\vec{p}_2} \rangle\}}, \tag{10}$$

where $n_{\vec{p}}$ is the number of particles of momentum $\vec{p}$ measured in a single event, $\vec{q} = \vec{p}_1 - \vec{p}_2$, and the averages, denoted by $\langle \cdots \rangle$, are over an ensemble of events. The pairs in the numerator are taken from the same event, and the pairs in the denominator from different events. Usually one also averages the numerator and denominator separately over a range of center-of-mass momenta $\vec{P} = \vec{p}_1 + \vec{p}_2$ of the pair, an average denoted here by $\{\cdots\}$. As q becomes very large the correlations between the particles are lost, and the correlation function approaches unity.

One important difference between astronomical observations and high energy physics is that the stars stay fixed, while in a collision, the system evolves a time scale of 10^{-23} to 10^{-22} seconds, and thus one has to take into account the changing geometry. The correlation funciton in high energy physics measures not the Fourier transform of the distribution in space alone, but rather to good approximation the Fourier transform in both space and time. A second important difference is that in astronomy, in the absence of a knowledge of the distance to the source, one cannot measure the actual difference in direction of the wavevectors of the light in the two detectors, and thus one measures only the angular size of the source as seen from the detectors. In a high energy experiment, one can determine the wavevectors of the detected particles, and thus measure the absolute size of the source.

To find an enhanced correlation at detector separation $\leq \lambda/\theta$ it is not necessary for the two detectors to be wired together. One needs only to compare the data trains. Why do two independent nearby detectors produce extra signal? As Hanbury Brown initially realized,[2] if the amplitude varies randomly then a positive fluctuation of the amplitude will produce a correlated increase in both measured signals, and vice versa for a negative fluctuation.

Purcell, in a paper[19] that was very influential in convincing people of the reality of the HBT effect for chaotic sources of light, showed simply how to understand observed bunching of photons in terms of electric field fluctuations. For example, in black-body radiation, both the real and imaginary parts of the complex electric fields $E \sim \alpha e^{i\vec{k}\cdot\vec{r}-i\omega t}$ are Gaussianly distributed. For independent Gaussianly distributed real variables x and y, one finds simply that $\langle (x^2+y^2)^2\rangle = 2\langle x^2+y^2\rangle^2$, so that the electric field at a given point obeys

$$\langle |E|^2|E|^2\rangle = 2(\langle |E|^2\rangle)^2, \tag{11}$$

The extra factor of two is precisely the source of the HBT correlations, the enhancement that arises from the cosine term in Eqs. (4) and (6). More generally, the time correlations of the electric field obey

$$\langle |E(t)|^2|E(t')|^2\rangle = \langle |E(t)|^2\rangle\langle |E(t')|^2\rangle + |\langle E^*(t)E(t')\rangle|^2. \tag{12}$$

Note that for a coherent source, e.g., a laser, one has instead $\langle |E(t)|^2|E(t')|^2\rangle \simeq \langle |E(t)|^2\rangle\langle |E(t')|^2\rangle$.

3 Quantum mechanics of HBT

The simple derivation of intensity interferometry in Sec. 2 is entirely classical. How can one understand the effect from a quantum mechanical viewpoint? If we think of the sources a and b in Fig. 1 as emitting photons, we can identify four different processes, shown in Fig. 2: *i*) source a emits two photons, one detected in each detector, *ii*) source b emits the two photons instead; *iii*) source a emits a photon detected in 1 and b emits a photon detected in 2, and finally, *iv*) source a emits a photon detected in 2 and b emits a photon detected in 1, the exchange of the previous process, *iii*. The first two processes are distinguishable, and do not produce any interferometry. They simply correspond to detection of the sources independently (the $|\alpha|^4$ and $|\beta|^4$ terms in Eq. (4)). Only the latter two processes, *iii* and *iv*, which are quantum-mechanically coherent, give rise to interferometry. Quantum mechanically, the HBT effect is a consequence of ordinary boson exchange, an effect included in the symmetry of the wave function of the pair of particles, e.g., for a pair of independent bosons in orthogonal states ϕ_α and ϕ_β, $\phi(1,2) = (\phi_\alpha(1)\phi_\beta(2) + \phi_\alpha(2)\phi_\beta(1))/\sqrt{2}$. The effect is present for all pairs of identical bosons, including pions and kaons produced in high energy collisions.

Formally one should describe HBT measurements in terms of the two-particle correlation functions of the emitted identical particles. It is more intuitive, however, first to study the problem in terms of particle wave functions. To be specific we focus on pions, although the discussion is quite general.

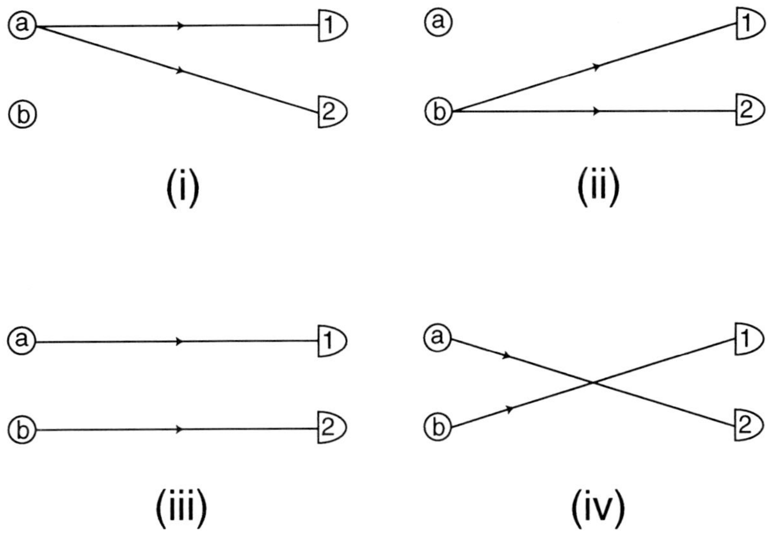

Fig. 2. The four independent photon emission and detection processes included in Eq. (4).

How does one describe quantum mechanically the set of pions emitted emitted in a nuclear collision? Even in the best of all possible worlds – where one knows the exact wave function of the two colliding nuclei, and knows exactly how the quantum mechanical evolution operator produces the final system as a coherent superposition of well defined pure quantum mechanical states $\Psi(1, 2, ..., N)$ of N particles – the subset of pions emitted is described by a *mixed* quantum state. Quite generally, any subset of particles in a pure state is described by a mixed state, even, e.g., for the electron in the ground state of a freely moving hydrogen atom. The single particle density matrix for pions of a given charge at equal time is given by

$$\langle \pi^\dagger(r,t)\pi(r',t)\rangle = \int d^3r_2 \cdots d^3r_n \Psi^*(r, r_2, \cdots, r_N, t)\Psi(r', r_2, \cdots, r_N, t), \tag{13}$$

where $\pi(x)$ is the operator destroying a pion of the given charge at point $x = (\vec{r}, t)$.

If Ψ is a product of single particle wave functions, then $\langle \pi^\dagger(x)\pi(x')\rangle$ factors into a product of single particle states, $\phi^*(x)\phi(x')$. In general the single-pion correlation function does not factor, even when the pions are no longer interacting, but it can be represented as a sum over a collection of single particle states ϕ_i:

$$\langle \pi^\dagger(x)\pi(x')\rangle = \sum_i f_i \phi_i^*(x)\phi_i(x'), \tag{14}$$

where the f_i give the probability of finding the pion in single particle state i. For example, the probability of finding a pion at point 1 is given by $\sum_i f_i|\phi_i(1)|^2$. Only if the pion part of the state Ψ factors out in the form of a product of the same single particle states for all the pions – a Bose-Einstein condensate – will single pions be in a pure state. The mixed single pion state always has finite entropy, $-\sum_i (f_i \ln f_i - (1+f_i)\ln(1+f_i))$.

The closest one can come to describing pions as little bullets is to picture the single particle states making up the mixed ensemble of pions as wavepackets with almost well defined momenta and energies, limited by the uncertainty principle. We picture the collision volume as made up of many sources of pions; whether the sources are fragmentation of strings, or in the language of low energy nuclear physics individual nucleon-nucleon collisions, the sources are localized to within a distance R which is less than the size of the entire collision volume, and the emission process is temporally localized to within a time τ. Thus the individual components of momentum, $\vec{p}$, and energy, ε_p, of the emitted particles are uncertain to within

$$\begin{aligned} \Delta p_a &\gtrsim \hbar/R, \quad a = x, y, z \\ \Delta \varepsilon_p &\gtrsim \hbar/\tau. \end{aligned} \tag{15}$$

A pion nominally of momentum $\vec{p}$ emitted from a source at the origin in space and time would have an amplitude to have four-momentum $q = (\varepsilon_q, \vec{q})$ that is roughly Gaussian,

$$A(q) \sim e^{-(\vec{q}-\vec{p})^2R^2/2}e^{-(\varepsilon_q-\varepsilon_p)^2\tau^2/2}, \tag{16}$$

and in space-time the particle would be described by a wavepacket

$$\phi_{\vec{p}}(x) = \int \frac{d^3q}{(2\pi)^3}\frac{e^{iqx}}{2\varepsilon_q}A(q). \tag{17}$$

How does the packet evolve in time after leaving the source? Assume that the collision takes place, and the particles emerge, into vacuum. (In real life secondary scattering on other atoms in the target and also scattering in air are both very important effects.) The transverse spatial spread, perpendicular to

$\vec{p}$, is determined by the uncertainty in transverse velocity, which for relativistic particles ($\varepsilon_p \approx p$) is:

$$\Delta v_\perp = \frac{\Delta p_\perp}{\varepsilon_p} \sim \frac{1}{pR}. \tag{18}$$

The spread in longitudinal velocity is

$$\Delta v_L = \Delta\left(\frac{p_L}{\varepsilon_p}\right) = \frac{\Delta p_L}{\gamma^2 \varepsilon p} \sim \frac{1}{\gamma^2 pR}, \tag{19}$$

where $\gamma = \varepsilon_p/m$ is the Lorentz factor, and m is the pion mass. The presence of the factor γ^2 reflects the fact that the more relativistic the particles become the closer to the speed of light the packet moves, with vanishing spread in longitudinal velocity. The transverse size of the wavepacket after travelling a distance L is thus $\sim L/pR$, while its thickness is $\sim L/\gamma^2 pR$.

To be specific, consider a 1 GeV pion ($\gamma = 7$) produced within an initial radius of 10 fm travelling to a detector 10 meters from the source. Then the transverse spread of the wavepacket is about 20 cm, while the thickness of the packet grows to half a centimeter. Pions emerge in rather extended pancake-like states. The characteristic time for such a pion wavepacket to cross a point at the distance of the detector is $\sim 10^{-11}$ sec. Smaller source sizes lead to even larger spreads in the pion wavepackets. Photons would have a similar transverse spread; however, the thickness of the wavepacket would not grow.

4 Detector response

Let us now turn to the question of the detection of particles in such wavepacket states by a magnetic spectrometer. In the "bullet" picture, the particle travels along a classical trajectory, excites an atom in the detector, which determines the direction of its momentum, and begins to make a track (or in emulsion makes a spot on photographic film); the particle continues on, producing a curved track in the spectrometer, from which one deduces the magnitude of its momentum. But, in fact, the particle has a distribution of probability amplitudes, wherever its wavepacket ϕ is non-zero, of where it makes the first spot in the spectrometer. A particle of mean momentum $\vec{p}$ can, because of the transverse momentum uncertainty, be detected anywhere within the range of momenta $\Delta p_\perp$ around $\vec{p}$ contained in its wavepacket. After the initial collision in the detector, a much more narrowly focused wave packet emerges. (If one does not actually do the measurement, the state that emerges is a mixed state corresponding to all possible points where the incident packet can excite a detector atom.) The narrowed packet continues through the magnetic field, and the subsequent collisions it makes selects its energy. One measures momentum by measuring a sequence of positions in the detection.

Consider the measurement of the momentum of an incident particle in state $\phi_i(x)$. The probability that it makes the first single excitation of an atom in the detector at point A and continues to produce a track corresponding to measuring its momentum to be $\vec{k}$ is then

$$P_{\vec{k}}(A;i) = \int dx dx' e^{ikx} \phi_i^*(x) s_A(x,x') e^{-ikx'} \phi_i(x'). \tag{20}$$

Here the detector atom response function, s, is localized in r and r' about the atom at A; it also extends over an interval in $t - t'$ which is the characteristic time that the detector analyzes the amplitude and phase variation of the incident wave, of order 10^{-16} sec $\sim \hbar/(10 \text{ eV})$, the inverse of a characteristic atomic excitation energy. This time is much shorter than the typical nanosecond scale resolution time of a detector, the time it takes to build up a million-fold cascade of electrons.[b] More generally, the probability of detection of a pion of given charge at A is given by

$$P_{\vec{k}}(A) = \int dx dx' e^{ik(x-x')} s_A(x,x') \langle \pi^\dagger(x) \pi(x') \rangle, \tag{21}$$

where $\langle \pi^\dagger(x)\pi(x')\rangle$ is the single-pion correlation function.

Consider next the detection of two independent particles. The question is why, if one detector lights up, do nearby detectors tend to have greater probability to light up than detectors further away – the HBT effect. Suppose first that the two particles are incident in orthogonal wavepackets ϕ_i and ϕ_j. The symmetrized two-particle wave function is $\phi(\vec{r},\vec{r}',t) = [\phi_i(\vec{r},t)\phi_j(\vec{r}',t) + \phi_i(\vec{r}',t)\phi_j(\vec{r},t)]/\sqrt{2}$. The probability of a joint detection by a detector atom at A of a pion that continues to make a track corresponding to momentum $\vec{k}$, and by a detector atom at B of a second pion that continues to make a track corresponding to momentum $\vec{k}'$, is then given by

$$\begin{aligned} & P_{\vec{k},\vec{k}'}(A,B;i,j) \\ = & \int dx dx'' e^{ik(x-x'')} s_A(x,x'') \int dx' dx''' e^{ik'(x'-x''')} s_B(x',x''') \\ & \qquad \times (\phi_i(x)\phi_j(x') + \phi_i(x')\phi_j(x))^* (\phi_i(x'')\phi_j(x''') + \phi_i(x''')\phi_j(x'')) \\ = & P_{\vec{k}}(A;i)P_{\vec{k}'}(B;j) + P_{\vec{k}}(A;j)P_{\vec{k}'}(B;i) \\ & + \mathcal{A}_{\vec{k}}(A;i,j)\mathcal{A}_{\vec{k}'}(B;j,i) + \mathcal{A}_{\vec{k}}(A;j,i)\mathcal{A}_{\vec{k}'}(B;i,j), \end{aligned} \tag{22}$$

[b] A detailed discussion of the role of the detectors is given in Refs. 16 and 17.

where

$$\mathcal{A}_{\vec{k}}(C;i,j) = \int dx dx'' e^{ik(x-x'')} s_C(x,x'')\phi_i^*(x)\phi_j(x''). \tag{23}$$

The first term in Eq. (22) is the probability of the particle in state i being detected at A times the probability of the particle in state j being detected at B, and the second is the same with i and j interchanged. These are the normal terms.

The final two terms in Eq. (22) are the enhancement of the detection probability – the HBT effect. As we see, in order to have enhancement, it is necessary that both wavepackets, i and j, overlap in the detector at A during the time that the detector is doing quantum mechanics on the incoming system, and similarly that they also must overlap in the detector at B (but note that the wavepackets do not have to be present in both detectors simultaneously). The presence of the wavepackets simultaneously in each of the detectors is the reason the detectors produce more signal when they are close to each other.

The maximum transverse separation d of the detectors that will produce an HBT signal is essentially the transverse size of a given wavepacket, L/pR, the wavelength divided by the angular size of the source as seen from the detectors, where again L is the distance from the source to the detector, R is the size of the source, and p is the average particle momentum. The scale of relative momenta q for which one finds a signal is $q/p = d/L \sim 1/pR$, and thus $q \sim 1/R$, the standard HBT result. As we see from this argument, the HBT effect directly measures the width of the wavepackets at the first detection. This width is in turn determined by the uncertainties in the momentum distribution at the time that the wavepacket is no longer affected by strong interactions with the other particles in the collision.

Generally the amplitudes, Eqs. (16) and (17), of the wavepackets, $A(q)$, vary slowly in q on an atomic scale, $\sim$ 1 KeV/c. In this case the overlap integral (23) is not sensitive to the detection time scales, $\sim 10^{-16}$ sec, and one finds an HBT correlation function:

$$C(p,p') \equiv 1 + \frac{|\sum_i f_i A_i^*(p) A_i^*(p')|^2}{\sum_i f_i |A_i(p)|^2 \sum_j f_j |A_j(p')|^2}, \tag{24}$$

where f_i is the single pion probability in the ensemble, Eq. (14).

However, if a particle is delayed in emission by more than the detection time scale, the HBT correlations between that particle and one produced directly will be suppressed. A simple example in heavy-ion collisions of this effect is in the correlation of π^- produced in Λ decay, $\Lambda \to \pi^- + p$, with directly produced pions. Because the Λ travels more slowly than a directly produced

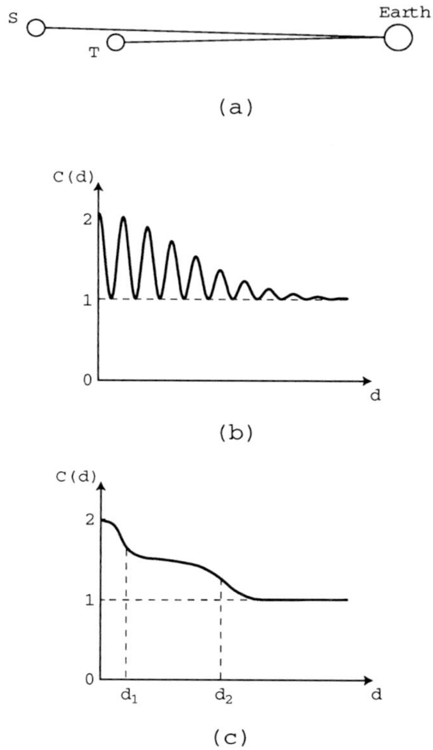

Fig. 3. a) Two stars, S, and T, along nearby lines of sight from the earth; b) schematic of HBT measurement of correlated intensity from the two stars; c) schematic of HBT measurement a bright star surrounded by a halo of dim stars.

pion of the same rapidity as one emitted in the decay, the pion from decay will lag the directly produced one by a time Δt. To estimate this effect, we note that a π^- emitted in the forward direction has rapidity $y_\pi^0 \approx 0.67$ in the Λ frame, and that a Λ of rapidity y travels on average a distance $\tau_\Lambda \sinh y$ before decaying, where τ_Λ is the Λ lifetime. Thus $\Delta t = \tau_\Lambda/(\cosh y + \sinh y/\tanh y_\pi^0)$, which for a Λ of typical rapidity 3 is $\sim 0.037\tau_\Lambda = 9.7 \times 10^{-12}$ sec, much longer than the detector timescale. Pions emitted in other than the forward direction will have an even greater time lag.

As the above discussion makes apparent, the enhanced signal is not a consequence of special preparation at the source, such as a particle of a given momentum inducing emission of other particles of similar momentum, as in a laser (which does not give an HBT signal.) Clearly, in the stellar case there can be no such connection between emission processes on opposite sides of the star, and yet photons from opposite sides give an HBT enhancement. The effect is a property of the detection. Furthermore, if two stars, S and T, which are at very different distances from the Earth, are approximately along the same line of sight, Fig. 3a, one will see HBT correlations between a photon from S and a photon from T. Of course the emission times have to be different in order that the two photons arrive at each of the detectors at the same time. (Even in measuring a single star one correlates earlier emitted photons from the stellar rim with later emitted photons from the front surface.) The HBT correlations between photons from the two sources as a function of detector separation, d, sketched in Fig. 3b, would have an oscillatory term characteristic of the angular separation of the two stars (cf. Eq. (6), modulated by a broader Gaussian characteristic of the angular size of an individual star. By comparison, a single bright star, T, surrounded by a halo of dim stars (an average over a distribution of stars S), would yield a correlation function, Fig. 3c, with a rise proportional, for $d \lesssim d_2$, to the inverse of the angular size of the central star λ/d_2, plus a much more narrow rise, for $d \lesssim d_1$, inversely proportional to the angular size of the halo.

That the HBT effect is not a property of production, but rather is a property of detection is well illustrated by the following experiment. Imagine, as in Fig. 4, that RHIC, the BNL heavy-ion collider, sends a pion into a far away particle detector array, and that by using the pionic analogue of a half-silvered mirror, we reflect a pion from CERN into the same detector array, so that it arrives at the same time as the pion from RHIC. As long as the wavepackets of the pions overlap sufficiently that the interference amplitudes $\mathcal{A}$ at the detector atoms are non-zero, one will see an HBT enhancement! As this experiment illustrates, the HBT effect does not depend on the history of the particles. What matters is the form of the wavepackets when they arrive at the detectors.

5 HBT in atomic and condensed matter systems

Studies of Hanbury Brown–Twiss correlations in atomic and condensed matter systems are at a very early stage. The first detection of intensity correlations among atoms was carried out only recently, by Yasuda and Shimizu[20] who observed the time correlations in laser-cooled ultracold (but not yet Bose-Einstein condensed) beams of bosonic ^{20}Ne atoms. The experiment is similar in principle to the original Hanbury Brown–Twiss tabletop experiment on photon

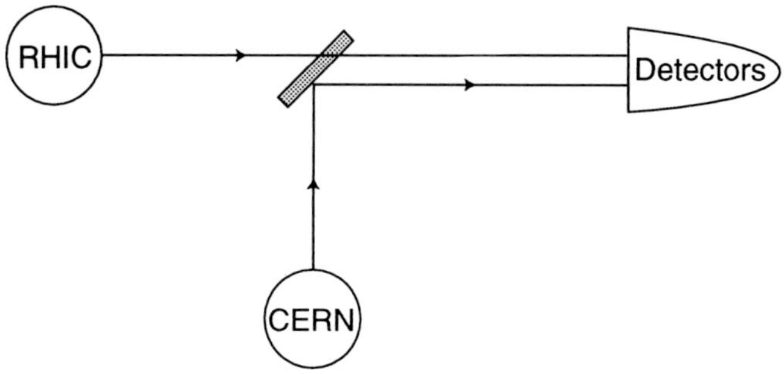

Fig. 4. Detection of the HBT effect between a pion from RHIC and a pion from CERN.

bunching in the beam of a Hg vapor lamp. The correlations in the beam are those expected from a thermal source, where the correlation time is the inverse of the temperature of the beam. Indeed the HBT correlation function begins to rise at time separations less than $\sim 0.5 \times 10^{-6}$ sec to a value a factor of two larger than at large time, corresponding to a beam temperature $\sim 10^2 \mu$K.

By contrast, measurement of HBT correlations in a beam emerging from a Bose-Einstein condensed system, e.g., the MIT atomic laser,[21] in which the atoms macroscopically occupy a single mode, $\psi(r)$, would yield no such atomic bunching, but rather the correlation function would remain flat. Lack of an HBT enhancement is an indicator of coherence of the beam. In general, loss of HBT correlations would probe the onset of Bose-Einstein condensation, not only in atomic systems, but in condensed matter systems such as the observed Bose-condensed paraexcitons in cuprous oxide.[22,23]

While single particles can undergo ordinary quantum interference with themselves and produce a diffraction pattern, as in a two-slit interferometer, similar interference between different particles results, as we have seen, from the Hanbury Brown–Twiss effect. This phenomenon is very well illustrated in the interference between two independent atomic Bose condensates, an experiment first carried out by Ketterle's group at MIT.[24] In this experiment, two condensates, A and B, of identical (^{23}Na) atoms are independently created initially spatially separated from each other, and released. The clouds expand and overlap; the resulting spatial distribution of particles, which is imaged by resonant absorption of a laser beam, shows a characteristic interference pattern.

The interpretation of the result is straightforward in the case that the condensates are mutually coherent. Then the atoms each occupy a coherent state $\psi \sim \sqrt{N_A}\psi_A(r,t) + \sqrt{N_B}\psi_B(r,t)$, with amplitude to be in either condensate, where N_A and N_B are the number of condensed particles in each condensate. During expansion, the waves $\sqrt{N_A}\psi_A(r,t)$ and $\sqrt{N_B}\psi_B(r,t)$ from the two condensates interfere, producing a particle density pattern

$$\langle\rho(r,t)\rangle = |\sqrt{N_A}\psi_A(r,t) + \sqrt{N_B}\psi_B(r,t)|^2, \tag{25}$$

the result of ordinary quantum interference of the individual particles with themselves.

For example, for initial Gaussian sources of free particles, individually of width R_0 and separated by $\vec{a}$, the time dependent waves are

$$\psi \sim e^{-(\vec{r}\pm\vec{a}/2)^2(1+it/mR_0^2)/2\langle\vec{r}(t)^2\rangle}, \tag{26}$$

where $\langle\vec{r}(t)^2\rangle = R_0^2 + (t/mR_0)^2$. The interference pattern thus exhibits a linear structure in space, $\sim \cos^2(\vec{r}\cdot\vec{a}m/2t)$ at large time, as is in fact observed.

However, if the condensates are prepared independently, as in the MIT experiment, the individual particles do not have a coherent wave function spreading over the two condensates. Rather a given particle is either in one condensate or the other. Nonetheless, the expanding system exhibits a similar interference pattern. Measurement of the two particle correlation function via two photon absorption processes would yield the expected Hanbury Brown and Twiss result, Eq. (4), for the intensity correlations from two independent sources:

$$\begin{aligned}\langle\rho(x)\rho(x')\rangle = \;& \left(N_A|\psi_A(x)|^2 + N_B|\psi_B(x)|^2\right)\left(N_A|\psi_A(x')|^2 + N_B|\psi_B(x')|^2\right)\\ & + N_A N_B \psi_A(x)\psi_B(x')\psi_B^*(x)\psi_A^*(x') + c.c..\end{aligned} \tag{27}$$

The various processes that contribute to the two particle correlation function are the same as in Fig. 2, only now read from right to left, corresponding to the initial laser photons being absorbed.

The crucial feature of the interference of two condensates is that in the limit of a large number of particles in the condensates, the system behaves in any single run in which the condensates are prepared and released, as if the two condensates are coherent with a relative phase 2ϕ, which varies randomly from run to run.[25,26] This behavior is analogous to that of two independent lasers. The density measured through multiphoton absorption in a single run has the form $\langle\rho(x)\rangle_\phi = |e^{i\phi}\sqrt{N_A}\psi_A(x) + e^{-i\phi}\sqrt{N_B}\psi_B(x)|^2$, where $x = (r,t)$, and thus shows interference. However, when measured over many runs and thus averaged over the random phase, the expectation value of the particle density is given by $\langle\rho(\dot{x})\rangle = N_A|\psi_A(x)|^2 + N_B|\psi_B(x)|^2$, which shows no interference.

Similarly, the two particle correlation function measured in a single run has the form

$$\begin{aligned}\langle\rho(x)\rho(x')\rangle_\phi = \; & |e^{i\phi}\sqrt{N_A}\psi_A(x) + e^{-i\phi}\sqrt{N_B}\psi_B(x)|^2 \\ & \times|e^{i\phi}\sqrt{N_A}\psi_A(x') + e^{-i\phi}\sqrt{N_B}\psi_B(x')|^2. \qquad (28)\end{aligned}$$

When averaged over the phase ϕ, the two particle correlation function becomes the HBT result (27). Underlying the measurement of apparent interference between independent condensates, through two (or more) photon absorption processes, is the Hanbury Brown–Twiss effect.

A further application of HBT has been in light scattering from atoms trapped in optical lattices. Jurczak et al.[27] have created an optical lattice with an arrangement of four lasers in which they trap atomic rubidium at a density $\sim 2 \times 10^9$ cm^{-3}, filling about 10^{-4} of the lattice sites. The lasers also scatter from the rubidium, and the time correlations in the scattered light (of two different polarizations) effectively measure the atom-atom correlation functions in the lattice. From these measurements they are able to extract the diffusion of the loosely packed atoms in the optical lattice.

I thank the members of my group in Urbana – Alejandro Ayala, James Popp, and Benoit Vanderheyden – who are responsible for much of the material reported here, Chris Pethick, Tony Leggett and Erich Mueller for insightful discussions of coherence in Bose-Einstein condensates, and Michael Baym for preparing the graphics. This work was supported in part by U.S. National Science Foundation Grant Nos. PHY94-21309 and PHY-9800978.

References

1. R. Hanbury Brown and R. Q. Twiss, *Phil. Mag., Ser. 7* 45, 663 (1954).
2. R. Hanbury Brown, *Boffin* (Adam Hilger, Bristol, 1991).
3. R. Hanbury Brown and R. Q. Twiss, *Nature* 178, 1046 (1956).
4. R. Hanbury Brown and R. Q. Twiss, *Nature* 177, 27 (1956).
5. G. Goldhaber, S. Goldhaber, W. Lee, and A. Pais, *Phys. Rev.* 120, 300 (1960).
6. Private communication, July 1997.
7. B. V. Jacak, *Nucl. Phys.* A590, 215c (1995).
8. W. Bauer, C. Gelbke and S. Pratt, *Ann. Rev. Nucl. Part. Sci.* 42, 77 (1992), and references therein.
9. U. Heinz, *Nucl. Phys.* A610, 264c (1996).
10. S. Pratt, *Proc. 13th Int. Conf. on Nucleus-Nucleus Collisions, QM97, Nucl. Phys.* A in press.
11. B. V. Jacak, *Rev. Mod. Phys.* (1998), in press.

12. N.M. Agababyan et al., NA22 collaboration, *Z. Phys.* C71, 409 (1996).
13. P.D. Acton et al. (OPAL), *Phys. Lett.* B267, 143 (1991), *Z. Phys.* C58, 207 (1993); P. Abreu et al. (DELPHI), *Phys. Lett.* B286, 201 (1992); D. Decamp et al. (ALEPH), *Z. Phys.* C54, 75 (1992).
14. R. Ghetti et al., *Nucl. Inst. and Meth.* A335, 156 (1993).
15. G. Baym, *Acta Physica Polonica* 29, 1 (1998).
16. J. Popp, Ph.D. thesis, University of Illinois, 1997 (unpublished).
17. A. Ayala, G. Baym, and J. Popp, *Nucl. Phys.* A, to be published.
18. G. Baym, *Lectures on Quantum Mechanics* (W. A. Benjamin, Inc., New York, 1969), pp. 431-434.
19. E. Purcell, *Nature* **178**, 1449 (1956).
20. M. Yasuda and F. Shimizu, *Phys. Rev. Lett.* 77, 3090 (1996).
21. M.-O. Mewes, M.R. Andrews, D.M. Kurn, D.S. Durfee, C.G. Townsend, and W. Ketterle, *Phys. Rev. Lett.* 78, 582 (1997).
22. D.W. Snoke, J.P. Wolfe, and A. Mysyrowicz, *Phys. Rev. Lett.* 59, 827 (1987); J.-L. Lin and J.P. Wolfe, *Phys. Rev. Lett.* 71, 1223 (1993).
23. G.M. Kavoulakis, G. Baym, and J.P. Wolfe, *Phys. Rev.* B53, 1 (1996).
24. M.R. Andrews, C.G. Townsend, H.-J. Miesner, D.S. Durfee, D. M. Kurn, and W. Ketterle, *Science* 275, 637 (1997).
25. Y. Castin and J. Dalibard, *Phys. Rev.* A, 55, 4330 (1997).
26. G. Baym, A.J. Leggett, and C.J. Pethick, to be published.
27. C. Jurczak, B. Desruelle, K. Sengstock, J.-Y. Courtois, C.I. Westbrook, and A. Aspect, *Phys. Rev. Lett.* 77, 1727 (1996); C.I. Westbrook, C. Jurczak, G. Birkl, B. Desruelle, W.D. Phillips, and A. Aspect, *J. Mod. Opt.* 44, 1837 (1997).

PROBING THE UNIVERSE WITH THE HUBBLE SPACE TELESCOPE

N. PANAGIA[a]

Space Telescope Science Institute, 3700 San Martin Drive, Baltimore, MD 21218, USA

E-mail: panagia@stsci.edu

I present a selection of the most exciting and spectacular results obtained with the Hubble Space Telescope in the last few years, such as (1) the Hubble Deep Field, *i.e.* galaxies in the far Universe, and studies of colliding galaxies in the near Universe, (2) the complete view of SN 1987A environment and the first measurements of the collision of SN 1987A ejecta with its inner circumstellar ring, (3) the observations of planetary nebulae that have revealed new aspects of stellar evolution and wonders of hydrodynamics, and (4) the studies of transient γ-ray sources, whose characteristics may imply the existence of enormously energetic stellar explosions.

1 Introduction

The Hubble Space Telescope (*HST*) is a cooperative program of the European Space Agency (ESA) and the National Aeronautics and Space Administration (NASA) to operate a long-lived space-based observatory for the benefit of the international astronomical community. *HST* is an observatory first conceived in the 1940s, designed and built in the 1970s and 80s, and become operational only in the 1990s. *HST* was designed to be a different type of mission for NASA – a long term space-based observatory. To accomplish this goal and protect the spacecraft against instrument and equipment failures, NASA had planned on regular servicing missions. A three year cycle of on-orbit servicing was adopted.

HST was deployed in low-Earth orbit ($\sim 600\ km$) by the crew of the space shuttle Discovery (STS-31) on 25 April 1990. The mission was troubled soon after launch by the discovery that the primary mirror was spherically aberrated. In addition, problems with the solar panels flexing as the spacecraft passed from the Earth's shadow into sunlight caused problems with the pointing stability. The first maintenance mission (STS-61 Endeavour, December 1993), succesfully fixed all of these problems and fully restored the functionality of *HST*. In fact, both *HST* servicing missions, *i.e.* both December 1993 and February 1997, were enormous successes. Future servicing missions are planned for early 2000 and mid-2002. Contingency flights could still be added to the shuttle manifest to perform specific tasks that cannot wait for the next

[a]On assignment from Astrophysics Division, Space Science Department of ESA

regularly scheduled servicing mission (and/or required tasks that were not completed on a given servicing mission).

HST's current science instruments include three cameras, ESA's Faint Object Camera (*FOC*), the Wide Field - Planetary Camera 2 (*WFPC*2) and the Near Infrared and Multi-Object Spectrometer (*NICMOS*), two spectrographs, the Space Telescope Imaging Spectrograph (*STIS*) and the near-IR *NICMOS*, and the Fine Guidance Sensors (*FGS*, primarily used for astrometric observations). Because of *HST*'s location above the Earth's atmosphere, these science instruments can produce high resolution images of astronomical objects. While ground-based telescopes can seldom provide resolution better than an arcsecond, *HST* can customarily achieve a 100 *mas* resolution and easily reach its diffraction limit of about 50 *mas* in the optical.

In the following sections I shall present my personal selection of the telescope's most compelling results obtained in recent years.

2 Galaxies

2.1 The Hubble Deep Field

One of the most ambitious and succesful projects carried out by *HST* is the study of the so-called *Hubble Deep Field* (*HDF*) (Williams *et al.* 1996). For 12 consecutive days in December 1995, the telescope was pointed at a field of the sky in the Ursa Major constellation, selected for being most devoid of nearby objects so as to probe the distant Universe, thus the distant past. The WFPC2 was used to image the field in four broad bands, covering the spectrum from near UV to near IR. A composite of the images obtained in various passbands is shown in Fig. 1. The *HDF* images show that nearly a third of very distant galaxies, which existed early in the history of the Universe, appear to be interacting galaxies (just like the Antennae, see section 2.2). In particular, the Hubble Deep Field has uncovered quite a number of odd-shaped, disrupted-looking galaxies. They offer direct visual evidence that galaxy collisions were more the rule than the exception in the early days of the Universe.

This analysis has led to the conclusion that the Big Bang may have been followed by a stellar "baby boom". The early Universe may have had an active, dynamic youth where stars formed out of clouds of dust and gas at incredible rates. Consequently, most of the stars the Universe will ever make may have already been formed, and the Universe now contains largely "mid-life" stars. An excellent summary of the first results from studies of the *HDF* can be found in Livio, Fall & Madau (1998).

Figure 1 - The Hubble Deep Field image obtained combining the U, B, V and I images (Williams *et al.* 1996).

2.2 Merging Galaxies: the Antennae

The *HST* resolution and sensitivity have allowed Whitmore *et al.* (1998) to detect over 1,000 exceptionally bright, young star clusters within the prototypical interacting galaxy system NGC 4038/4039, which has been dubbed the *Antennae* because the long tidal tails resemble an insect's antennae. Ground-based observations were able to detect only the brightest of these star clusters, and even in these cases were not able to show that the clusters were very compact, with the sizes of classical globular clusters.

Observing these and other interacting galaxies, Whitmore and collaborators discovered the presence of many young, compact star clusters in the case of ongoing collisions, whereas they found that older merger remnants had only fainter and redder clusters. This allowed them to place the "snapshots" of galaxy collisions into a chronological sequence.

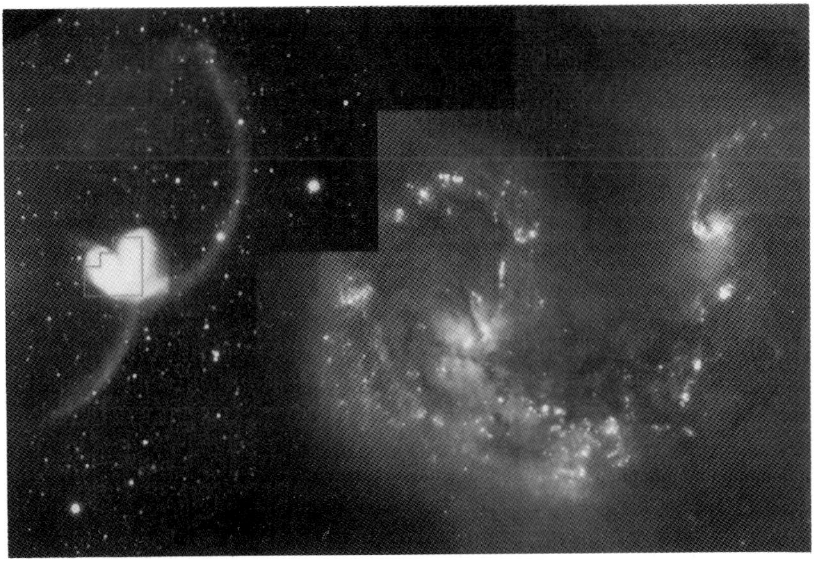

Figure 2 - WFPC2 image (right hand side) of the central region of the interacting galaxies NGC 4038/4039 (left hand side) (Whitmore *et al.* 1998).

The images of so many young globular star clusters in the Antennae suggest, for example, that they are not necessarily relics of the earliest generations of stars formed in a galaxy, as was once commonly thought, but may be the fossils of much more recent collisions. These spectacular images will help understanding how globular star clusters formed from giant hydrogen clouds in space. The Antennae system is an excellent laboratory for studying the formation of stars and star clusters since it is one of the nearest and youngest examples of a pair of colliding galaxies.

3 SN 1987A

Supernova 1987A exploded on 23 February, 1987, in the Large Magellanic Cloud. Because of its relative proximity to us (a mere 51,000 parsecs) SN 1987A is by far the best studied supernova of all times. *HST* did not miss its opportunity and the first images, taken with the FOC in August 23 and 24, 1990, revealed the inner circumstellar ring in all its "glory" and detail (*cf.* Jakobsen *et al.* 1991), showing that, despite spherical aberration, *HST* was not a complete disaster, after all.

Figure 3 - Combined B, V and I image of the central 35" field centered on SN 1987A (Panagia *et al.* 1998).

Since those early times, *HST* has assiduously kept an eye on SN 1987A, obtaining both imaging (*e.g.*, Fig. 3) and spectrographic observations at least once a year, accumulating valuable data and revealing quite a number of interesting results, both on the supernova itself (*e.g.*, Jakobsen *et al.* 1994, 1995, Pun *et al.* 1996), its circumstellar rings (*e.g.*, Panagia *et al.* 1991, Burrows *et al.* 1995, Plait *et al.* 1995, Panagia *et al.* 1996), and its companion stars (Scuderi *et al.* 1996).

3.1 Stellar Population around SN 1987A

SN 1987A is located at the SW edge of the Tarantula Nebula, some 20 arcminutes away from its center. The whole area contains a large number of early type stars interspersed with HII regions and SNR shells. Obviously, the study of SN 1987A neighborhood offers a unique opportunity to place the supernova explosion in the proper context of stellar evolution and the evolution of stellar populations. Thus, taking advantage of multiband observations made with the $WFPC2$ over the course of the years, Panagia, Romaniello *et al.* have carried out a systematic study of the stellar population within 30 *pc* of SN 1987A (Romaniello *et al.* 1998, Panagia *et al.* 1998, Romaniello 1998). An analysis of the stellar parameters accurately derived for about 10,000 stars reveals the presence of several generations of stars, with ages between 1 and 150 *Myr*, superposed on a much older field population. The youngest stars in the field appear to be *T Tau* stars, *i.e.* intermediate mass stars $(1\text{–}3\ M_{\odot})$ which are easily recognized because of their strong Hα excesses: this is the first time that such young stars are identified outside our Galaxy. It is clear that SN 1987A is associated with a region in which star formation has been active over a long time interval and is still very active at present.

3.2 Ejecta-Ring Interaction

The highest velocity material expelled in SN 1987A explosion has been detected for the first time by the Space Telescope Imaging Spectrograph (STIS). The spectrograph has found the first direct evidence for material from SN 1987A colliding with its inner circumstellar ring. The fastest debris, moving at 15,000 $km\ s^{1}$ or 1/20th the speed of light, are colliding with the slower moving gas which was ejected when the progenitor was a red supergiant (Sonneborn *et al.* 1998).

The top image, taken with Hubble's WFPC2 in 1995, shows the circumstellar rings surrounding Supernova 1987A. The glowing debris of the supernova explosion, is at the center of the inner ring. The small white square indicates the location of the STIS aperture used for the ultraviolet observation. The open square in this image is where the spectrum was masked to remove the emission from hydrogen in the Earth's outer atmosphere.

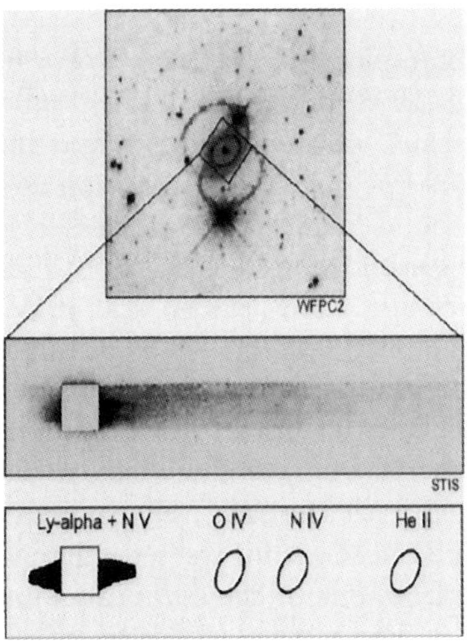

Figure 4 - The field of view centered on SN 1987A which was observed with STIS (top), and the resulting ultraviolet spectrum (Sonneborn *et al.* 1998).

The *STIS* data in the middle panel (and a schematic representation in the bottom panel) show the presence of ionized hydrogen expanding at a speed of 15,000 $km\ s^{1}$ coming from an extended area inside the inner ring. In addition to hydrogen emission, *STIS* also detected emission from high-velocity ionized nitrogen, as well as emission lines from oxygen, nitrogen, and helium ions, produced in the inner ring itself.

4 Planetary Nebulae

Although the final outbursts of ordinary stars are not as powerful as supernova explosions, they do create quite a show. A star with original mass lower than about 8 $M_{\odot}$ at the end of its nuclear burning life ejects a fraction of their mass (typically, say, 10%). The envelope gas is ionized by the remnant star, thus creating a *planetary nebula* (PN) which will shine for about 10,000 years, until either the central star fades away and/or the expanding gas becomes too tenuous to absorb a significant fraction of the ionizing radiation. *HST* observations of planetary nebulae have provided images of both great scientific interest and unsurpassed beauty. Here I shall limit myself to illustrating the case of M2-9.

M2-9 is a striking example of a "butterfly" or a bipolar planetary nebula (see Fig. 5; Balick *et al.* 1998) which justifiably could be dubbed the "Twin Jet Nebula". Indeed, because of the nebula's shape and the measured velocity of the gas, in excess of 300 $km\ s^{1}$, the description as a supersonic jet exhaust is quite appropriate. Ground-based studies have shown that the nebula's size increases with time, suggesting that the stellar outburst that formed the lobes occurred just 1,200 years ago. The central star in M2-9 is known to be the primary star in a close binary system. It is possible that one star is being engulfed by the other and that the gravity of one star pulls weakly bound gas from the surface of the other, thus forming a thin, dense disk which surrounds both stars and extends well into space. The disk can actually be seen in shorter exposure images obtained with the *HST*. Hydrodynamical models show that such a disk can successfully account for the jet-exhaust-like appearance of M2-9. The high-speed wind from one of the stars rams into the surrounding disk, which serves as a nozzle. The wind is deflected in a perpendicular direction and forms the pair of jets that we see in the nebula's image.

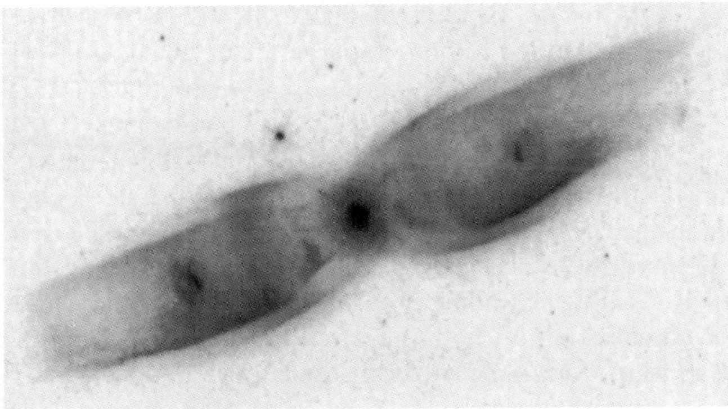

Figure 5 - WFPC2 image of the planetary nebula M 2-9 (Balick *et al.* 1998).

5 Gamma-Ray Bursts

Gamma-ray bursts are "mysterious" flashes of high-energy radiation that appear from random directions in space and typically last a few seconds. They were first discovered by U.S. Air Force Vela satellites in the 1960s. Since then, numerous theories of their origin have been proposed, but the causes of γ-ray bursts remain unknown. NASA's Compton Gamma-Ray Observatory (CGRO) satellite has detected several thousand bursts so far.

The principal limitation in understanding the bursts was the difficulty in pinpointing their direction on the sky: unlike visible light, γ-rays are exceedingly difficult to observe with a telescope, and the bursts' short duration exacerbates the problem. The Italian/Dutch satellite BeppoSAX, launched in 1996, had the ability to localize the bursts on the celestial sphere with a sufficient precision to permit follow-up observations with the world's most powerful ground-based telescopes. This breakthrough led to the discovery of long-lived "afterglows" of bursts in X-rays, visible and infrared light, and radio waves. While γ-ray bursts last only a few seconds, their afterglows can be studied for several months, allowing observers to probe the immediate environment of γ-ray burst sources and get clues to their nature.

The first γ-ray burst source studied with *HST* was GRB 970228 which Sahu *et al.* (1997) demonstrated to be associated with a faint (thus probably distant) late type galaxy and, by inference, with a region of active star formation. A few months later Pian *et al.* (1998) imaged the afterglow of another γ-ray burst, GRB 970508, with the $WFPC2$, again finding this source to be associated with a late type galaxy, at a redshift $z = 0.835$.

The most energetic one, the γ-ray burst GRB 971214, was detected on

December 14, 1997, by the BeppoSAX and CGRO satellites. BeppoSAX and NASA's Rossi X-ray Timing Explorer spacecraft detected an X-ray afterglow. BeppoSAX precision led to the detection of a visible light afterglow, found by Halpern *et al.*(1998) using the KPNO 2.4 m telescope, but no distance could be measured from those observations. As the visible light from the burst afterglow faded, an extremely faint galaxy ($R = 25.6 \pm 0.2$) was detected at its location, using the W.M. Keck Observatory 10-meter Keck II telescope on Mauna Kea, Hawaii (Kulkarni *et al.*1998) for which a redshift of z=3.42 was measured. Subsequent images taken with the Hubble Space Telescope confirmed the association of the burst afterglow with this faint galaxy. (Note that such a redshift correponds to a distance of about 12 billion light years distant if we assume the Universe to be about 14 billion years old). From the distance and the observed brightness of the burst, one can estimate the amount of energy released in the flash: $E \simeq 3 \times 10^{53}\ erg$.

At this stage the scenario of GRB sources being very energetic explosions (hypernovae? *e.g.*, Paczynski 1998) occurring at cosmological distances appeared "well established".

And then, GRB 980425 was discovered and part of the certainties were shaken and models had to be revised. In fact, a bright supernova, SN 1998bw, was found within the error box of the γ-ray burst source (Galama *et al.* 1998). The probable explosion date corresponded very well with the time of the γ-ray burst and, at the same time, the supernova presented a number of peculiarities to"justify" the suspicion that it may be the counterpart of GRB 980425: it has been classified as a type Ib/c SN (Sadler *et al.* 1998) but its absolute magnitude is more than 3 magnitudes brighter than expected for such supernova type, its radio emission was quite unusual in that it has reached unexpectedly high luminosities (almost $10^{29}\ erg\ s^{-1}\ \ Hz^{-1}$, *i.e.* 4 times higher than any other radio supernova; Weiler *et al.*1998) at unexpectedly early times (a few days after explosion). Worse of all, SN 1998bw is located in ESO 184-G82, a galaxy at an estimated distance of "only" $\sim$ 44 Mpc.

Clearly, these findings do not agree with the "simple" picture of ultra powerful sources at cosmological distances and more sophisticated models appear to be needed to explain this phenomenon. A possible way out is to attribute to beaming the apparently high energy deduced for distant GRBs, so that one could see them at large distances only if their beams were oriented in our direction, whereas one could see the nearby ones even under less favourable beaming conditions (Wang & Wheeler 1998).

Acknowledgments

Part of the material presented in this review was extracted from the invaluable set of STScI press releases which are produced with support to Space Telescope Science Institute, operated by the Association of Universities for Research in Astronomy, Inc., from NASA contract NAS5-2655 and are reproduced with permission from AURA/STScI.

References

1. B. Balick, *et al.*, 1998, in preparation.
2. C. Burrows *et al.*, 1995, ApJ, 452, 680.
3. T.J Galama *et al.*, 1998, IAUC 6895.
4. J.P Halpern, J.R. Thorstensen, D. Helfand, & E. Costa, 1998, Nature, 393, 41.
5. P. Jakobsen, *et al.*, 1991 ApJ, 369, L59.
6. S. Kulkarni, *et al.*1998, Nature, 393, 35.
7. M. Livio, S.M. Fall, & P. Madau, (editors) 1998, *The Hubble Deep Field*, Cambridge University Press, in press.
8. B. Paczynsky 1998, ApJ, 484, L45.
9. N. Panagia, *et al.*, 1991, ApJ, 380, L23.
10. N. Panagia, *et al.*, 1996, ApJ, 459, L17.
11. N. Panagia, *et al.*, 1998, ApJ, in press.
12. E. Pian, *et al.*, 1998, ApJ, 492, L103.
13. P. Plait, *et al.*, 1995, ApJ, 439, 730.
14. M. Romaniello, 1998, PhD Thesis, Scuola Normale Superiore, Pisa.
15. M. Romaniello, N. Panagia, S. Scuderi, S., & the SINS Collaboration, 1998, in *The Magellanic Clouds and Other Dwarf Galaxies*, eds. R. Dettmar and U. Klein, Springer Verlag, in press.
16. E.M. Sadler, R.A. Stathakis, B.J. Boyle, & R.D. Ekers, 1998, IAUC 6899.
17. K. Sahu, *et al.*, 1997, ApJ, 489, L127
18. S. Scuderi, *et al.*, 1996, ApJ, 465, 956.
19. G. Sonneborn *et al.*, 1998, ApJ, 492, L139
20. L. Wang & J.C. Wheeler, 1998, ApJ, 504, L87.
21. K.W. Weiler, S.D. Van Dyk, M.J. Montes, N. Panagia, & R.A. Sramek, 1998, ApJ, 500, 51.
22. B. Whitmore, *et al.*, 1998, in preparation.
23. R.E. Williams, *et al.*, 1996, AJ, 112, 1335.

TWO-NEUTRON INTERFEROMETRY IN LOW- AND INTERMEDIATE-ENERGY HEAVY ION REACTIONS

N. COLONNA[a], R. GHETTI[a,b], B. JAKOBSSON[b], E. DEFILIPPO[c], G. TAGLIENTE[a] V. BELLINI[c], L. CARLEN[b], S. CAVALLARO[c], L. CELANO[a], G. D'ERASMO[a], D. DI SANTO[a], E.M. FIORE[a], A. FOKIN[b], M. GERACI[C], J. HELGESSON[d], A. KUZNETSOV[e], G. LANZANO'[c], J. MAARTENSSON[b], D. MAHBOUB[c], A. PAGANO[c], F. PALAZZOLO[c], M. PALOMBA[a], A. PANTALEO[a], V. PATICCHIO[a], R. POTENZA[C], G. RIERA[c], V. RUSSO[c], L. SPERDUTO[c], C. SUTERA[c], M. URRATA[c], L. WESTERBERG[f]
The CHIC collaboration
[a]*INFN and Dip. Fisica, V. Amendola 173, 70126 Bari, Italy*
E-mail:colonna@bari.infn.it
[b]*Dept. of Cosmic and Subatomic Physics, Lund Univ. Sweden*
[c]*INFN (LNS and Sez.) and Dip. Fisica, Catania, Catania, Italy*
[d]*ECT*, Trento, Italy*
[e]*Khlopin Radium Inst., St. Petersburg, Russia*
[f]*The Svedberg Laboratory, Uppsala, Sweden*

Two-nucleon interferometry can be used in low- and intermediate-energy heavy ion reactions to probe the space-time evolution of hot nuclear systems. While angle- and momentum-gated correlation functions measured in a Tandem experiment have allowed to successfully extract compound nucleus lifetime, at intermediate energy n-n interferometry can provide information on the spacial extension of a hot and possibly compressed nuclear system.

1 Introduction

Since it was first proposed by Hanbury-Brown and Twiss [1], the use of intensity interferometry has been extended from astrophysics, where it can be applied to determine the dimension of stellar objects, to nuclear and particle physics, where it can provide information on the space-time evolution of excited nuclear systems [2]. The relative wave function of identical nucleons is influenced by the Quantum-Statistical Interference (QSI), the short range Final State nuclear Interaction (FSI) and, for charged pairs, the long-range repulsive Coulomb Interaction (CI). Since the magnitude of these effects depends on the separation between particles, two-nucleon interferometry can provide an estimate of the radius and lifetime of a decaying system, with the space-time ambiguity resolved by means of a selection on the angle between the relative and the total momenta of the particle pair [3,4,5].

In low energy heavy ion reactions, both two-proton and two-neutron interferometry can be applied to study the lifetime of compound nuclei [6,7]. However, the presence of strong distortions introduced by the Coulomb interaction between the proton pair and the evaporation residue complicates the theoretical interpretation of two-proton interferometry. On the other hand, two-neutron correlation functions are easier to interpret, but are affected by several experimental problems, such as detector cross-talk, in- and out-scattering and low detection efficiency. An accurate study of these problems, and the availability of complex neutron detection systems, has recently opened the way to measurements of reliable angle- and momentum-gated two-neutron correlation functions at low energy [8,9].

At intermediate energy, where the dynamical evolution of the reaction is more complicated and still somewhat unknown, two-neutron interferometry can be applied to verify processes, such as expansion or the formation of hot fireball-like regions, whose occurrence is predicted by theoretical models or suggested by indirect experimental evidences [10]. In this energy region, however, the clarification power of the correlation functions may be hampered by the complicated dynamical evolution of the reaction, which depends on the impact parameter and often results in sources of vastly different characteristics. Therefore, a selection of the events and a characterization of the emitting sources is mandatory in the Fermi-energy domain to disentangle the contribution of the different emission sources.

In this talk, we present the results of a study of two-neutron interferometry at Tandem energy, together with the first, preliminary results of an experiment in the Fermi energy domain, recently performed at the Superconducting Cyclotron of Laboratorio Nazionali del Sud, Catania. In both experiments, a modular neutron apparatus has been employed, alone or in conjunction with other detector systems. The individual neutron detector consists of a BC501A liquid scintillator cell, 12.7 cm in diameter by 12.7 cm thickness. To minimize background, the detectors are mounted on a light, aluminum support structure, with a thin-walled aluminum spherical scattering chamber 1 m in diameter. More details on the individual detectors, as well as on the performance of the whole apparatus can be found in ref. [11].

2 Two-neutron correlation functions and compound nucleus lifetime

A study of two-neutron interferometry at low energy was performed by the Bari Nuclear Physics group at the Tandem Accelerator of Laboratori Nazionali di Legnaro, Italy. Correlation functions were measured for neutrons emitted in the 130 MeV ^{18}O + ^{26}Mg reaction. For this system, a large part of the total reaction cross-section ($\sim$ 90 %) goes into fusion, with the formation of ^{44}Ca

compound nucleus at 100 MeV excitation energy. The formation of a unique source with well defined velocity is a fundamental requisite for the analysis of angle-gated correlation functions, since transforming the neutron momenta into the frame of a well-characterized source is crucial for detecting lifetime effects [5]. The experimental apparatus consisted of 22 neutron detectors, 18 of which mounted in a close-packed configuration, with angular separation between detectors of $\sim$ 6 degrees, so as to allow the measurement of small-angle correlation functions. The cluster was positioned at 45 degrees, at a distance of 2 meters from the target. In this condition, only neutrons emitted from the compound nucleus were detected.

When studying two-neutron correlation functions, particular attention has to be paid to the problem of cross-talk, that is neutron scattering from one detector into a neighboring one, which gives rise to a false coincidence. For low multiplicity and for small separation between detectors, cross-talk might produce large distortions of the measured correlation functions. Several methods can be used to identify and reject cross-talk events. The one used in this analysis, described in more details in ref. [11], relies purely on the time-of-flight measurement. A very restrictive energy conservation condition is applied which takes into account the time resolution, the thickness of the detectors and their finite transverse dimension. According to this procedure, a coincidence between two detectors is attributed to cross-talk, and therefore rejected, if it fulfills the following relation:

$$E_{diff} = E_1 - \frac{1}{2}m\left(\frac{d_{min}}{\Delta t}\right)^2 > 0. \tag{1}$$

with E_1 the energy of the first (or fastest) neutron, d_{min} the minimum distance between the struck detectors and Δt the difference between the time signals from the two detectors. Monte Carlo simulations and measurements with the ^{241}Am/^{9}Be source indicate that this method is very efficient in rejecting cross-talk events. However, along with cross-talk neutrons a large number of true two-neutron coincidences is also rejected, with a consequent distortion of the neutron energy distribution which has to be considered in the analysis of two-neutron correlation functions.

Experimental angle-gated correlation functions for neutrons emitted in the 130 MeV ^{18}O + ^{26}Mg reaction are shown in fig. 1. Cross-talk events, as well as real events identified as cross-talk, were singled out and rejected in the numerator of the correlation function. The singles technique was used to construct the denominator. Although single spectra are not strongly affected by cross-talk, the rejection procedure (1) was also applied to the uncorrelated pairs of the denominator, to compensate for the real coincidences rejected in the numerator.

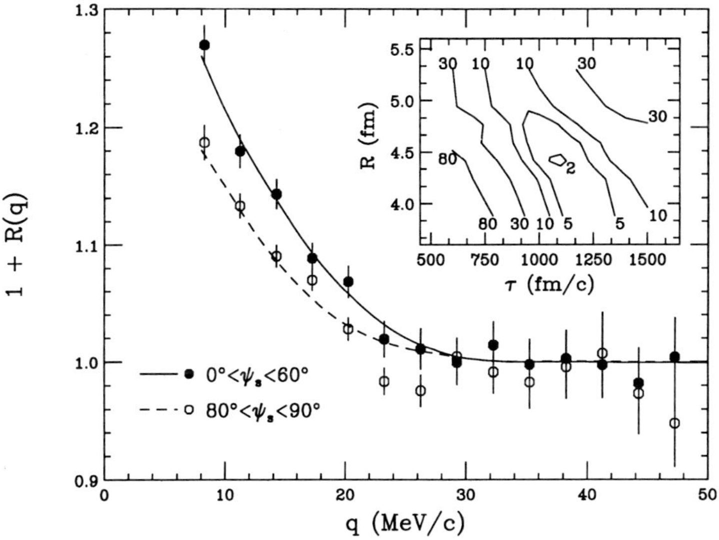

Figure 1: Experimental longitudinal (solid symbols) and transverse (open symbols) correlation functions. The solid and dashed curves depict calculations (see text for details) with a source radius R = 4.4 fm and lifetime of τ = 1100 fm/c. The panel in the figure is a contour diagram of the reduced χ^2 between the experimental and theoretical correlation functions in the region of q = 0 - 20 MeV/c.

The solid and empty symbols in the figure represent respectively longitudinal and transverse correlation functions, constructed by gating on the angle ψ between the relative and the total momentum of the neutron pair, the latter evaluated in the rest frame of the ^{44}Ca compound nucleus. As expected, transverse correlation functions are suppressed relative to the longitudinal ones, because the fermionic anticorrelation from quantum statistics is stronger and reduces the positive correlation from the attractive final state interaction. (see ref. [5] for more details). The magnitude of both correlation functions, together with the difference between them, were exploited to obtain information on the source dimension and on the neutron emission time-scale. Quantitative information was obtained by comparing the measured correlation functions with theoretical calculations performed with a formalism which incorporates both quantum-statistical interference and final-state nuclear interaction [3,12]. In the calculations, neutrons are assumed to be emitted from a hard sphere source with a characteristic radius R and emission time scale τ [13]. All resolutions were included in the calculations, and the cross-talk rejection procedure described above was applied to the theoretical neutron pairs, so as to take into account the bias in the neutron energy spectrum introduced by the procedure. The χ^2 between experimental and theoretical correlation functions was computed for different combinations of source radii and emission time scales (panel in fig. 1). The minimum value of χ^2 is obtained for a radius of 4.4 fm and an emission time scale of $\sim$ 1100 fm/c. The corresponding correlation functions are shown by the solid and dashed curves in the figure.

The extracted value of 1100 fm/c should be considered only as an average value of the emission time scale. As the compound nucleus evaporates nucleons, its temperature decreases and, consequently, its lifetime is expected to increase. A direct evidence of this cooling effect on the compound nucleus lifetime can be obtained by constructing correlation functions gated on the total momentum of the neutron pair. The results, compared with Koonin-Pratt calculations, are shown in fig. 2. Neutron pairs with high total momentum, presumably emitted early in the decay chain, when the system still retains much of its initial excitation energy, are characterized by an emission time scale sensibly shorter than the average value. On the contrary, pairs with low total momentum, emitted along the whole decay chain, show an emission time slightly larger than the overall emission time, as expected for an averagely cooler nucleus. This trend represents a clear experimental evidence of the cooling effects on the lifetime of compound nuclei.

3 Two-nucleon interferometry in the Fermi energy domain

While at low energy the reaction picture is sufficiently clear, at intermediate energies ($10 < E/A < 100$ MeV) the evolution of the reaction is far from being understood. Most studies suggest a strong dependence of the reaction features on the incident energy as well as on the impact parameter and mass asymmetry of the reacting nuclei. In general, particle emission is characterized by a fast pre-equilibrium stage followed by a slower emission from an equilibrated system, but neither the initial stages of the reaction, nor the evolution of the system on its way to deexcitation have been consistently described. Some aspects of the reaction dynamics in the Fermi-energy domain might be clarified with the help of two-nucleon interferometry. In fact, several studies have already been performed with p-p, and to a lesser extent, with n-n and n-p correlation functions [5,14,15,16], but an exhaustive picture of the reaction mechanism is yet to come. In particular, a unique interpretation of the interferometry studies performed with different particle pairs in this energy region

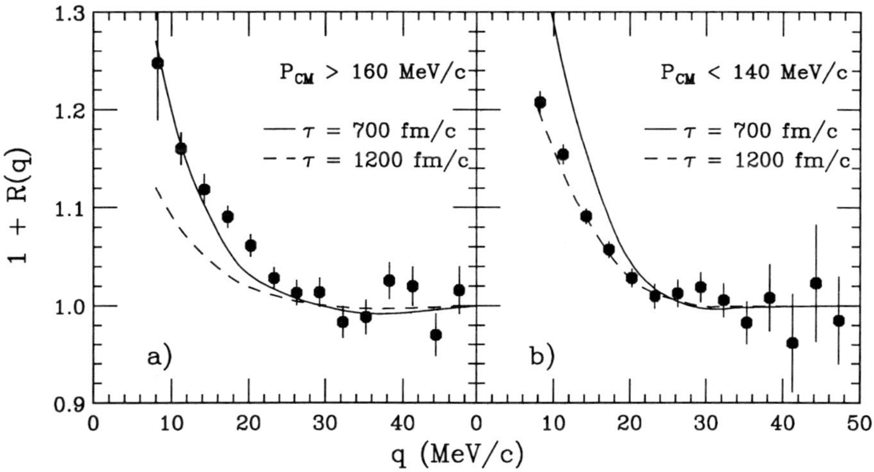

Figure 2: Angle-integrated correlation functions for two cuts on the total neutron pair momentum in the compound nucleus frame. The solid and dashed curves are results of theoretical calculations with the indicated emission time scales.

has not been fully achieved, although the different sensitivity of n-n, n-p and p-p correlation functions to the QSI, FSI and CI should allow to extract more reliable information on the space-time evolution of excited nuclear systems [16]. For a consistent understanding of the correlation functions, however, it is fundamental in this energy region to select and identify the different classes of events, since the emission sources produced in the reaction present characteristics largely dependent, for example, on the impact parameter. A measurement of the fragment decay channel, in coincidence with the nucleon pair, may help correlate the results of interferometry studies to the properties of the emission sources, in particular to their excitation energies.

With the aim of studying the space-time evolution of the reaction zone in intermediate energy heavy ion reactions, a collaboration between the the nuclear physics group in Bari, a group from INFN-Catania and the CHIC collaboration [17], has recently undertaken an experimental investigation of 45 MeV/u ^{58}Ni-induced reactions at the Superconducting Cyclotron of Laboratori Nazionali del Sud, Catania, Italy. A complex apparatus has been put together in order to simultaneously measure n-n, p-p and n-p correlation functions in coincidence with the fragment decay channel. In brief, the setup is here described:

- The neutron apparatus consists of 48 liquid scintillator cells, of the same type used for the measurement at low energy. The detectors are mounted in 4 clusters placed at 2.7 m from the target at angular positions of 90 and 45 degrees on one side of the beam, and 45 and 25 degrees on the other side. Each cluster is made of 12 detectors arranged in a matrix of 5x5 cells, with two consecutive detectors separated by an empty cell. A minimum separation of 5 degrees between adjacent detectors is achieved in the setup.

- The proton apparatus used in the experiment, the EMRIC array [18], consists of 13 CsI crystals, with face of 4x4 cm^2 and 10 cm thickness. The array is positioned at 45 degrees and at 70 cm from the target, just outside a thin mylar window mounted on the scattering chamber. Each detector is placed in correspondence of an empty cell of one neutron cluster, so to allow a measurement of n-p correlation functions, together with two-proton interferometry.

- A fundamental addition to the apparatus is represented in these measurements by the ARGOS [19] detector, a forward wall of 32 high-resolution phoswich detectors made of a thin plastic scintillator foil (up to 700 μm) optically coupled to a 10 cm thick BaF_2 crystal. The detectors, with face of 25 cm^2 and 10 cm thickness, are placed at 2 m from the target and cover between 1 and 8 degrees around the beam, thus allowing the measurement of fragments emitted in projectile fragmentation reactions. In reverse kinematic reactions, the array can also be employed to characterize central collisions.

Fig. 3 shows the integrated two-neutron correlation function measured in the 45 MeV/u ^{58}Ni + ^{27}Al reaction, in coincidence with at least one fragment in the forward wall. The strong correlation measured for small relative momenta indicates a source of small space-time extent. To extract more detailed informations, we have compared the results with theoretical predictions from models used to generate the neutron phase-space distribution. The solid line in the figure represents the results of calculations assuming time-dependent evaporative emission according to the Friedman-Lynch formalism [20], from a compound nuclear system of size A=80, Z=39 and initial excitation energy of 10 MeV/u. A reasonable agreement is obtained between the data and the calculations, which predict emission on a time-scale of $\sim$ 100 fm/c from a system close to normal density. This result is not surprising, considering that the available energy in the center-of-mass for this system is not extremely high, and only a moderate thermal and mechanical pressure may be produced in the reaction. Furthermore, since a selection of the emitting source has not yet been performed, the shown correlation function is the result of an averaging

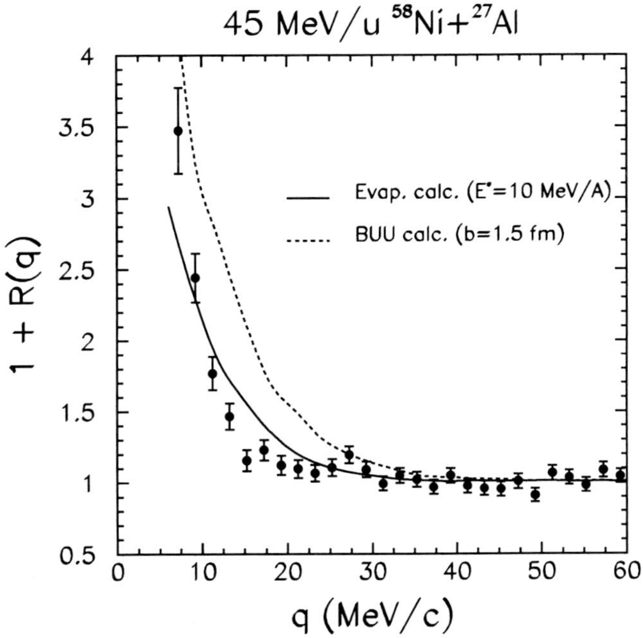

Figure 3: Experimental integrated two-neutron correlation function (solid symbols), measured in coincidence with the forward fragment array in the 45 MeV/u ^{58}Ni + ^{27}Al reaction. The curve represents calculations with two complementary approaches (see text). The theoretical predictions have not been corrected for energy and angular resulution nor for cross-talk effects.

over the whole range of inpact parameters, and may be affected by the less violent peripheral collisions.

Although a good agreement is obtained with the evaporative model both for the correlation function and the neutron energy spectra, the choice of the input parameters may mask a shortcoming of the evaporative approach, that is the absence of preequilibrium effects, which are expected to play an important role in this energy region. To check the influence of preequilibrium emission on the correlation functions, we have simulated the early stages of the reaction (up to 200 fm/c) in within the BUU approach [21]. The results for an impact parameter of 1.5 fm are shown by the dashed line in the figure. The calculations reproduce the general features of the correlation function, considering that no impact parameter averaging was performed in the simulations. Most probably, however, the overestimate of the correlation function (as well as of the energy distribution) in the BUU approach may simply reflect the importance of slower evaporation.

Although the comparison of two complementary approaches indicates that a combination of preequilibrium and evaporative process is necessary to account for the measured correlation functions and energy spectra, the results obtained so far do not provide enough information to discriminate between the different models. A comparison with p-p and n-p correlation functions and, expecially, a more accurate event selection, will probably help clarify the theoretical interpretation. Furthermore, correlation functions gated on the total momentum of the neutron pair may help disentangle the relative contribution of preequilibrium and equilibrium emission.

4 Conclusions

Two-nucleon interferometry can be used in low- and intermediate-energy heavy ion reactions to study the space-time properties of excited nuclear matter. At low energy multidimensional two-neutron correlation functions have been successfully used to extract the lifetime of a moderately excited compound nucleus, which in turn can be used to obtain direct information on the total decay width of the compound nucleus. Experimental evidences of cooling effects on particle emission timescale have been also obtained by means of momentum-gated correlation functions.

At intermediate energy, two-neutron correlation functions measured in coincidence with fragments detected in the forward direction indicate emission from a source at normal density on a fast timescale ($\sim$ 100 fm/c). Preliminary comparisons to theoretical calculations suggest that a careful analysis of both single particle spectra and correlation functions with different cuts on the total pair momentum and accurate event identification is needed in order to disentangle the different contributions of mechanisms such as pre-equilibrium, flow, evaporation, temperature gradient, impact parameter averaging etc. The analysis of identical and mixed-pairs correlation functions, made possible by the complete apparatus used at LNS, Catania, will hopefully allow to extract more reliable information on these aspects of the reaction dynamics.

Acknowledgments

The authors wish to thank the technical staff of INFN-Bari and Catania for their efficient support in mounting and testing of the neutron apparatus. The authors acknowledge the excellent accelerator staff of LNS for providing a high-intensity, high-quality ^{58}Ni beam.

References

1. R. Hanbury-Brown and R.Q. Twiss, Nature **178**, 1046 (1956).
2. D.H. Boal, C.K. Gelbke, and B.K. Jennings, Rev. Mod. Phys. **62**, 553 (1990), and references therein.
3. S. Pratt and M.B. Tsang, Phys. Rev. C **36**, 2390 (1987).
4. W.G. Gong et al., Phys. Rev. C **43**, 1804 (1991).
5. M.A. Lisa et al., Phys. Rev. Lett. **71**, 2863 (1993).
6. W. Dunnweber et al., Phys. Rev. Lett. **65**, 297 (1990).
7. A. Elmani et al., Phys. Rev. C **49**, 284 (1994).
8. N. Colonna et al., Phys. Rev. Lett. **75**, 4190 (1995).
9. P.A. DeYoung et al., Nucl. Phys. A **597**, 127 (1996).
10. N. Marie et al., Phys. Lett. B **391**, 15 (1997).
11. N. Colonna et al., Nucl. Instr. and Meth. A **381**, 472 (1996).
12. W. Bauer, C.K. Gelbke, and S. Pratt, Ann.Rev. Nucl. Part. Sci. **42**, 77, (1992), and references therein.
13. M.A. Lisa et al., Phys. Rev. C **49**, 2788 (1994).
14. B. Jakobsson et al., Phys. Rev. C **44**, R1238 (1991).
15. D. Rebreyend et al., Phys. Rev. C **46**, 2387 (1992).
16. M. Cronqvist et al., Phys. Lett. B **317**, 505 (1993).
17. The CHIC collaboration (spokeperson B. Jakobsson, Lund U., Sweden) includes institutions from Sweden, Russia, Japan, France and Italy.
18. F. Merchez et al., Nucl. Instr. and Meth. A **275**, 133 (1989).
19. G. Lanzano et al., Nucl. Instr. and Meth. A **323**, 694 (1992).
20. W.A. Friedman and W.G. Lynch, Phys. Rev. C **28**, 16 (1983).
21. B. Li and W. Bauer, Phys. Rev. C **44**, 450 (1991).

Two-neutron HBT interferometry to measure time-scales in the reaction ^{40}Ar + ^{165}Ho at E/A = 25 MeV

S.J. Gaff
Triangle Universities Nuclear Laboratory
Duke University
Box 90308
Durham, NC 27708-0308 USA
E-mail: gaff@tunl.duke.edu

As a test for the time scale of neutron emission, two-neutron HBT interferometry was applied to the reaction ^{40}Ar+ ^{165}Ho at E/A = 25 MeV. After appropriate corrections for neutron scattering, the correlation function showed a lifetime that was about a factor of two shorter than the lifetime predicted for the statistical decay of an equilibrated nucleus. Preequilibrium emission may explain some of the difference.

1 Introduction

In nuclear physics an important topic is understanding how nuclear reactions proceed and the motion of the nucleons in these reactions. One important characteristic of a reaction is the time it takes for individual steps in the evolution of the nucleus. HBT intensity interferometry provides a useful technique to be able to describe the lifetime of a compound nucleus as it evaporates particles.

Two-neutron intensity interferometry was applied to the system ^{40}Ar + ^{165}Ho at E/A = 25 MeV. [3,4,5,6,7] In this system, for more central events, a compound nucleus is expected to be formed which will decay by evaporating neutrons. Two neutron intensity interferometry eliminates the Coulomb interaction with the residual nucleus which might be significant in this type of reaction if charged particles were used.

2 Experimental Details

The experiment was carried out at the National Superconducting Cyclotron Laboratory (NSCL) at Michigan State University (MSU).[13] The Ar beam with E/A = 25 MeV impinged on the Ho target. Si telescopes surrounded the target at 30° and 45° which provided a start time and a rough centrality filter, as a fragment with $Z > 2$ was required for each event.[8] The neutrons were detected in the NSCL Neutron Walls which are each made up of 25 glass cells filled with liquid scintillator.[9] Photomultiplier tubes at each end of the cells provide time and pulse height signals. Pulse shape discrimination is used to select the neutron events.

The neutron data had to be corrected for two main scattering backgrounds. One correction is for crosstalk where one neutron interacts and produces a light signal in multiple detectors. The other correction is for background scattering. Since this experiment was done in a relatively small room some neutrons hit the detector after scattering from material in the room.

The correction for crosstalk is based on the technique suggested in Wang.[10] This technique assumes that all crosstalk comes from neutrons that scatter elastically from protons which is the case for most of the low energy neutrons (E $<$ 40 MeV) measured. Elastic scattering kinematics give the expected energy, $E_{n'}$,and angle, θ',for the scattered neutron as:

$$E_{n'} = E_n - E_p, \tag{1}$$

$$\cos^2\theta' = \frac{E_{n'}}{E_n} \tag{2}$$

where E_n is the incident neutron energy, and E_p is the scattered proton's energy. The neutron energy is calculated from the time-of-flight, and the proton energy is found from the amplitude of the measured light signal. Then these calculated values are compared to the measured values and criteria set to determine the most likely crosstalk events. Comparisons are made between the measured and calculated angle, $\cos\theta$ vs $\cos\theta'$, the measured and calculated time between hits, $(t_2 - t_1)$ vs $t_{n'}$, and the measured and calculated light signal in the second detector, L_2 vs $L_{n'}^{max}$. Where $L_{n'}^{max}$ is the most light a neutron with energy $E_{n'}$ can leave in a detector. A cut of $|\vec{P}_{tot}| < 250$ MeV/c is applied to the correlation function because higher energy neutrons are more likely to include crosstalk.

The correlation function is constructed from the data using a singles technique:

$$\sum Y_2(\mathbf{p_1}, \mathbf{p_2}) = C[1 + R(q)] \sum Y_1(\mathbf{p_1})Y_1(\mathbf{p_2}) \tag{3}$$

where $Y_2(\mathbf{p}_1, \mathbf{p}_2)$ is the coincident neutron yield and $Y_1(\mathbf{p})$ is the singles yield. Both sums are over all detector combinations for a given relative momentum,$q = \frac{1}{2}|\mathbf{p_1} - \mathbf{p_2}|$. When the crosstalk cuts are made the lowest relative momentum bins are not changed, but the correlation function around q = 10 has a large, anomalous peak that is removed as shown in Figure 1.

In order to understand how the crosstalk cuts effect the data, we looked at a GEANT simulation.[11] The simulation includes the detector geometry and all possible interactions of the neutrons with the scintillator. The neutron energies are sampled form the experimental singles spectra. The results from the simulation, especially the effects of crosstalk on various distributions are very similar to the data.

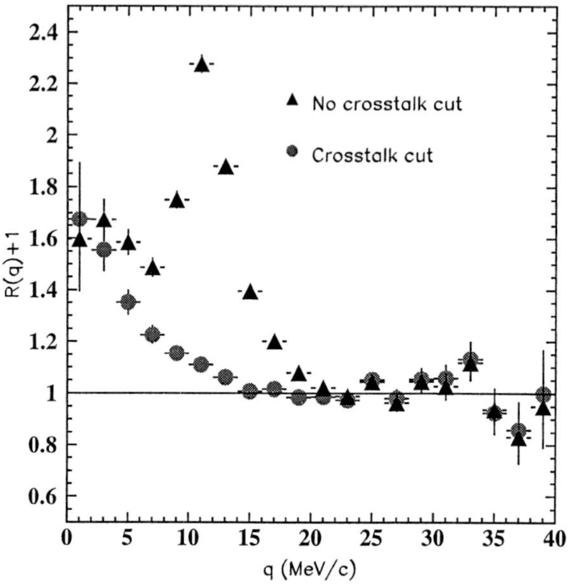

Figure 1: Correlation function before and after crosstalk corrections. Events are selected with $|\vec{P}_{tot}| < 250$ MeV/c and neutron energies E > 5 MeV.

Corrections are necessary for the background scattering. The background is reduced by requiring E > 5 MeV for each neutron and eliminating events where the light measured is greater than the largest response possible from the reconstructed energy. Even with these cuts about 10% of the singles are still background, as measured during shadow bar runs. These background neutrons suppress the true correlation. A corrected correlation function can be defined as

$$1 + R'(q) = \frac{\sum Y_2(\mathbf{p_1}, \mathbf{p_2}) - \alpha \sum Y_1(\mathbf{p_1}) Y_1(\mathbf{p_2}) b(\mathbf{p_1}, \mathbf{p_2})}{\alpha \sum Y_1(\mathbf{p_1}) Y_1(\mathbf{p_2})(1 - b(\mathbf{p_1}, \mathbf{p_2}))}. \tag{4}$$

Here $Y_2(\mathbf{p_1}, \mathbf{p_2})$, $Y_1(\mathbf{p})$ are the same as in equation (3).[12] The function $b(\mathbf{p_1}, \mathbf{p_2})$ is the fraction of times that an event measuring neutrons with momenta $\mathbf{p_1}$ and $\mathbf{p_2}$ would be a background event. The function $(1 - b(\mathbf{p_1}, \mathbf{p_2}))$ is the probability an event is a good event, and α is the normalization chosen so that $\langle R'(q) \rangle = 0$ for $16 \leq q \leq 40$ MeV/c. This correction enhances the correlation function by about 10% as shown in Figure 2.

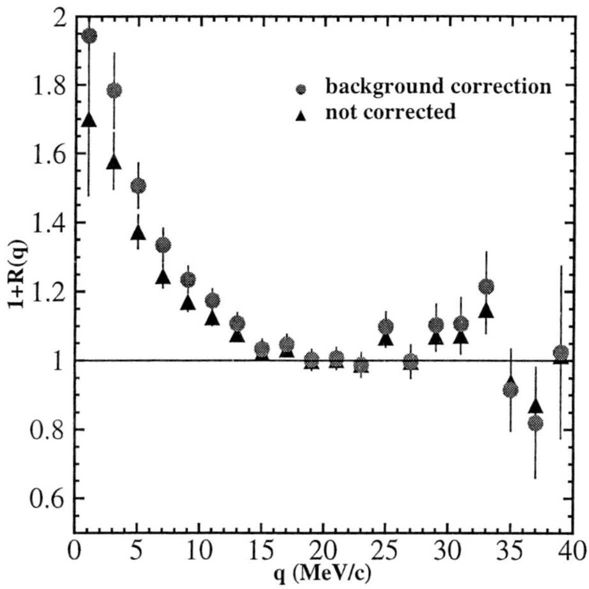

Figure 2: Correlation functions corrected for crosstalk, before and after background correction. Cuts are the same as in Figure 1.

3 Interpretation of Results

After making all the corrections it is necessary to compare the result to a model in order to extract a lifetime. Results are shown in Figure 3 for a model with surface emission from a sphere with a radius r = 7 fm. The energy for each neutron is chosen from the efficiency-corrected energy spectrum and is independent of the time which is chosen from an exponential distribution with a lifetime τ. The model events are filtered through a simulation of detector effects and the crosstalk cuts are applied. A correlation function is produced from the simulated singles events by the Koonin-Pratt formalism. [14,15] The best fit is for the case where τ = 700 fm/c.

As statistical models are often used to predict lifetimes for this type of reaction, one based on Friedman and Lynch[16] is compared to the data. The level density parameter is a=A/10 and the number of nucleons in the compound source is equal to the total number available in the reaction. The total possible excitation energy of the system E^*/A = 3.93 MeV is used as an upper limit. As this value gives an energy spectrum which is too shallow, as shown in Figure 4, a lower limit for the excitation energy is determined by fitting the lower energy

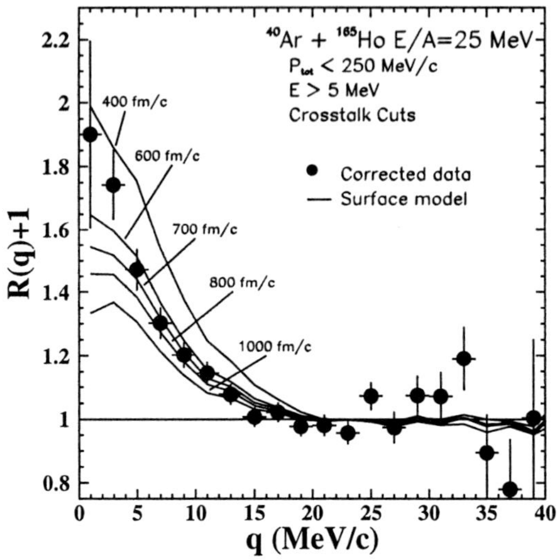

Figure 3: Experimental correlation function compared to results for different lifetimes from a source emitting neutrons from the surface of a r = 7 fm sphere. The model calculations include the detector response and crosstalk corrections.

part of the energy spectrum. The correlation function in both cases is smaller than the measured correlation function as shown in Figure 5.

One reason the measured correlation function may be larger, show a shorter lifetime, is because of preequilibrium neutrons. To study the effects of preequilibrium neutrons a second component was added to the evaporation model. Each component is defined by a temperature, or energy slope parameter. The compound-source component is taken from the evaporation model while the preequilibrium component has an energy sampled from an exponential distribution with temperature T_p and source points from the surface of a sphere. The fraction of preequilibrium neutrons is defined between 5 and 20 MeV,

$$f_p = \frac{\int_{5MeV}^{20MeV} Ne^{-E/T_p} dE}{\int_{5MeV}^{20MeV} (e^{-E/T_c} + Ne^{-E/T_p}) dE}, \tag{5}$$

where T_c, the temperature of the compound-source component, determines the excitation energy. Considering fits to the energy spectrum which have similar χ^2 values, the best fit to the experimental correlation function is for $f_p = 25\%$. See Figure 6.

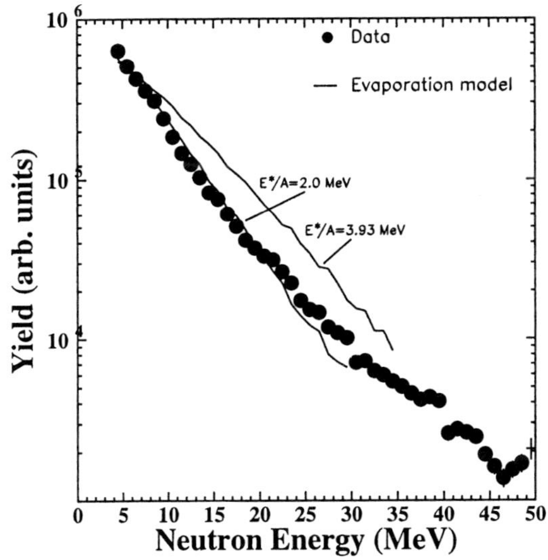

Figure 4: Experimental energy spectrum corrected for efficiency compared to the energy spectra from an evaporation model for different excitation energies.

To predict a reasonable fraction for the preequilibrium contribution, calculations based on the Boltzman-Uehling-Uhlenbeck equations were used to find the slope parameter for the preequilibrium temperature.[17,18,19] This slope combined with the experimental energy spectrum predicts 18% preequilibrium. Constructing the correlation function for this amount of preequilibrium produces a correlation function that is too small as shown in Figure 6. Assuming 18% preequilibrium neutrons, the evaporation lifetime may be scaled by some factor. The correlation between energy and emission time from the model is maintained, but the time for each evaporation neutron is scaled by a factor s, $t = st_e$. The best fit to the experimental correlation function is for s = 0.4 as seen in Figure 7. In the model if the neutrons are emitted about a factor of two slower then the correlation function agrees with the data.

4 Conclusion

The correlation function was measured for the reaction ^{40}Ar + ^{165}Ho at E/A = 25 MeV. Crosstalk was effectively removed for the low momentum pairs. Surface emission from a sphere with a radius of 7 fm gives a lifetime of $\tau =$

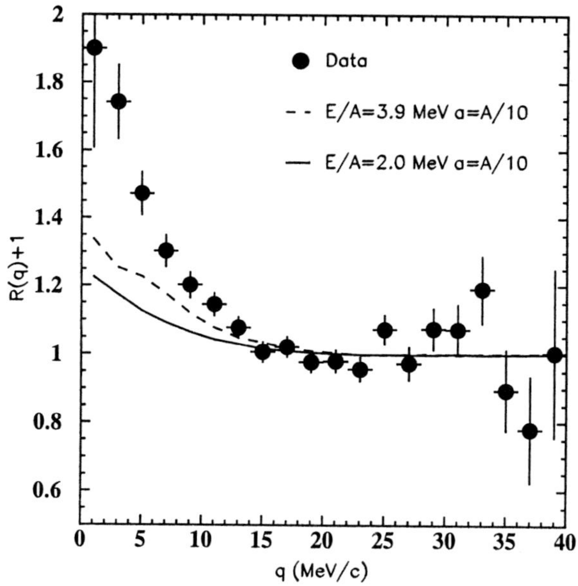

Figure 5: Experimental correlation function compared to the results from an evaporation model. The level density parameter is a=A/10.

700± 200 fm/c. The evaporation model results are within a factor of two of the measured lifetime.

Acknowledgments

This work was supported by the National Science Foundation under Grant No. PHY-95-28844.

References

1. W.G. Gong, Y.D. Kim and C.K. Gelbke, Phys. Rev. C **45**, 863 (1992).
2. M.A. Lisa, W.G. Gong, C.K. Gelbke, N. Carlin, R.T. deSouza, Y.D. Kim,W.G. Lynch, T.Murakami, G. Poggi, M.B. Tsang, and H.M. Xu, Phys. Rev. C **49**, 2788 (1994).
3. P.A. DeYoung, M.S. Cordon, Xiu qin Lu, R.L. McGrath, J. M. Alexander, D.M. de Castro Rizzo, and L.C. Vaz, Phys. Rev. C **39**, 128 (1989).
4. B. Jakobsson, B. Norén, A. Oskarsson, M. Westenius, M. Cronqvist, S. Mattson, M. Rydehell, Ö. Skeppstedt, J.C. Gondrand, B. Khelfaoui, S.

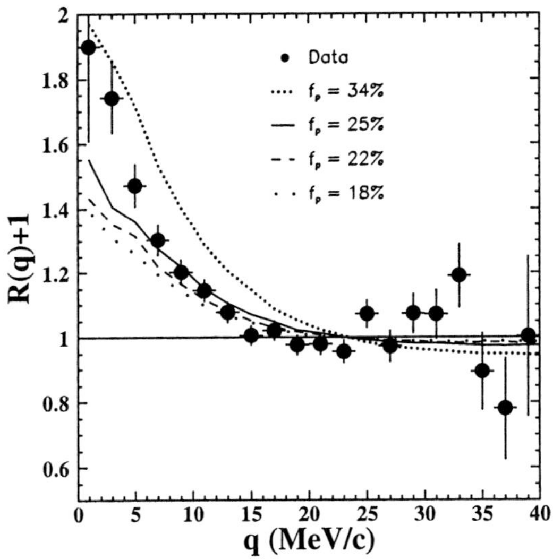

Figure 6: Experimental correlation function compared to the results from a hybrid model with evaporation plus a preequilibrium component. Various fractions of preequilibrium neutrons are shown including 18% which is what the BUU calculations predict.

Kox, F. Merchez, C. Perrin, D. Rebreyend, L. Westerberg and S. Pratt, Phys. Rev. C **44**, R1238 (1991).

5. R. Ghetti, L. Carlén, M. Cronqvist, B. Jakobsson, F. Merchez, B. Norén, D. Rebreyend, M. Rydehell, Ö. Skeppstedt and L. Westerberg, NIM **335**, 156 (1993).
6. N. Colonna, D.R. Bowman, L. Celano, G. D'Erasmo, E.M. Fiore, A. Pantaleo, V. Paticchio, G. Tagiente, and S. Pratt, Phys. Rev. Lett. **75**, 4190 (1995).
7. D.A. DeYoung, R. Bennink, T. Butler, W. Chung, C. Dykstra, G. Gilfoyle, J. Hinnefeld, M. Kaplan, J.J. Kolata, R.A. Kryger, J. Kugi, C. Mader, M. Nimchek, P. Santi, A. Snyder, Nuc Phys. A **597**, 127 (1996).
8. Y.D. Kim, M.B. Tsang, C.K. Gelbke, W.G. Lynch, N. Carlin, Z. Chen, R. Fox, W.G. Gong, T. Murakami, T.K. Nayak, R.M. Ronningen, H.M. Xu, F. Zhu, W. Bauer, L.G. Sobotka, D. Stracener, D. G. Sarantites, Z. Majka, And V. Abnante, H. Griffin, Phys. Rev. Lett. **63**, 494 (1989).
9. P.D. Zecher, A. Galonsky, J.J. Kruse, S.J. Gaff, J. Ottarson, J. Wang, F. Deák, Á. Horváth, Á. Kiss, Z. Seres, K. Ieki, Y. Iwata, and H. Schelin,

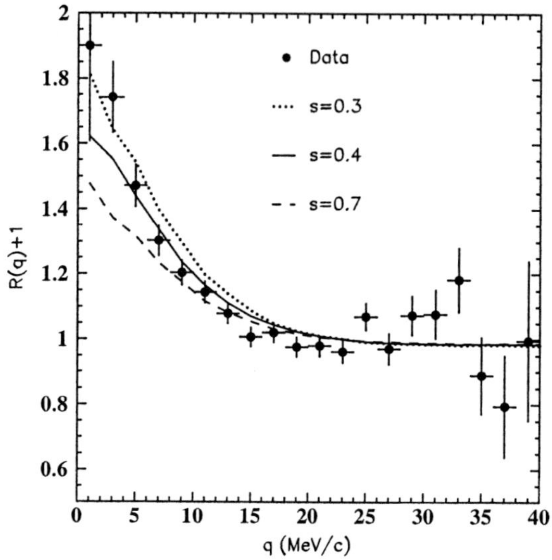

Figure 7: Experimental correlation function compared to a hybrid model with 18% preequilibrium and the time for each neutron from the compound-source component scaled by s.

NIM **401**, 329 (1997).

10. J. Wang, A. Galonsky, J.J. Kruse, P.D. Zecher, F. Deák, Á. Horváth, Á. Kiss, Z. Seres, K. Ieki, and Y. Iwata, NIM **397**, 380 (1997).
11. GEANT Manual, CERN Program Library Long Writeup W5013. Found at WWW site: http://wwwinfo.cern.ch/asdoc/geant_html3/geantall.html.
12. S.J. Gaff, Ph.D. thesis, Michigan State University, 1997.
13. S.J. Gaff, A. Galonsky, C.K. Gelbke, T. Glasmacher, M. Huang, J.J. Kruse, G.J. Kunde, R. Lemmon, W.G. Lynch, M.B. Tsang, J. Wang, P.D. Zecher, F. Deák, Á. Horváth, Á. Kiss, Z. Seres, K. Ieki, Y. Iwata, submitted Phys. Rev. C. (1998).
14. S.E. Koonin, Phys. Lett. **70B**, 43 (1977).
15. S. Pratt and M.B. Tsang, Phys. Rev. C **36**, 2390 (1987).
16. W.A. Friedman, W.G. Lynch, Phys. Rev. C **28**, 16 (1983).
17. W. Bauer, Nucl. Phys. A **471**, 604 (1987).
18. B.A. Li and W. Bauer, Phys. Rev. C **44**, 450 (1991).
19. W. Bauer, C.K. Gelbke, and S. Pratt, Annu. Rev. Nucl. Part. Sci. **42**, 77 (1992).

Recent progress in Three-particle Correlation Study

J. Schmidt-Sørensen
Department of Elementary Particle Physics,
Box 118, Lund University
221 00 Lund, Sweden
E-mail: janus@quark.lu.se

Three-particle correlations can reveal new information on the coherence of the particle emitting source. We present evidence of a partially coherent particle emitting source in Sulphur-Lead collisions at 200 GeV/c per nucleon measured by NA44. A brief summery of other experimental three-particle correlation results are presented.

1 The NA44 Experiment

The NA44 is a second generation experiment optimized for measuring identified particles in the low four-momentum region where the interferometry signal exist. In order to increase the acceptance, three superconducting quadrupoles are used in either an Horizontal or Vertical mode - whereas in this analysis data are sampled only in Horizontal setting. Two dipole magnets select the momentum range of the secondary particles into the acceptance, why only one sign of charges can be measured at the time. In the presented analysis we measure positive pions of momentum 4 GeV/c $\pm$ 20%. The momentum resolution is about 5 MeV.

The trigger is identical to the one used in our two-particle Sulphur-Lead analysis [1]. This trigger is constructed by requiring signal in a cherenkov firing on particles with masses including pions, and rejecting lighter particles by vetoing electrons in another cherenkov detector. Further, a multiplicity trigger is constructed requiring two or more hits in the three hodoscopes, which also provide TOF (time-of-flight) information in the offline analysis. The NA44 acceptance in terms of transverse momentum and rapidity can be studied elsewhere [1]. The mean p_T is 121 MeV/c and mean rapidity is 3.8.

2 Correlation Function in theory

Let us assume that the particle emitting source is fully chaotic, i.e. that there exists no correlation between the particles of when and where they are being emitted from the source, then the amplitude of detecting pions (bosons) with momentum p_i emitted at r_i and detected at r_I is given by:

$$A_{12} = \frac{1}{\sqrt{2}} \left\{ e^{ip_1(r_1 - r_A)} e^{ip_2(r_2 - r_B)} + e^{ip_1(r_1 - r_B)} e^{ip_2(r_2 - r_A)} \right\} \tag{1}$$

if the wave-functions are described by plane-waves. The second term is required due to the identity of the pions. The probability of such an event is thus given by the square of A_{12} integrated over all source points:

$$P_{12} = \int dr_A dr_B |A_{12}|^2 \rho(r_A)\rho(r_B) = 1 + |F_{12}|^2 \equiv C_{12} \tag{2}$$

where F_{ij} is the Fourier transform of the source density function, $\rho(r)$:

$$F_{ij} = \int dr \rho(r) e^{-iQ_{ij}r} \tag{3}$$

where $Q_{ij} = p_i - p_j$ is the four-momentum difference of a pair. The source density function $\rho(r)$ provides information on where and when the particles are being emitted from the source. Normally ρ is taken to be a Gaussian of width R and then Eq. 2 can be parametrized as:

$$C_2 = 1 + \lambda e^{-R^2 Q^2} \tag{4}$$

where λ is introduced due to the experimental fact that the correlation functions never reach the full value of 2. Possible explanations for this are misidentification of particles, partially coherent particle emission and the effect of long-lived resonnances which will all decrease the magnitude of the seen correlation function.

The definition leading to Eq. 2 can easily be extended to the three-particle case:

$$C_3 = 1 + |F_{12}|^2 + |F_{23}|^2 + |F_{31}|^2 + 2Re\{F_{12}F_{23}F_{31}\} \tag{5}$$

which again is only valid for a totally chaotic source. For a Gaussian source density function and by using plane-waves, C_3 can be parametrized as:

$$C_3 = 1 + \lambda \left\{ e^{-R^2 Q_{12}^2} + e^{-R^2 Q_{23}^2} + e^{-R^2 Q_{31}^2} \right\} + 2\lambda^{3/2} e^{-\frac{1}{2} R^2 Q_3^2} \tag{6}$$

where $Q_3^2 \equiv \Sigma Q_{ij}^2$ where $ij = 12, 23, 31$. We note that R and λ are the same in Eq. 4 and Eq. 6, i.e. within the assumptions mentioned above, the three-particle correlations provide no extra information than ordinary two-particle correlations.

We normally simplify the three-particle correlation function to a 1-dimensional function by using this parametrization:

$$C_3 = 1 + \lambda_3 e^{-R_3^2 Q_3^2} \tag{7}$$

The Q_{ij} parameter has been subject to a lot of criticism, and in ordinary two-particle correlation measurement Q_{ij} is usually decomposed into three components, namely Q_{long} in longtudinal direction along the beam-axis and two in transverse direction, Q_{tside} and Q_{tout}, for more details see [2,3]. Such a parametrization is not developed for three-particle systems which would also result in a 9-dimensional three-particle correlation function for which the statistics in current experiments are too poor to perform a proper analysis.

3 Correlation Function in practice

Experimentally we construct the correlation function by:

$$C = \frac{Real}{Back} \tag{8}$$

where *Real* is a two- or three-particle distribution in some invariant parameter like Q, and *Back* is a reference distribution without any HBT effect. The *Back* is constructed by eventmixing where pairs/triplets are constructed from different events. This ensures that the *Back* has the same trigger bias and experimental acceptance as *Real*. However, due to small acceptance all the events from which the particles are taken are affected by HBT and for this we need to correct.

3.1 Corrections

We apply the following corrections to our raw correlation function:

$$C_{final} = C_{raw} \times K_{SPC} \times K_{ACC} \times K_{Coulomb} \tag{9}$$

The background spectrum is distorted with respect to the true uncorrelated many-particle spectrum owing to the effect of the many-particle correlations on the single-particle spectrum. This is iteratively corrected by K_{SPC}.

The data are also corrected for the momentum resolution of the spectrometer and the many-particle acceptance. This correction factor, K_{ACC}, is calculated by a Monte Carlo programme with a full simulation of the tracking detectors and multiple scattering.

The two-particle correlation function has been Coulomb corrected, $K_{Coulomb}$, either by the standard Gamow correction or by a Coulomb wave-function integration method[3]. The three-particle correlation function is Coulomb corrected by using the product of the three two-particle Coulomb correction factors[4]:

$$G = G_{12}G_{23}G_{31} \tag{10}$$

This method is an approximation since no analytic solution to the three-body Coulomb problem exists. Alternative methods[5] have also been applied, and they all give consistent results.

4 Results

A raw sample of 24k triplets have been analysed. TOF cuts have been applied and only events which are separated by one slat in all the hodoscopes are accepted. The last requirement is to ensure a high purity. After cuts we end up with about 8k triplet events. The four-momentum resolution is about 15 MeV/c which determines the binsize used to study the correlation function.

4.1 Two-particle correlations in a three-particle sample

We have constructed a two-particle correlation function by using the three combinations of a pair in the three-particle sample. In Figure 1 we compare this function with the ordinary two-particle correlation function from published NA44 data[1]. We observe a very good agreement which indicate that the genuine three-body correlation is weak, i.e. that a third particle within the NA44 acceptance has no or very little effect on the two-particle correlation.

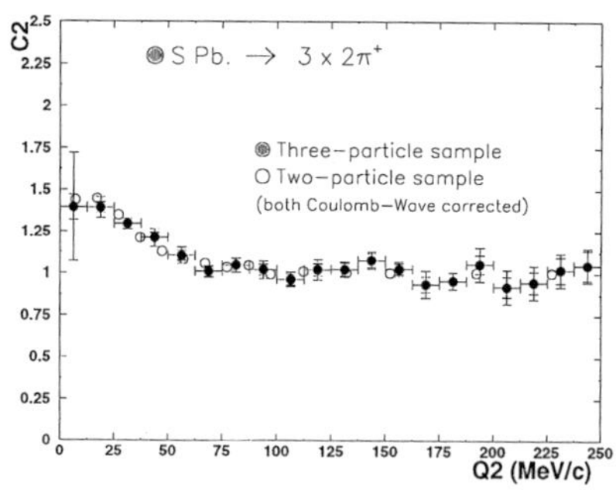

Figure 1: The two-particle correlation function from this three-particle sample and from an ordinary two-particle sample. (NA44 PRELIMINARY)

System	λ	R (fm)	χ^2/N_{dof}	Coulomb-correction
$3\pi^+ \to 2\pi^+$	0.51 ± 0.04	4.08 ± 0.28	19.4/19	Gamow
$2\pi^+$	0.52 ± 0.04	4.12 ± 0.20	-	Gamow
$3\pi^+ \to 2\pi^+$	0.45 ± 0.04	4.21 ± 0.32	17.8/19	Coulomb-Wave
$2\pi^+$	0.46 ± 0.04	4.5 ± 0.31	18.1/16	Coulomb-Wave

Table 1: Results from two-pion analysis from the three-particle data set compared to two-particle analysis measured by NA44. Both datasets are minimum bias, horizontal setting and errors are statistical only.

One should keep in mind that the trigger is identical for the two datasets and the difference in the cuts has no effect on the correlation function. The extracted fit parameters are listed in Table 1.

4.2 Three-particle correlation

We have measured the three-particle correlation, see Figure 2, for both Gamow and Coulomb-Wave corrected data. The extracted fit parameters are listed in Table 2.

The genuine three-body correlation, i.e. the last term in Eq. 6, has been estimated, see the dashed-dotted line in Figure 2. The genuine three-body correlation is only expected to be present in the first 3-4 bins.

Ingenuously, we would expect a three-particle correlation function with a λ_3 of about 2.0-2.3 as $\lambda_3 = 3\lambda + 2\lambda^{3/2}$ at $Q_3 = 0$. Our measured three-particle correlation function shows a decrease in magnitude. In fact, C_3 can be made of the sum of the three C_2 factors alone.

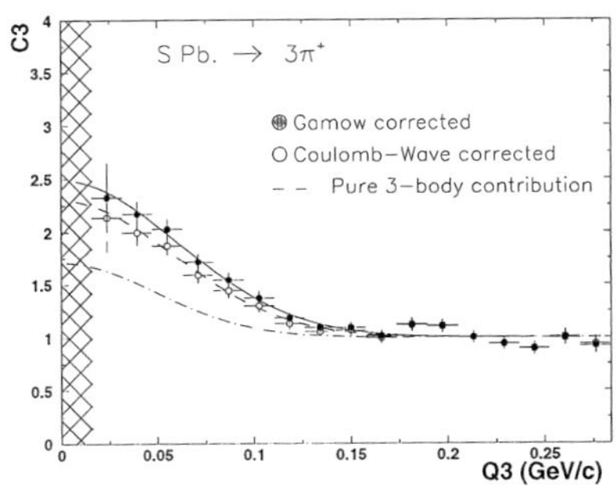

Figure 2: The three-particle correlation function for both Gamow and Coulomb-Wave corrected data. Errors are statistical only. (NA44 PRELIMINARY)

System	λ_3	R_3 (fm)	χ^2/N_{dof}	Coulomb-Correction
$3\pi^+$	1.50 ± 0.12	2.32 ± 0.13	17.5/15	Gamow
$3\pi^+$	1.30 ± 0.12	2.39 ± 0.15	15.8/15	Coulomb-Wave

Table 2: Fitted results from three-pion analysis for both kinds of Coulomb correction methods. The given errors are statistical only.

4.3 Phases in three-particle correlations

As already mentioned, the three-particle correlation function provides no additional information than ordinary two-particle correlations when assumming a totally chaotic source and using a Gaussian parametrization of the source density function. However, in case of an asymmetric source density function, F_{ij} in Eq. 3 will become complex, i.e. a phase is introduced. This phase can be isolated in Eq. 5 and the real part looks like:

$$\cos\phi = \frac{(C_3 - 1) - (C_2^{12} - 1) - (C_2^{23} - 1) - (C_2^{31} - 1)}{2\sqrt{(C_2^{12} - 1)(C_2^{23} - 1)(C_2^{31} - 1)}} \tag{11}$$

This formula is valid for a totally chaotic source. In case of a partial coherent source more terms must be included in the expressions for two- and three-particle correlation functions - hence we end up with a different and more complicated expression for $\cos\phi$ [7].

For a totally chaotic and symmetric source we would expect $\cos\phi = 1$. For a chaotic but asymmetric source theoretical calculations [6,7] have shown that only a minor decrease in the value of $\cos\phi$ is expected, i.e. $\cos\phi \simeq 1$. In the case that $\cos\phi \ll 1$, this could be an evidence for coherence [7]. Here one should note that Eq. 11 is no longer valid and the exact level of $\cos\phi$ carries no real information on the phase.

We can calculate $\cos\phi$ for each event and the result is shown in Figure 3. The result shows $\cos\phi \simeq 0$. This means that C_3 can be made upon C_2 only, i.e. the last term in Eq. 5, the genuine three-body correlation, is zero.

5 Results from other experiments

In the following sections we have a brief summary of those high-energy experiments which have been able to measure three-particle correlations.

5.1 AFS

The AFS collaboration have measured [8] three-particle correlations at ISR energies in proton-proton collisions, see Figure 4. The data-points following the

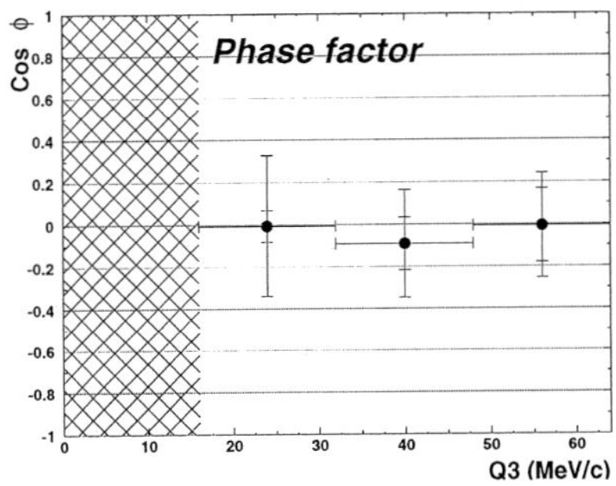

Figure 3: The phase factor as a function of Q3 calculated for the Gamow corrected data. The Coulomb-Wave corrected gives consistent results. The statistical and systematic errors are shown as a double error-bar. (NA44 PRELIMINARY)

flat line arise when all known correlation is extracted, using Eq. 6 and assuming $\cos\phi = 1$. Their conclusion is that no extra information is provided by three-particle correlations not already seen in two-particle correlations.

Based on their two-particle correlation measurement we have estimated the genuine three-body correlation, which is indicated on Figure 4 as the thick black line. In the region $Q > 0.2$ GeV/c, when errors are small, the non-correlated data-points are below unity. This would be the case if one extracts too much correlation, i.e. an indication that $\cos\phi < 1$. When the genuine three-body correlation vanish, i.e. in the region $Q > 0.4$ GeV/c, the non-correlated data-points are back on unity.

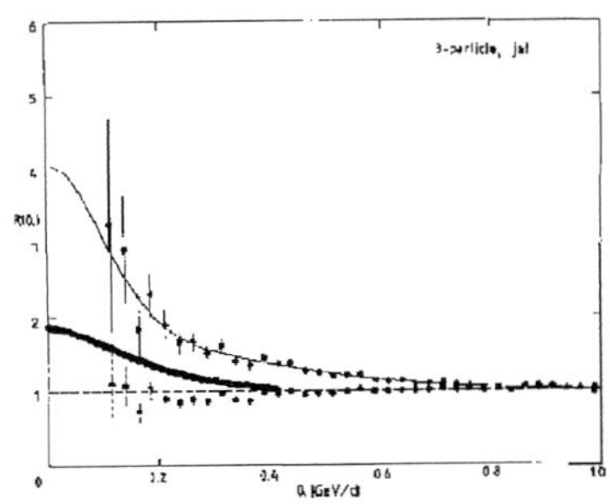

Figure 4: The three-particle correlation function from pp collisions at ISR energies. The data-points following the flat line is when all known correlations, using Eq. 6, are extracted. The thick black line indicates the genuine three-body correlation for $\cos\phi = 1$.

5.2 *OPAL and Delphi*

Both OPAL [10] and DELPHI [11] collaborations have measured three-particle correlations at the Z^0 pole in e^+e^- collisions at LEP. They report a clear three-body signal, see Figure 5, where the solid line represents a fit to $C_3 = 1 + 2\lambda exp(-R_3^2 Q_3^2)$. The extracted fitparameters are consistent with fitparameters from two-particle correlations, i.e. $\cos\phi = 1$.

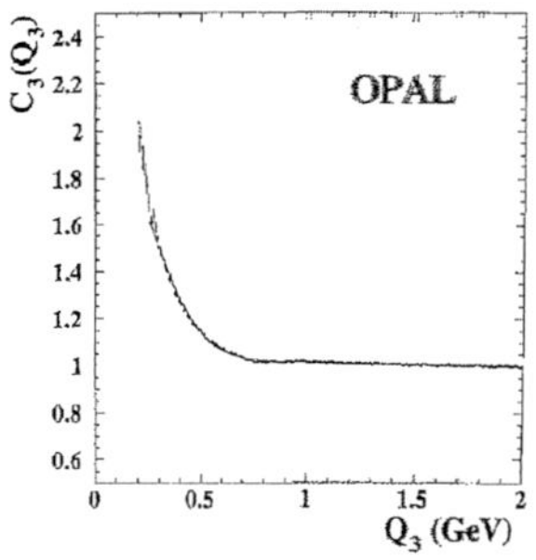

Figure 5: The genuine three-particle correlation term in e^+e^- collisions measured by OPAL. (OPAL PRELIMINARY)

5.3 *NA22*

The NA22 collaboration have measured [9] the genuine three-body correlation function in $\pi^+ p$ and $K^+ p$ collisions at 250 GeV/c. In Figure 6 all known correlations have been extracted the three-particle correlation function using Eq. 6 and assuming $\cos\phi = 1$. They claim to have observed a small three-particle interference effect not described in terms of two-particle correlations.

Except for the region with very large errors ($Q < 0.2$ GeV/c), the data-points are below unity up to $Q = 0.5$ GeV/c. This will happen if too much correlation has been extracted, i.e. an indication that $\cos\phi < 1$.

6 Summary and conclusion

The NA44 collaboration have measured the three-particle correlation function and find that the genuine three-body correlation is suppresed. This is seen partly in the similarity of two-particle correlation function from a three- and two-particle datasample, partly in the three-particle correlation function whose magnitude is decreased, and partly from the $\cos\phi \ll 1$ determination. The interpretation is that we observe partial coherence.

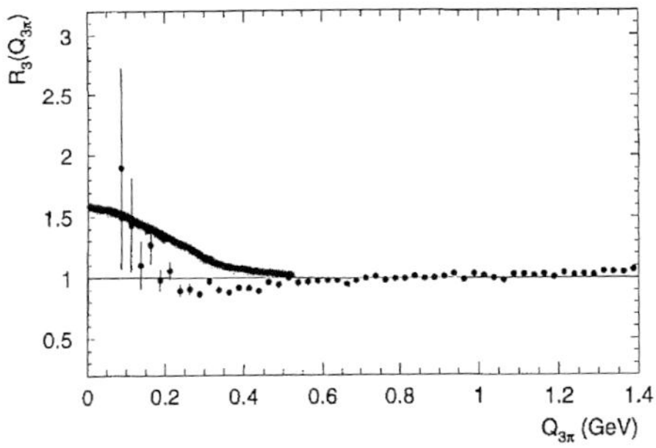

Figure 6: The three-particle correlation function measured by NA22 when all known correlation as been extracted. The enhancement at $Q < 0.2$ GeV/c has rather large errors and the data-points are below unity in the region $0.2 < Q < 0.5$ GeV/c. The thick black line is an estimate of the genuine three-body correlation for $\cos\phi = 1$, based on information from the two-particle correlation function.

We have looked at other experimental results on three-particle correlations and we find indications that the genuine three-body correlation is suppressed in heavy-ion collisions, but not in e^+e^- collisions where a clear signal is observed.

This is the first measurement of the $\cos\phi$ phase factor in three-pion correlations. It is a very sensitive direct measurement of the importance of the genuine three-body correlation.

References

1. H. Bøggild et al., (NA44 Coll.) *Phys. Lett.* B **302**, 510-516 (1993)
2. G. Bertsch, M. Gong, and M. Tohyama, *Phys. Rev.* C **37**, 1896 (1988)
3. S. Pratt, *Phys. Rev.* D **33**, 1314 (1986)
4. Y. M. Liu, D. Beavis, S. Y. Chu, S. Y. Fung, D. Keane, G. Van Galen, and M. Vient, *Phys. Rev.* C **34**, 1667 (1986)
5. J. G. Cramer, *Phys. Rev.* C **43**, 2798 (1991)
6. H. Heiselberg and A. P. Vischer, **NBI-97-32** (1997)
7. U. Heinz and Q. Zhang, *Phys. Rev.* C **56**, 426 (1997)
8. T. Åkesson et al., (AFS Coll.) *Z. Phys.* C **36**, 517 (1987)
9. N. M. Agababyan et al., (NA22 Coll.) *Z. Phys.* C **68**, 229 (1995)
10. K. Ackerstaff et al., (OPAL Coll.), hep-ex/9806036, Submitted to European Physical Journal **C**, (1998)
11. P. Abreau et al., (Delphi Coll.) *Phys. Lett.* B **355**, 415 (1995)

HANBURY BROWN – TWISS INTERFEROMETRY IN HIGH ENERGY NUCLEAR AND PARTICLE PHYSICS

Ulrich Heinz
Institut für Theoretische Physik, Universität Regensburg, D-93040 Regensburg, Germany

I review recent applications of two-particle intensity interferometry in high energy physics, concentrating on relativistic heavy ion collisions. By measuring hadronic single-particle spectra and two-particle correlations in hadron-hadron or heavy-ion collisions, the size and dynamical state of the collision fireball at freeze-out can be reconstructed. I discuss the relevant theoretical methods and their limitations. By applying the formalism to recent pion correlation data from Pb+Pb collisions at CERN we demonstrate that the collision zone has undergone strong transverse growth before freeze-out (by a factor 2 in each direction), and that it expands both longitudinally and transversally. From the thermal and flow energy density at freeze-out the energy density at the onset of transverse expansion can be estimated from conservation laws. It comfortably exceeds the critical value for the transition to color deconfined matter.

1 Introduction

The principle of two-particle intensity interferometry, developed by R. Hanbury Brown and R.Q. Twiss in the mid 1950's for radio and optical astronomy,[1] was independently rediscovered by particle physicists just a few years later.[2] It has since become known under the names "HBT interferometry", "GGLP effect", and "Bose-Einstein correlations". The method exploits the effects on the phase-space density of Bose-Einstein symmetrization (or Pauli antisymmetrization) of multiparticle states of identical particles. In astronomy one studies photon pairs, in nuclear and particle physics one typically uses pion, kaon or nucleon pairs (recently also pion triplets.[3,4]) Strictly speaking, the method is used oppositely in astronomy and high energy physics: in astronomy one measures the two-photon correlation function as a function of the *space-time* distance of the two detected photons and extracts from it information about the size of the emitter in momentum space (specifically in the opening angle between the two photon momentum vectors). By knowing the distance of the emitter (star, radio source) this "angular size" in momentum space can then be translated into a spatial radius of the source. In particle physics one measures the correlation function as a function of the *momentum* difference between the two pions and extracts from it information about the space-time extension of the emitting source. These two opposite ways of looking at intensity interferometry emphasize its generic nature as a phase-space effect which can only be understood if both the space-time and momentum-space structure of the emitter are taken into account simultaneously.

Another big difference between HBT interferometry in astronomy and nuclear physics is that the sources studied in astronomy are static on the timescale of the observation, while the sources created in high energy collisions are highly dynamical and very shortlived. As I will show this has dramatic implications for the HBT formalism to be used for the analysis of measured correlation functions which were only gradually realized during the last 10 years and for which a common understanding was developed only quite recently.

Relativistic heavy ion collisions are performed in order to create color deconfined strongly interacting matter, the "quark-gluon plasma". This plasma is hot and generates a huge pressure which drives a strong collective expansion of the reaction zone into the surrounding vacuum. As a consequence of expansion, the quark-gluon matter cools and dilutes until it can no longer remain in the color deconfined state: it hadronizes. If the reaction zone is sufficiently large and locally equilibrated, this process manifests itself as a bulk phase transition, very similar to the quark-hadron phase transition in the very early universe (about 20-40 μs after the Big Bang). Two crucial questions are therefore (a) whether the initial energy density in the reaction region is large enough to make color deconfined quark-gluon matter, and (b) whether this partonic system equilibrates sufficiently to undergo a thermodynamic phase transition while expanding. While the longitudinal expansion of the reaction zone may be simply due to incomplete stopping of the two colliding nuclei, *transverse* collective expansion flow can only be generated by the buildup of a locally isotropic pressure component. Since this requires a certain degree of local equilibration of the momentum distributions, the identification of collective *transverse* flow, superimposed by random thermal motion, plays an essential role in any attempt to answer these two questions. As we will see, two-particle interferometry features prominently in such an endeavour.

Over the last years ample evidence was accumulated that the hot and dense collision region in relativistic heavy ion collisions indeed thermalizes and shows collective dynamical behaviour. Most of this evidence is based on a comprehensive analysis of the hadronic single particle spectra. It was shown that all available data on hadron production in heavy ion collisions at the AGS and the SPS can be understood within a simple model which assumes locally thermalized momentum distributions at freeze-out, superimposed by collective hydrodynamical expansion in both the longitudinal and transverse directions.[5,6] The collective dynamical behaviour in the transverse direction is reflected by a characteristic dependence of the inverse slope parameters of the $m_\perp$-spectra ("effective temperatures") at small $m_\perp$ on the hadron masses.[5]

New data from the Au+Au and Pb+Pb systems [7] support this picture and show that the transverse collective dynamics is much more strongly exhibited in larger collision systems than in the smaller ones from the first rounds of experiments. The amount of transverse flow also appears to increase monotonically with collision energy from GSI/SIS to AGS energies, but may show signs of saturation at the even higher SPS energy.[7]

The extraction of flow velocities and thermal freeze-out temperatures from the measured single particle spectra relies heavily on model assumptions.[5] The single-particle spectra are ambiguous because they contain no direct information on the space-time structure and the space-momentum correlations induced by collective flow. In terms of the phase-space density at freeze-out ("emission function") $S(x,p)$ the single-particle spectrum is given by $E\,dN/d^3p = \int d^4x\, S(x,p)$; the space-time information in S is completely washed out by integration. Thus, on the single-particle level, comprehensive model studies are required to show that a simple hydrodynamical model with only a few thermodynamic and collective parameters can fit all the data, and additional consistency checks are needed to show that the extracted fit parameter values lead to an internally consistent theoretical picture. The published literature abounds with examples demonstrating that without such consistency checks the theoretical ambiguity of the single particle spectra is nearly infinite.

At this point Bose-Einstein correlations between the momenta of identical particle pairs provide crucial new input. They give direct access to the space-time structure of the source *and* its collective dynamics. In spite of some remaining model dependence, the set of possible model sources can thus be reduced dramatically. The two-particle correlation function $C(\boldsymbol{q}, \boldsymbol{K})$ is usually well approximated by a Gaussian in the relative momentum q whose width parameters are called "HBT (Hanbury Brown-Twiss) radii". It was recently shown [8--10] that these radius parameters measure certain combinations of the second central space-time moments of the source. In general they mix the spatial and temporal structure of the source in a nontrivial way,[9] and the remaining model dependence enters when trying to unfold these aspects.

Collective dynamics of the source leads to a dependence of the HBT radii on the pair momentum K.[11,12] This feature was recently quantitatively reanalyzed, both analytically [9,10,13,14] and numerically.[16,17] The velocity gradients associated with collective expansion lead to a dynamical decoupling of different source regions, and the HBT radii measure the size of the resulting "space-time regions of homogeneity" of the source [12,13] around the point of maximum emissivity for particles with the measured momentum K. The velocity gradients are smeared out by random thermal motion of the emitters around the fluid velocity.[9] Due to the exponential decrease of the Maxwell distribution, this thermal smearing factor shrinks with increasing transverse pair momentum $K_\perp$; this is the basic reason for the $K_\perp$-dependence of the HBT radii.

Unfortunately, other gradients in the source (for example spatial and temporal temperature gradients) can also generate a K-dependence of the HBT radii.[9,14,15] Furthermore, the pion spectra in particular are affected by resonance decay contributions, but only at small $K_\perp$. This may also affect the HBT radii in a $K_\perp$-dependent way.[18,19] The isolation of collective flow, in particular transverse flow, from the $K_\perp$-dependence of the HBT radii thus requires a careful investigation of these different effects.

Our group studied this K-dependence of the HBT radii within a simple analytical model for a finite thermalized source which expands both longitudinally and transversally. For presentation I use the Yano-Koonin-Podgoretskii (YKP) parametrization of the correlator which, for sources with dominant longitudinal expansion, provides an optimal separation of the spatial and temporal aspects of the source.[10,17] The YKP radius parameters are independent of the longitudinal velocity of the observer frame. Furthermore, in all thermal models without transverse collective flow, they show perfect $M_\perp$-scaling (in the absence of resonance decay contributions). Only the transverse gradients induced by a non-zero transverse flow can break this $M_\perp$-scaling, causing an explicit dependence on the particle rest mass. This allows for a rather model-independent identification of transverse flow from accurate measurements of the YKP correlation radii for pions and kaons. High-quality data should also allow to control the effects from resonance decays.

A comprehensive and didactical discussion of the formalism and a more extensive selection of numerical examples can be found in the lecture notes [20] to which I refer the interested reader for more details.

2 Spectra and emission function

2.1 Single-particle spectra and two-particle correlations

The covariant 1- and 2-particle momentum spectra are defined by

$$P_1(\boldsymbol{p}) = E\,\frac{dN}{d^3p} = E\,\langle \hat{a}^+_{\boldsymbol{p}} \hat{a}_{\boldsymbol{p}} \rangle\,, \tag{1}$$

$$P_2(\boldsymbol{p}_a, \boldsymbol{p}_b) = E_a\,E_b\,\frac{dN}{d^3p_a d^3p_b} = E_a\,E_b\,\langle \hat{a}^+_{\boldsymbol{p}_a} \hat{a}^+_{\boldsymbol{p}_b} \hat{a}_{\boldsymbol{p}_b} \hat{a}_{\boldsymbol{p}_a} \rangle\,, \tag{2}$$

where $\hat{a}^+_{\boldsymbol{p}}$ ($\hat{a}_{\boldsymbol{p}}$) creates (destroys) a particle with momentum $\boldsymbol{p}$. They are normalized to $\langle N \rangle$ and $\langle N(N-1) \rangle$ (i.e.the average number of pions or of pions in pairs per event), respectively. The angular brackets denote an ensemble average $\langle \hat{O} \rangle = \mathrm{tr}\,(\hat{\rho}\hat{O})$ where $\hat{\rho}$ is the density operator associated with the ensemble. The two-particle correlation function is defined as

$$C(\boldsymbol{p}_a, \boldsymbol{p}_b) = \frac{P_2(\boldsymbol{p}_a, \boldsymbol{p}_b)}{P_1(\boldsymbol{p}_a) P_1(\boldsymbol{p}_b)} \,. \tag{3}$$

If the two particles are emitted independently and final state interactions are neglected one can prove [20] a generalized Wick theorem

$$C(\boldsymbol{p}_a, \boldsymbol{p}_b) = 1 \pm \frac{|\langle \hat{a}^+_{\boldsymbol{p}_a} \hat{a}_{\boldsymbol{p}_b} \rangle|^2}{\langle \hat{a}^+_{\boldsymbol{p}_a} \hat{a}_{\boldsymbol{p}_a} \rangle \langle \hat{a}^+_{\boldsymbol{p}_b} \hat{a}_{\boldsymbol{p}_b} \rangle} \,. \tag{4}$$

Note that the second term is positive definite, i.e. the correlation function cannot, for example, oscillate around unity. This is no longer true if final state interactions are included (see below). I will here assume that the emitted particles are bosons which I will call pions.

2.2 Source Wigner function and spectra

In the language of the covariant current formalism [21] the source of the emitted pions can be described in terms of classical currents $J(x)$ which act as classical sources of freely propagating pions. They parametrize the last collision from which the free outgoing pion emerges. Very helpful for the following will be the so-called "emission function" $S(x, K)$: [11,22]

$$S(x, K) = \int \frac{d^4 y}{2(2\pi)^3}\, e^{-iK\cdot y} \left\langle J^*(x + \tfrac{1}{2}y) J(x - \tfrac{1}{2}y) \right\rangle \,. \tag{5}$$

It is the Wigner transform of the density matrix associated with the classical source amplitudes $J(x)$. This Wigner density is a quantum mechanical object defined in phase-space (x, K); it is real but not always positive definite. Textbooks on Wigner functions show that their non-positivity is a genuine quantum effect resulting from the uncertainty relation and is concentrated at short phase-space distances; when the Wigner function is averaged over phase-space volumes large compared to the volume $(2\pi\hbar)^3$ of an elementary phase-space cell, the result is real and positive definite and behaves exactly like a classical phase-space density.

The emission function $S(x, K)$ is thus the quantum mechanical analogue of the classical phase-space distribution which gives the probability of finding at point x a source which emits free pions with momentum K. It allows to express the single-particle spectra and two-particle correlation function via the following fundamental relations:[11,22,23]

$$E_p \frac{dN}{d^3p} = \int d^4x\, S(x,p)\,, \tag{6}$$

$$C(\boldsymbol{q},\boldsymbol{K}) = 1 + \frac{\left|\int d^4x\, S(x,K)\, e^{iq\cdot x}\right|^2}{\int d^4x\, S(x,K+\frac{1}{2}q)\ \int d^4x\, S(x,K-\frac{1}{2}q)}\,. \tag{7}$$

For the single-particle spectrum (6), the Wigner function $S(x,p)$ on the r.h.s. must be evaluated on-shell, i.e. at $p^0 = E_p = \sqrt{m^2+\boldsymbol{p}^2}$. For the correlator (7) we have defined the relative momentum $\boldsymbol{q} = \boldsymbol{p}_a - \boldsymbol{p}_b$, $q^0 = E_a - E_b$ between the two particles in the pair, and the total momentum of the pair $\boldsymbol{K} = (\boldsymbol{p}_a + \boldsymbol{p}_b)/2$, $K^0 = (E_a+E_b)/2$. Of course, since the 4-momenta $p_{a,b}$ of the two measured particles are on-shell, $p_i^0 = E_i = \sqrt{m^2+\boldsymbol{p}_i^2}$, the 4-momenta q and K are in general off-shell. They satisfy the orthogonality relation

$$q \cdot K = m_a^2 - m_b^2 = 0\,. \tag{8}$$

Thus, the Wigner function on the r.h.s. of Eq. (7) is *not* evaluated at the on-shell point $K^0 = E_K$. This implies that for the correlator, in principle, we need to know the off-shell behaviour of the emission function, i.e. the quantum mechanical structure of the source! Fortunately, nature is nice to us: the interesting behaviour of the correlator (its deviation from unity) is concentrated at small values of $|\boldsymbol{q}|$. Expanding $K^0 = (E_a+E_b)/2$ for small q one finds

$$K^0 = E_K \left(1 + \frac{\boldsymbol{q}^2}{8E_K^2} + \mathcal{O}\left(\frac{\boldsymbol{q}^4}{E_K^4}\right)\right) \approx E_K\,. \tag{9}$$

Since the relevant range of q is given by the inverse "size" of the source (to be defined more exactly below), this approximation is valid as long as the Compton wavelength of the particles is small compared to this "source size". For pion, kaon, or proton interferometry in heavy-ion collisions this is the case, due to the particle rest masses. This is of enormous practical importance: it allows one to replace the source Wigner density by a classical phase-space distribution function for on-shell particles. This provides a necessary theoretical foundation for the calculation of HBT correlations from classical hydrodynamic or kinetic (e.g. cascade or molecular dynamics) simulations of the collision.

If one approximates the product of single-particle distributions in the denominator of (7) by the square of the spectrum at the average momentum K, one obtains the simpler result

$$C(\boldsymbol{q},\boldsymbol{K}) \approx 1 + \left|\frac{\int d^4x\, e^{iq\cdot x}\, S(x,K)}{\int d^4x\, S(x,K)}\right|^2 \equiv 1 + \left|\langle e^{iq\cdot x}\rangle\right|^2\,. \tag{10}$$

The deviations from this approximation are proportional to the curvature of the single-particle distribution in logarithmic representation.[9] They are small in practice because the measured single-particle spectra are usually more or less exponential. In the second equality of (10) we defined $\langle \ldots \rangle$ as the average taken with the emission function; due to the K-dependence of $S(x,K)$ this average is a function of K. This notation will be used extensively below.

The fundamental relations (6,7,10) show that *both the single-particle spectrum and the two-particle correlation function can be expressed as simple integrals over the emission function.* The emission function thus is *the* basic ingredient in the theory of HBT interferometry: once it is known, the calculation of one- and two-particle spectra is straightforward (even if the evaluation of the integrals may in some cases be technically involved). More interestingly, measurements of the one- and two-particle spectra provide access to the emission function and thus to the space-time structure of the source. This latter aspect is, of course, the motivation for exploiting HBT in practice. In my talk I will concentrate on the question to what extent this access to the space-time structure from only momentum-space data really works, whether it is complete, and (since we will find it is not and HBT analyses will thus necessarily be model-dependent) what can be reliably said about the extension and dynamical space-time structure of the source anyhow, based on a minimal set of intuitive and highly suggestive model assumptions.

2.3 Final state interactions (FSI)

Equation (7),via the plane wave factor $e^{iq\cdot x}$ under the integral of the exchange term, reflects the absence of final state interactions:

$$P_2(\boldsymbol{p}_a,\boldsymbol{p}_b) = \int_{x_1,x_2} \left[S\left(x_1, K+\tfrac{q}{2}\right)\, S\left(x_2, K-\tfrac{q}{2}\right) \pm e^{iq\cdot(x_1-x_2)} S(x_1,K) S(x_2,K)\right].$$

In practice particle interferometry is done with charged particle pairs which suffer long-range Coulomb final state repulsion on their way out to the detector. In addition, there may be strong final state interactions, e.g. in proton-proton interferometry where there is a strong s-wave resonance just above the two-particle threshold. In this case Eq. (11) must be replaced by [24]

$$\begin{aligned} P_2(\boldsymbol{p}_a,\boldsymbol{p}_b) = & \int_{x,y} S\left(x+\tfrac{y}{2}, p_a\right)\, S\left(x-\tfrac{y}{2}, p_b\right) \\ & \times \left[\theta(y^0)\left|\phi_{\boldsymbol{q}/2}(\boldsymbol{y}-\boldsymbol{v}_b y^0)\right|^2 + \theta(-y^0)\left|\phi_{\boldsymbol{q}/2}(\boldsymbol{y}-\boldsymbol{v}_a y^0)\right|^2\right] \\ \pm & \int_{x,y} S\left(x+\tfrac{y}{2}, K\right)\, S\left(x-\tfrac{y}{2}, K\right)\, \phi^*_{-\boldsymbol{q}/2}(\boldsymbol{y}-\boldsymbol{v}y^0)\, \phi_{\boldsymbol{q}/2}(\boldsymbol{y}-\boldsymbol{v}y^0). \end{aligned} \quad (11)$$

Here $\boldsymbol{v}=\boldsymbol{K}/E_K$, $\boldsymbol{v}_a=\boldsymbol{p}_a/E_K$, $\boldsymbol{v}_b=\boldsymbol{p}_b/E_K$ are (to quadratic accuracy in q) the velocities of the particles with momentum $\boldsymbol{K}$, $\boldsymbol{p}_a$, $\boldsymbol{p}_b$, respectively, and $\phi_{\boldsymbol{q}/2}(\boldsymbol{r})$ is an FSI distorted wave with asymptotic relative momentum $\boldsymbol{q}/2$, evaluated at the two-particle relative distance $\boldsymbol{r}$. Upon replacing the latter by plane waves (11) turns into (11). The FSI distorted waves can be calculated by solving a non-relativistic Schrödinger equation for the relative motion which includes the FSI potential *in the rest system of the pair* (where $\boldsymbol{K} = \boldsymbol{v} = 0$). Eq. (11) represents a non-relativistic Galilei-transformation of the result from the pair rest frame to the frame in which $\boldsymbol{p}_a$ and $\boldsymbol{p}_b$ are measured; therefore it is only valid in observer frames in which the pair moves non-relativistically. In order to evaluate Eq. (11) one must therefore first transform the 4-momenta $p_{a,b}$ to such a frame (best directly into the pair rest frame). The momentum argument $\boldsymbol{q}$ of the FSI distorted waves ϕ is then the difference between the two spatial momenta *in that frame*, and their space-time argument $\boldsymbol{y} - \boldsymbol{v}_i y^0$ is the relative distance of the two particles in that frame *at the time when the second particle is emitted*.[24] Since the latter depends not only on the time difference y^0 between emission points, but also on the velocity of the first emitted particle, these arguments depend on the momentum argument of the emission function associated with the first emitted particle. The two terms $\sim \theta(\pm y^0)$ in the direct term reflect the two possible time orderings between the emission points.

Equations (11) and (11) can be implemented into event generators, following the procedure given elsewhere.[24,25]

2.4 The redundance of wavepackets

In the last two years it has been repeatedly suggested that, due to the smallness of the sources, the theory of particle interferometry in high energy physics should be based on a finite-size wave-packet description rather than on plane wave propagation from the source to the detector.[25--29] I will argue that, if done correctly, it doesn't matter whether you start from wave packets or not, and that the free parameter σ (the wave packet width) occurring in these approaches is essentially unmeasurable (except, perhaps, in elementary e^+e^- or pp collisions).

To the extent that the detector really measures the momenta of the particles (which it is supposed to do with highest possible accuracy), the measurement process can be described as a projection at $t = \infty$ of the emitted pion state on a plane wave momentum eigenstate, irrespective of the actual localization properties of the emitted states; the latter will be reflected in the momentum spectrum resulting from this projection. The usefulness of HBT interferometry rests exactly on the fact that free propagation after emission from the source does not change this overlap integral, i.e. that the 1- and 2-particle momentum spectra remain unchanged on their way from source to detector. Otherwise data measured meters away from the collision could not be used to extract information about the reaction zone. Final state interactions spoil this feature; therefore they must be accounted for analytically or numerically via Eq. (11) before source information can be extracted from the measured correlator. For free propagation, however, the spectrum calculated from the overlap between the emitted wavefunction and the momentum eigenstates is identical if calculated at the time of particle freeze-out or at the detector time several nanoseconds later.[24] In both cases, the relation between the spectra and the source distribution $S(x,K)$ is given by Eqs. (6) and (7).

While wave packets are a physically intuitive concept, one must take care not to exaggerate their role in particle interferometry. I will show that, to first order, their intrinsic structure can be completely absorbed into the emission function, leaving no measurable trace of the wave packet width. Statements to the contrary[27,29] are at best misleading. To prove my point, let me anticipate from Sec. 3.1 that the primary feature of the two-particle correlator, its Gaussian width in $\boldsymbol{q}$, determines only the r.m.s. width of the source in space-time; the extraction of finer structures of the source requires a quantitative study of the deviations from Gaussian behaviour of the correlator and therefore at least an order of magnitude more accurate data. The dominant features of the correlator can thus be reproduced by replacing the true source function $S(x,K)$ by a Gaussian with the same r.m.s. width in space-time.

Let us now follow custom[25--29] and assume that the source emits at times τ_i Gaussian wave packets of spatial width σ, centered at points $\boldsymbol{\xi}_i$ in coordinate space and $\boldsymbol{\pi}_i$ in momentum space:

$$\psi_i(\boldsymbol{x},\tau_i) = \frac{1}{(\pi\sigma^2)^{3/4}} \exp\left(-\frac{(\boldsymbol{x}-\boldsymbol{\xi}_i)^2}{2\sigma^2} + i\boldsymbol{\pi}_i\cdot\boldsymbol{x}\right) . \tag{12}$$

The Wigner density corresponding to such an individual wave packet is given by

$$S_{\text{w.p.}}(x,\boldsymbol{p}) = \frac{E_p}{\pi^3}\delta(x^0-\tau_i)\ \exp\left(-\frac{(\boldsymbol{x}-\boldsymbol{\xi}_i)^2}{\sigma^2} - \sigma^2(\boldsymbol{p}-\boldsymbol{\pi}_i)^2\right) ; \tag{13}$$

it is normalized to $\int(d^3p/E_p)\int d^4x\, S_{\text{w.p.}}(x,\boldsymbol{p}) = 1$. Its r.m.s. width parameters $\Delta x = \sigma/\sqrt{2}$ and $\Delta p = 1/(\sqrt{2}\sigma$ in the three Cartesian directions saturate the uncertainty relation $\Delta x \Delta p \geq \hbar/2$.

Let us distribute the wave packet emission points $(\tau_i,\boldsymbol{\xi}_i,\boldsymbol{\pi}_i)$ according to the *classical* Gaussian phase-space distribution

$$\rho(\tau,\boldsymbol{\xi},\boldsymbol{\pi}) = \langle N\rangle \frac{\exp\left(-\frac{(\tau-\tau_0)^2}{2(\Delta\tau)^2}\right)}{\sqrt{2\pi(\Delta\tau)^2}} \frac{\exp\left(-\frac{\boldsymbol{\xi}^2}{2R_0^2}-\frac{\boldsymbol{\pi}^2}{2\Delta_0^2}\right)}{(2\pi R_0\Delta_0)^3}, \tag{14}$$

normalized to the average pion multiplicity $\langle N\rangle$ per event. The r.m.s. widths $\Delta\tau, R_0, \Delta_0$ of this classical probability density are not constrained by quantum mechanics. It can be shown [25--28] that the spectra and correlation functions are then given by Eqs. (6) and (7) with an emission function $S(x,K)$ which is obtained by folding the classical phase-space distribution (14) with the intrinsic Wigner density (13) of the wave packets:

$$\begin{aligned} S(x,\boldsymbol{K}) &= \int d\tau\, d^3\xi\, d^3\pi\, \rho(\tau,\boldsymbol{\xi},\boldsymbol{\pi})\, S_{\rm w.p.}(x^0-\tau,\boldsymbol{x}-\boldsymbol{\xi},\boldsymbol{K}-\boldsymbol{\pi}) \\ &= \langle N\rangle \frac{\exp\left(-\frac{(x^0-\tau_0)^2}{2(\Delta\tau)^2}\right)}{\sqrt{2\pi(\Delta\tau)^2}} \frac{\exp\left(-\frac{\boldsymbol{x}^2}{2R^2}-\frac{\boldsymbol{K}^2}{2\Delta^2}\right)}{(2\pi R\Delta)^3}, \qquad (15) \\ R^2 &= R_0^2+\frac{\sigma^2}{2}, \qquad \Delta^2 = \Delta_0^2+\frac{1}{2\sigma^2}. \qquad (16) \end{aligned}$$

Only the combinations R,Δ from (16) enter the 1- and 2-particle spectra; at the Gaussian level there exists no measurement which allows to disentangle R_0, Δ_0 and σ. As long as $R\Delta \gg \hbar/2$, there are infinitely many combinations of R_0, Δ_0, σ which describe the same source. Only if R and Δ saturate the uncertainty limit, $R\Delta = \hbar/2$, one may conclude $R_0 = \Delta_0 = 0$ and $\sigma = \sqrt{2}R$.

The wave packet size σ is thus generally not measurable; as long as the effective emission function $S(x,\boldsymbol{K})$ (i.e. R,Δ) is not changed, variations of σ have no influence on the momentum distributions. Final state interactions do not affect this reasoning either – instead of (7) one must then evaluate (11) with the same emission function (15). The effect of 2-particle final state interactions on the time evolution of the wave packets [29] is fully taken into account by weighting the effective emission function $S(x,K)$ with the distorted waves $\phi_{\boldsymbol{q}/2}$ in (11); a generalization which includes also the FSI with the electric charge of the central fireball is known [30,31]. There is no need to describe the time evolution of the wave packets explicitly as done by Merlitz and Pelte [29] using a sophisticated numerical algorithm. Finally, the above statement remains true [32] even if multi-boson symmetrization effects [27,28] are included.

I have presented the argument using Gaussian wave packets and a Gaussian parametrization for the distribution of their centers. As explained above this is sufficient since, in leading order, HBT interferometry gives only access to the r.m.s. widths of the source such that one cannot distinguish Gaussians from other source shapes. At the next level of accuracy these statements are surely subject to (small) corrections. So far, however, I cannot see how in heavy-ion collisions the wave packet size might be determined experimentally. For this reason I prefer avoiding the concept of wave packets, thereby eliminating the poorly controlled free parameter σ from the theory of HBT interferometry in nuclear physics.

From here on I will neglect final state interactions, assuming that the measured correlators have been or can be corrected for them.

2.5 The mass-shell constraint

Expressions (7,10) show that the correlation function is related to the emission function by a Fourier transformation. At first sight this might suggest that one should easily be able to reconstruct the emission function from the measured correlation function by inverse Fourier transformation, the single particle spectrum (6) providing the normalization. This is, however, not correct. The reason is that, since the correlation function is constructed from the on-shell momenta of the measured particle pairs, not all four components of the relative momentum q occurring on the r.h.s. of (10) are independent. They are related by the "mass-shell constraint" (8) which can, for instance, be solved for q^0:

$$q^0 = \boldsymbol{\beta} \cdot \boldsymbol{q} \qquad \text{with} \qquad \boldsymbol{\beta} = \frac{\boldsymbol{K}}{K^0} \approx \frac{\boldsymbol{K}}{E_K}\,. \tag{17}$$

$\boldsymbol{\beta}$ is (approximately) the velocity of the c.m. of the particle pair. The Fourier transform in (10) is therefore not invertible, and the reconstruction of the space-time structure of the source from HBT measurements will thus always require additional model assumptions.

It is instructive to insert (17) into (10):

$$C(\boldsymbol{q},\boldsymbol{K}) \approx 1 + \left| \frac{\int d^4x \, \exp\big(i\boldsymbol{q}\cdot(\boldsymbol{x} - \boldsymbol{\beta}\,t)\big)\, S(x,K)}{\int d^4x \, S(x,K)} \right|^2 . \tag{18}$$

This shows that the correlator $C(\boldsymbol{q},\boldsymbol{K})$ actually mixes the spatial and temporal information on the source in a non-trivial way which depends on the pair velocity $\boldsymbol{\beta}$. Only for a pulsed source $\sim \delta(t-t_0)$ things are simple: the correlator then just measures the Fourier transform of the spatial source distribution.

It is instructive to look at the problem also in the following way: If one rewrites Eq. (18) in the pair rest frame where $\boldsymbol{K} = 0$ and hence $q^0 = 0$, one obtains

$$C(\boldsymbol{q},\boldsymbol{K}) - 1 = \int d^3r \, \cos(\boldsymbol{q}\cdot\boldsymbol{r})\, S_{\rm rel}(\boldsymbol{r};K) \tag{19}$$

where

$$S_{\rm rel}(\boldsymbol{r};K) = \int d^3R\; \bar{s}_K(\boldsymbol{R}+\tfrac{1}{2}\boldsymbol{r})\; \bar{s}_K(\boldsymbol{R}-\tfrac{1}{2}\boldsymbol{r})\,, \tag{20}$$

with

$$\bar{s}_K(\boldsymbol{x}) = \int dt\; s(\boldsymbol{x},t;K) = \int dt\; \frac{S(x,K)}{\int d^4x'\, S(x',K)}\,, \tag{21}$$

is the *time-integrated* normalized relative distance distribution in the source. The latter can, in principle, be uniquely reconstructed from the measured correlator [33] by inverting the cosine-Fourier transform (19). But since it gives only the time integral of the relative distance distribution for fixed pair momentum $\boldsymbol{K}$ in the pair rest frame, no direct information on the time structure of the source is obtainable! Only by looking at the result as a function of $\boldsymbol{K}$, which, as I will show, brings out the collective dynamical features of the source, can one hope to unfold the time-dependence of the emission function. It is clear that this will be only possible within the context of specific source parametrizations.

2.6 *K-dependence of the correlator*

We have seen that in general the correlator is a function of *both* $\boldsymbol{q}$ and $\boldsymbol{K}$. Only if the emission function factorizes in x and K, $S(x,K) = F(x)\,G(K)$ (no "x-K-correlations", i.e. every point x in the source emits particles with the same momentum spectrum $G(K)$), the K-dependence in $G(K)$ cancels between numerator and denominator of (10), and the correlator seems to be K-independent. However, not even this is really true: even after the cancellation of the explicit K-dependence $G(K)$, there remains an implicit K-dependence via the pair velocity $\boldsymbol{\beta} \approx \boldsymbol{K}/E_K$ in the exponent on the r.h.s. of Eq. (18)! Only if both conditions, factorization of the emission function in x and K *and* instantaneous emission $\sim \delta(t-t_0)$, apply simultaneously, the correlation function is truely K-independent. Even for stars which are time-independent sources there remains a K-dependence: the correlator is then [34] $\sim \delta(\boldsymbol{\beta}\cdot\boldsymbol{q})$, i.e. there are only transverse ($\perp \boldsymbol{\beta}$), but no longitudinal ($\parallel \boldsymbol{\beta}$) correlations.

It is hard to believe that this complication in the application of the original HBT idea to high-energy collisions went nearly unnoticed for more than 20 years. It was only brought to light in 1984 by Scott Pratt [11] in his pioneering work on HBT interferometry for heavy-ion collisions.

If one parametrises the correlator by a Gaussian in q (see below) this means that in general the parameters ("HBT radii") depend on K. Typical sources of x-K correlations in the emission function are a collective expansion of the emitter and/or temperature gradients in the particle source: in both cases the momentum spectrum $\sim \exp[-p{\cdot}u(x)/T(x)]$ of the emitted particles (where $u^\mu(x)$ is the 4-velocity of the expansion flow) depends on the emission point. In the case of collective expansion, the spectra from different emission points are Doppler shifted relative to each other. If there are temperature gradients, e.g. a high temperature in the center and cooler matter at the edges, the source will look smaller for high-momentum particles (which come mostly from the hot center) than for low-momentum ones (which receive larger contributions also from the cooler outward regions).

We thus see that collective expansion of the source induces a K-dependence of the correlation function. But so do temperature gradients. The crucial question is: does a careful measurement of the correlation function, in particular of its K-dependence, permit a separation of such effects, i.e. can the collective dynamics of the source be quantitatively determined through HBT experiments? We will see that this is not an easy task; however, with sufficiently good data, it should be possible. In any case, the K-dependence of the correlator is a decisive feature which puts the HBT game into a completely new ball park. Two-particle correlation measurements which are not able to resolve the K-dependence of the HBT parameters are, in high energy nuclear and particle physics, only of very limited use.

3 Model-independent discussion of HBT correlation functions

3.1 The Gaussian approximation – HBT radii as homogeneity lengths

The most interesting feature of the two-particle correlation function is its half-width. Actually, since the relative momentum $\boldsymbol{q} = \boldsymbol{p}_1 - \boldsymbol{p}_2$ has three Cartesian components, the fall-off of the correlator for increasing q is not described by a single half-width, but rather by a (symmetric) 3×3 tensor [10] which describes the curvature of the correlation function near $\boldsymbol{q} = 0$. We will see that in fact nearly all relevant information that can be extracted from the correlation function resides in the 6 independent components of this tensor. This in turn implies that in order to compute the correlation function C it is sufficient to approximate the source function S by a Gaussian in x which contains only information on its space-time moments up to second order.

Let us write the arbitrary emission function $S(x,K)$ in the following form:

$$S(x,K) = N(K)\; S(\bar{x}(K),K)\; e^{-\frac{1}{2}\tilde{x}^\mu(K)\,B_{\mu\nu}(K)\,\tilde{x}^\nu(K)} + \delta S(x,K)\,, \qquad (22)$$

where we adjust the parameters $N(K)$, $\bar{x}^\mu(K)$, and $B_{\mu\nu}(K)$ of the Gaussian first term in such a way that the correction term δS has vanishing zeroth, first and second order space-time moments:

$$\int d^4x\, \delta S(x,K) = \int d^4x\, x^\mu\, \delta S(x,K) = \int d^4x\, x^\mu x^\nu\, \delta S(x,K) = 0\,. \tag{23}$$

This is achieved by setting

$$N(K) = E_K \frac{dN}{d^3K} \frac{\det^{1/2} B_{\mu\nu}(K)}{S(\bar{x}(K),K)}\,, \tag{24}$$

$$\bar{x}^\mu(K) = \langle x^\mu \rangle\,, \tag{25}$$

$$\left(B^{-1}\right)_{\mu\nu}(K) = \langle \tilde{x}_\mu \tilde{x}_\nu \rangle \equiv \langle (x-\bar{x})_\mu (x-\bar{x})_\nu \rangle\,. \tag{26}$$

The (K-dependent) average over the source function $\langle \ldots \rangle$ has been defined in Eq. (10). The normalization factor (24) ensures that the Gaussian term in (22) gives the correct single-particle spectrum (6); it fixes the normalization on-shell, i.e. for $K^0 = E_K$, but as we discussed this is where we need the emission function also for the computation of the correlator. $\bar{x}(K)$ in (25) is the centre of the emission function $S(x,K)$ and approximately equal to its "saddle point", i.e. the point of highest emissivity for particles with momentum K. The second equality in (26) defines $\tilde{x}$ as the space-time coordinate relative to the centre of the emission function; only this quantity enters the further discussion, since, due to the invariance of the momentum spectra under arbitrary translations of the source in coordinate space, the absolute position of the emission point is not measurable in experiments which determine only particle momenta. Since $\bar{x}(K)$ is not measurable, neither is the normalization $N(K)$ [17] as its definition (24) involves the emission function at $\bar{x}(K)$. Finally, Eq. (26) ensures that the Gaussian first term in (22) correctly reproduces the second central space-time moments $\langle \tilde{x}_\mu \tilde{x}_\nu \rangle$ of the original emission function, in particular its r.m.s. widths in the various space-time directions.

Inserting the decomposition (22) into Eq. (10) we obtain

$$C(\boldsymbol{q},\boldsymbol{K}) = 1 + \exp\left[-q^\mu q^\nu \langle \tilde{x}_\mu \tilde{x}_\nu \rangle(\boldsymbol{K})\right] + \delta C(\boldsymbol{q},\boldsymbol{K})\,. \tag{27}$$

The Gaussian in q results from the Fourier transform of the Gaussian contribution in (22); the last term δC receives contributions from the second term δS in (22) which contains information on the third and higher order space-time moments of the emission function, like sharp edges, wiggles, secondary peaks, or non-Gaussian tails in the source. It is at least of order q^4; the second derivative of the full correlator at $q = 0$ is given *exactly* by the Gaussian in (27).

In the past it has repeatedly been observed that the correlation data appear to be better fit by exponentials than by Gaussians. As far as I know, however, this happened usually for 1-dimensional fits as a function of the single Lorentz invariant variable $Q_{\rm inv}^2 = (q^0)^2 - \boldsymbol{q}^2$ while, at least for heavy ion collisions, the 3-dimensional correlators look much more Gaussian. (Correlators from elementary collisions are more strongly affected by resonance decay contributions (Sec.6), giving rise to severe deviations from Gaussian behavior even in three q-dimensions.[35]) Contemplating the structure of Eq. (27) one realizes that a fit in $Q_{\rm inv}^2$ does not make sense: the generic structure of the exponent, $-q^\mu q^\nu \langle \tilde{x}_\mu \tilde{x}_\nu \rangle$, tells us that the term $(q^0)^2$ should come with the time variance of the source while the spatial components $(q^i)^2$ should come with the spatial variances of the source. Since all variances are positive semidefinite by definition, it does not make sense to parametrize the correlation function by a variable in which $(q^0)^2$ and $\boldsymbol{q}^2$ appear with the opposite sign! Such a fit might work if the time variance and all mixed variances vanished identically and all three spatial variances were equal, but this is certainly not generic and also not frame-independent. *The variable $Q_{\rm inv}$ should therefore* not *be used for fitting HBT data.*

Note that Eq. (27) has no factor $\frac{1}{2}$ in the exponent. If the measured correlator is fitted by a Gaussian as defined in (27), its q-width can be directly interpreted in terms of the r.m.s. widths of the source in coordinate space. Model comparisons are thus most easy if the latter are directly parametrized in terms of r.m.s. widths.

Eqs. (22) and (27) would not be useful if the contributions from δS and δC were not somehow small enough to be neglected. This requires a numerical investigation. It was shown [16] that in typical (and even in some not so typical) situations δS has a negligible influence on the half width of the correlation function. It contributes only weak, essentially unmeasurable structures in $C(\boldsymbol{q}, \boldsymbol{K})$ at large values of $\boldsymbol{q}$. The reader can easily verify this analytically for an emission function with a sharp box profile; the results for the exact correlator and the one resulting from the Gaussian approximation (22) differ by less than 5% in the half widths;[10] the exact correlator has, as a function of q, secondary maxima with an amplitude below 5% of the value of the correlator at $q = 0$. We have checked that similar statements remain even true for a source with a doughnut structure, i.e. with a hole in the middle, which was obtained by rotating the superposition of two 1-dimensional Gaussians separated by twice their r.m.s. widths around their center. The only situation where these statements require qualification is if the correlator receives contributions from the decay of long-lived resonances; unfortunately, this is of relevance for pion interferometry as will be discussed in Sec. 6.

Eq. (27) implies that the two-particle correlator measures the second central space-time moments of the emission function. That's it – finer features of its space-time structure (edges, wiggles, holes) cannot be measured with two-particle correlations, but require the analysis of three-, four-, ..., many-particle correlations.[4] It follows that possible rapid quantum oscillations of the source Wigner density are also essentially unmeasurable by 2-particle interferometry. The variances $\langle \tilde{x}_\mu \tilde{x}_\nu \rangle$ are in general *not* identical with our naive intuitive notion of the "source radius": unless the source is stationary and has no x-K-correlations at all, the variances depend on the pair momentum $\boldsymbol{K}$ and cannot be interpreted in terms of simple overall source geometry. Their correct interpretation[9,12,13] is in terms of "lengths of homogeneity" which give, for each pair momentum $\boldsymbol{K}$, the size of the region around the point of maximal emissivity $\bar{x}(\boldsymbol{K})$ over which the emission function is sufficiently homogeneous to contribute to the correlator. Thus HBT measures "regions of homogeneity" in the source and their variation with the momentum of the particle pairs. As we will see, the latter is the key to their physical interpretation.

3.2 YKP parametrization for the correlator and HBT radius parameters

A full characterization of the source in terms of its second order space-time variances requires knowledge of the 10 functions $\langle \tilde{x}_\mu \tilde{x}_\nu \rangle(\boldsymbol{K})$. However, due to the mass-shell constraint (17) which leaves only three independent components of q, only 6 linear combinations of theses functions are actually measurable.[10] For azimuthally symmetric sources three of these 10 functions vanish by symmetry[10], but again the mass-shell constraint permits to measure only 4 linear combinations of the remaining 7 functions of $\boldsymbol{K}$.

Before the correlator (27) can thus be fit to data, the redundant components of q must first be eliminated via (17). We use a Cartesian coordinate system with the z-axis along the beam direction and the x-axis along $\boldsymbol{K}_\perp$. Then $\boldsymbol{\beta} = (\beta_\perp, 0, \beta_l)$. We assume an azimuthally symmetric source (impact parameter ≈ 0) and eliminate from (27) q_x and q_y in terms of $q_\perp^2 = q_x^2 + q_y^2$, q_l and q^0. This yields the YKP parametrization[10,17]

$$C(\boldsymbol{q}, \boldsymbol{K}) = 1 + \exp\left[-R_\perp^2 q_\perp^2 - R_\parallel^2 \left(q_l^2 - (q^0)^2\right) - \left(R_0^2 + R_\parallel^2\right) (q \cdot U)^2\right]. \quad (28)$$

Here $R_\perp$, $R_\parallel$, R_0, U are four K-dependent parameter functions. $U(\boldsymbol{K})$ is a 4-velocity with only a longitudinal spatial component:

$$U(\boldsymbol{K}) = \gamma(\boldsymbol{K}) \left(1, 0, 0, v(\boldsymbol{K})\right), \quad \text{with} \quad \gamma = \frac{1}{\sqrt{1 - v^2}}. \quad (29)$$

Its value depends, of course, on the measurement frame. The "Yano-Koonin velocity" $v(\boldsymbol{K})$ can be calculated[17] in an arbitrary reference frame from the second central space-time moments of $S(x,K)$. It is, to a good approximation, the longitudinal velocity of the fluid element from which most of the particles with momentum $\boldsymbol{K}$ are emitted.[10,17] For sources with boost-invariant longitudinal expansion velocity the YK-rapidity associated with $v(\boldsymbol{K})$ is linearly related to the pair rapidity Y.[17]

The other three YKP parameters do not depend on the longitudinal velocity of the observer. (This distinguishes the YKP form (28) from the Pratt-Bertsch parametrization[8,9,11] which results from eliminating q^0 in (27).) Their physical interpretation is easiest in terms of coordinates measured in the frame where $v(\boldsymbol{K})$ vanishes. There they are given by[10]

$$
\begin{aligned}
R_\perp^2(\boldsymbol{K}) &= \langle \tilde{y}^2 \rangle\,, &(30)\\
R_\parallel^2(\boldsymbol{K}) &= \left\langle \left(\tilde{z} - (\beta_l/\beta_\perp)\tilde{x}\right)^2 \right\rangle - (\beta_l/\beta_\perp)^2 \langle \tilde{y}^2 \rangle \approx \langle \tilde{z}^2 \rangle\,, &(31)\\
R_0^2(\boldsymbol{K}) &= \left\langle \left(\tilde{t} - \tilde{x}/\beta_\perp\right)^2 \right\rangle - \langle \tilde{y}^2 \rangle/\beta_\perp^2 \approx \langle \tilde{t}^2 \rangle. &(32)
\end{aligned}
$$

$R_\perp$, $R_\parallel$ and R_0 thus measure, approximately, the (K-dependent) transverse, longitudinal and temporal regions of homogeneity of the source in the local comoving frame of the emitter. The approximation in (31,32) consists of dropping terms which for the model discussed below vanish in the absence of transverse flow and were found to be small even for finite transverse flow.[17] Note that it leads to a complete separation of the spatial and temporal aspects of the source. This separation is spoiled by sources with $\langle \tilde{x}^2 \rangle \neq \langle \tilde{y}^2 \rangle$. For our source this happens for non-zero transverse (in particular for large) transverse flow η_f, but for opaque sources where particle emission is surface dominated[37] this occurs even without transverse flow.[37,38]

4 A model for a finite expanding source

For our quantitative studies we used the following model for an expanding thermalized source:[10]

$$
S(x,K) = \frac{M_\perp \cosh(\eta - Y)}{8\pi^4 \Delta\tau} \exp\left[-\frac{K\cdot u(x)}{T(x)} - \frac{(\tau-\tau_0)^2}{2(\Delta\tau)^2} - \frac{r^2}{2R^2} - \frac{(\eta-\eta_0)^2}{2(\Delta\eta)^2} \right]. \quad (33)
$$

Here $r^2 = x^2 + y^2$, the spacetime rapidity $\eta = \frac{1}{2}\ln[(t+z)/(t-z)]$, and the longitudinal proper time $\tau = \sqrt{t^2 - z^2}$ parametrize the spacetime coordinates x^μ, with measure $d^4x = \tau\, d\tau\, d\eta\, r\, dr\, d\phi$. $Y = \frac{1}{2}\ln[(E_K + K_L)/(E_K - K_L)]$ and $M_\perp = \sqrt{m^2 + K_\perp^2}$ parametrize the longitudinal and transverse components of the pair momentum $\boldsymbol{K}$. $\sqrt{2}R$ is the transverse geometric (Gaussian) radius of the source, τ_0 its average freeze-out proper time, $\Delta\tau$ the mean proper time duration of particle emission, and $\Delta\eta$ parametrizes the finite longitudinal extension of the source. $T(x)$ is the freeze-out temperature; if you don't like the idea of thermalization in heavy ion collisions, you can think of it as a parameter that describes the random distribution of the particle momenta at each space-time point around their average value. The latter is parametrized by a collective flow velocity $u^\mu(x)$ in the form

$$u^\mu(x) = (\cosh\eta \cosh\eta_t(r),\, \sinh\eta_t(r)\, \boldsymbol{e}_r,\, \sinh\eta \cosh\eta_t(r)) \,, \tag{34}$$

with a boost-invariant longitudinal flow rapidity $\eta_l = \eta$ ($v_l = z/t$) and a linear transverse flow rapidity profile

$$\eta_t(r) = \eta_f \left(\frac{r}{R}\right) . \tag{35}$$

η_f scales the strength of the transverse flow. The exponent of the Boltzmann factor in (33) can then be written as

$$K \cdot u(x) = M_\perp \cosh(Y-\eta) \cosh\eta_t(r) - \boldsymbol{K}_\perp {\cdot} \boldsymbol{e}_r \sinh\eta_t(r) \,. \tag{36}$$

For vanishing transverse flow ($\eta_f = 0$) the source depends only on $M_\perp$, and remains azimuthally symmetric for all $K_\perp$. Since in the absence of transverse flow the β-dependent terms in (31) and (32) vanish and the source itself depends only on $M_\perp$, all three YKP radius parameters then show perfect $M_\perp$-scaling. Plotted as functions of $M_\perp$, they coincide for pion and kaon pairs (see Fig. 1, left column). For non-zero transverse flow (right column) this $M_\perp$-scaling is broken by two effects: (1) The thermal exponent (36) receives an additional contribution proportional to $K_\perp = \sqrt{M_\perp^2 - m^2}$. (2) The terms which were neglected in the second equalities of (31,32) are non-zero, and they also depend on $\beta_\perp = K_\perp/E_K$. Both effects induce an explicit rest mass dependence and destroy the $M_\perp$-scaling of the YKP size parameters.

5 *K*-dependence of YKP parameters and collective flow

Collective expansion induces correlations between coordinates and momenta in the source, and these result in a dependence of the HBT parameters on the pair momentum K. At each point in the source the local velocity distribution is centered around the average fluid velocity; two points whose fluid elements move rapidly relative to each other are thus unlikely to contribute particles with small relative momenta. Essentially only such regions in the source contribute to the correlation function whose fluid elements move with velocities close to the velocity of the observed particle pair.

5.1 *The Yano-Koonin velocity and longitudinal flow*

Fig. 1 shows (for pion pairs) the dependence of the YK velocity on the pair momentum $\boldsymbol{K}$. In Fig. 1a we show the YK rapidity $Y_{\rm YK} = \frac{1}{2}\ln[(1+v)/(1-v)]$ as a function of the pair rapidity Y (both relative to the CMS) for different values of $K_\perp$, in Fig. 1b the same quantity as a function of $K_\perp$ for different Y. Solid lines are without transverse flow, dashed lines are for $\eta_f = 0.6$. For large $K_\perp$ pairs, the YK rest frame approaches the LCMS (which moves with the pair rapidity Y); in this limit all pairs are thus emitted from a small region in the source which moves with the same longitudinal velocity as the pair. For small $K_\perp$ the YK frame is considerably slower than the LCMS; this is due to the thermal smearing of the particle velocities in our source around the local fluid velocity $u^\mu(x)$.[17] The linear relationship between the rapidity $Y_{\rm YK}$ of the Yano-Koonin frame and the pion pair rapidity Y is a direct reflection of the boost-invariant longitudinal expansion flow.[17] For a non-expanding source $Y_{\rm YK}$ would be independent of Y. Additional transverse flow is seen to have nearly no effect. The dependence of the YK velocity on the pair rapidity thus measures directly the longitudinal expansion of the source and cleanly separates it from its transverse dynamics.

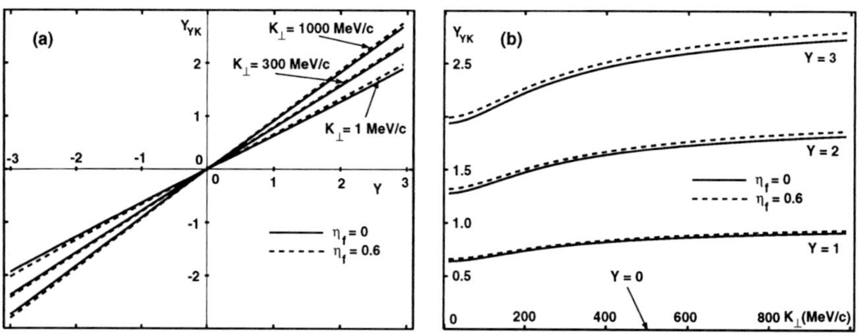

Figure 1: (a) The Yano-Koonin rapidity for pion pairs, as a function of the pair c.m. rapidity Y, for various values of $K_\perp$ and two values for the transverse flow η_f. (b) The same, but plotted against $K_\perp$ for various values of Y and η_f. Source parameters: $T = 140$ MeV, $\Delta\eta = 1.2$, $R = 3$ fm, $\tau_0 = 3$ fm/c, $\Delta\tau = 1$ fm/c.

The NA49 data for 160 A GeV Pb+Pb collisions[39] show very clearly such a more or less linear rise of the Yano-Koonin source rapidity with the rapidity of the pion pair. This confirms, in the most transparent way imaginable, their earlier conclusion[40] based on the Y-dependence of the longitudinal radius parameter R_l in the Pratt-Bertsch parametrization that the source created in 200 A GeV S+A collisions expands longitudinally in a nearly boost-invariant way.

Note that this longitudinal flow need not be of hydrodynamical (pressure generated) nature. Similar longitudinal position-momentum correlations arise in string fragmentation. This should cause a similar linear rise of the YK-rapidity with the pair rapidity in jet fragmentation (with the z-axis oriented along the jet axis). It would be interesting to confirm this prediction in e^+e^- or pp collisions.

5.2 $M_\perp$-dependence of YKP radii; transverse flow

If the source expands rapidly and features large velocity gradients, the "regions of homogeneity" contributing to the correlation function will be small. Their size will be inversely related to the velocity gradients, scaled by a "thermal smearing factor" $\sqrt{T/M_\perp}$ which characterizes the width of the Boltzmann distribution.[9] If one evaluates the expectation values (30-32) by saddle point integration one finds for pairs with $Y=0$

$$R_\perp^2 = R_*^2\,, \qquad R_0^2 = (\Delta t_*)^2\,, \qquad R_\parallel^2 = L_*^2\,, \tag{37}$$

with

$$\frac{1}{R_*^2} = \frac{1}{R^2} + \frac{1}{R_{\rm flow}^2}\,, \tag{38}$$

$$(\Delta t_*)^2 = (\Delta\tau)^2 + 2\left(\sqrt{\tau_0^2 + L_*^2} - \tau_0\right)^2\,, \tag{39}$$

$$\frac{1}{L_*^2} = \frac{1}{(\tau_0\Delta\eta)^2} + \frac{1}{L_{\rm flow}^2}\,, \tag{40}$$

where $R_{\rm flow}$ and $L_{\rm flow}$ are the transverse and longitudinal "dynamical lengths of homogeneity" due to the expansion velocity gradients:

$$R_{\rm flow}(M_\perp) = \frac{R}{\eta_f}\sqrt{\frac{T}{M_\perp}} = \frac{1}{\partial\eta_t(r)/\partial r}\sqrt{\frac{T}{M_\perp}}\,, \tag{41}$$

$$L_{\rm flow}(M_\perp) = \tau_0\sqrt{\frac{T}{M_\perp}} = \frac{1}{\partial{\cdot}u_l}\sqrt{\frac{T}{M_\perp}}\,, \tag{42}$$

where u_l is the longitudinal 4-velocity.

Thus, for expanding sources, the HBT radius parameters are generically decreasing functions of $M_\perp$. The slope of this decrease grows with the expansion rate[16,17] (this cannot be seen in the saddle point approximated expressions above). Longitudinal expansion affects mostly the longitudinal radius parameter $R_\parallel$ and the temporal parameter R_0;[17] the latter is a secondary effect since particles from different points are usually emitted at different times, and a decreasing longitudinal homogeneity length thus also leads to a reduced effective duration of particle emission (see lower panels in Fig. 2).

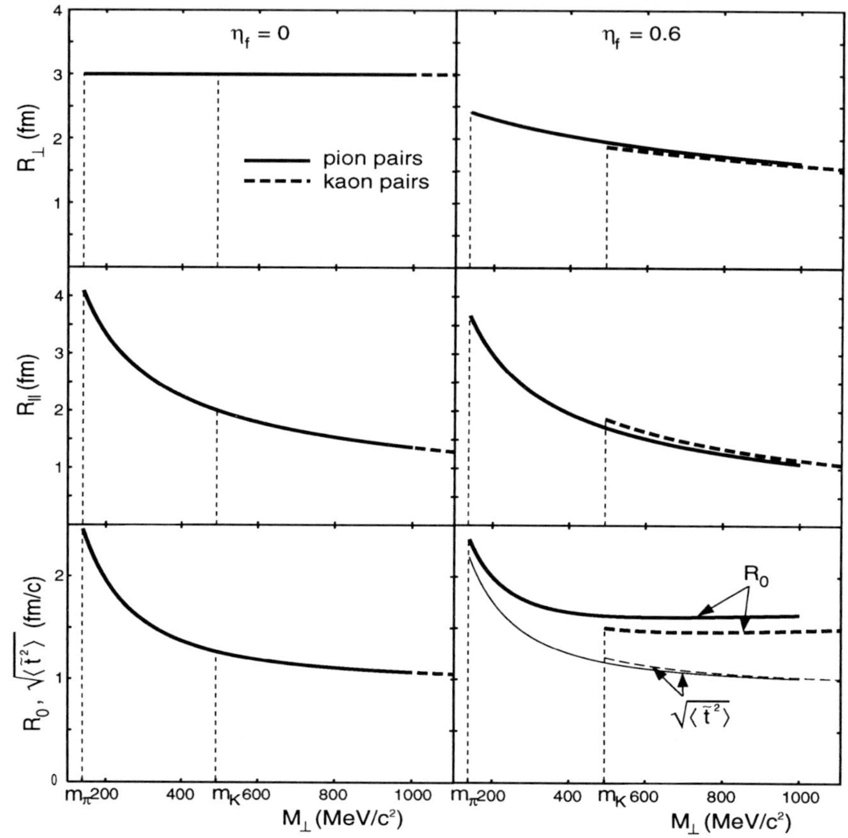

Figure 2: The YKP radii $R_\perp$, $R_\parallel$, and R_0 (top to bottom) for zero transverse flow (left column) and for $\eta_f = 0.6$ (right column), as functions of $M_\perp$ for pairs at $Y_{\rm cm} = 0$. Solid (dashed) lines are for pions (kaons). The breaking of the $M_\perp$-scaling by transverse flow is obvious in the right column. For nonzero transverse flow R_0 also does not agree exactly with the effective source lifetime $\sqrt{\langle \tilde{t}^2 \rangle}$ (lower right panel). Source parameters as in Fig. 1.

The transverse radius parameter $R_\perp$ is invariant under longitudinal boosts and thus not affected at all by longitudinal expansion (upper left panel in Fig. 2). It begins, however, to drop as a function of $M_\perp$ if the source expands in the transverse directions (upper right panel). Comparing the lower two left and right panels in Fig. 2 one sees that the sensitivity of $R_\parallel$ and R_0 to transverse flow is much weaker.[17] Transverse (longitudinal) flow thus mostly affects the transverse (longitudinal) regions of homogeneity.

While longitudinal "flow" is not necessarily a signature for nuclear collectivity but could be "faked" as discussed at the end of the previous subsection, transverse flow is much more generic in this respect: the ingoing channel has no transverse collective motion, and the only mechanism imaginable for the creation of transverse collective dynamics during the collision is multiple (re-)scattering among the secondaries, leading ultimately to hydrodynamic transverse flow.

Unfortunately, the observation of an $M_\perp$-dependence of $R_\perp$ by itself is not sufficient to prove the existence of radial transverse flow. It can also be created by other types of transverse gradients, e.g. a transverse temperature gradient.[9,14,15] To exclude such a possibility one must check the $M_\perp$-scaling of the YKP radii, i.e. the independence of the functions $R_i(M_\perp)$ $(i = \perp, \|, 0)$ of the particle rest mass (which is not broken by temperature gradients). Since different particle species are affected differently by resonance decays, such a check further requires the elimination of resonance effects.

6 Resonance decays

Resonance decays contribute additional pions at low $M_\perp$; these pions originate from a larger region than the direct ones, due to resonance propagation before decay. They cause an $M_\perp$-dependent modification of the HBT radii.

Quantitative studies [18,19] have shown that the resonances can be subdivided into three classes with different characteristic effects on the correlator:
(i) Short-lived resonances with lifetimes up to a few fm/c do not propagate far outside the region of thermal emission and thus affect $R_\perp$ only marginally. They contribute to R_0 and $R_\|$ up to about 1 fm via their lifetime; $R_\|$ is larger if pion emission occurs later because for approximately boost-invariant expansion the longitudinal velocity gradient decreases as a function of time.
(ii) Long-lived resonances with lifetimes of more than several hundred fm/c do not contribute to the measured correlation and thus only reduce the correlation strength (the intercept at $q = 0$), without changing the shape of the correlator.[a] Decaying at large distances from their production point, they simulate a very large source which contributes to the correlation signal only for unmeasurably small relative momenta.
(iii) Only the ω meson with its lifetime of 23.4 fm/c does not fall in either

[a] A reduced correlation strength in the two-particle sector could also arise from partial phase coherence in the source.[36] By comparing two- and three-particle correlations, the intercept-reducing effects of resonances can be eliminated, and the degree of coherence resp. chaoticity in the source can be unambiguously determined.[4]

of these two classes and can thus distort the form of the correlation function. It contributes a second bump at small q to the correlator, giving it a non-Gaussian shape and thus complicating [19] the extraction of HBT radii from a Gaussian fit. At small $M_\perp$ up to 10% of the pions can come from ω decays, and this fraction doubles effectively in the correlator since the other pion can be a direct one; thus the effect is not always negligible.

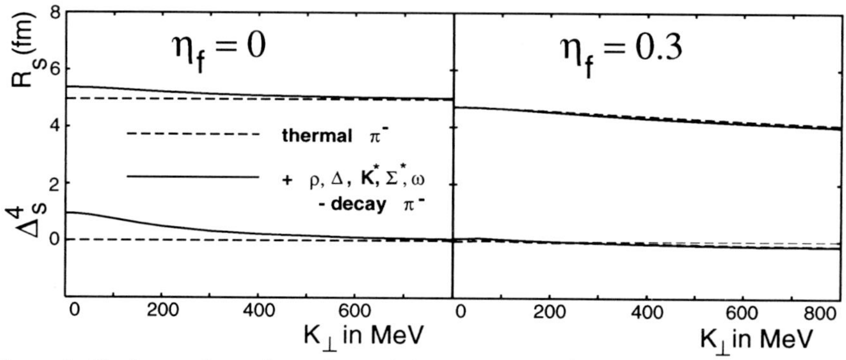

Figure 3: The inverted q-variance $R_\perp$ and the kurtosis $\Delta_\perp$ (the index s in the figure stands for "sideward") at $Y = 0$ as functions of $K_\perp$. Left: $\eta_f = 0$ (no transverse flow). Right: $\eta_f = 0.3$. The difference between dashed and solid lines is entirely dominated by ω decays. Source parameters as in Fig. 1, except for $R = 5$ fm.

In a detailed model study [19] we showed that resonance contributions can be identified through the non-Gaussian features in the correlator induced by the tails in the emission function resulting from resonance decays. To this end one computes the second and fourth order q-moments of the correlator.[19] The second order moments define the HBT radii, while the kurtosis (the normalized fourth order moments) provide a lowest order measure for the deviations from a Gaussian shape. We found [19] that, at least for the model (33), a positive kurtosis can always be associated with resonance decay contributions (Fig. 3, left panel). Strong flow also generates a non-zero, but small and apparently always negative kurtosis (Fig. 3, right panel). Any $M_\perp$-dependence of $R_\perp$ which is associated with a positive $M_\perp$-dependent kurtosis must therefore be regarded with suspicion; an $M_\perp$-dependence of $R_\perp$ with a vanishing or negative kurtosis, however, cannot be blamed on resonance decays.

In our model, the first situation is realized for a source without transverse expansion (left panel of Fig. 1): At small $M_\perp$ the ω contribution increases $R_\perp$ by up to 0.5 fm while for $M_\perp > 600$ MeV it dies out. The effect on $R_\perp$ is small because the heavy ω moves slowly and doesn't travel very far before decaying. The resonance contribution is clearly visible in the positive kurtosis (lower curve). For non-zero transverse flow (right panel) there is no resonance contribution to $R_\perp$; this is because for finite flow the effective source size for the heavier ω is smaller than for the direct pions, and the ω-decay pions thus always remain buried under the much more abundant direct ones. Correspondingly the kurtosis essentially vanishes; in fact, it is slightly negative, due to the weak non-Gaussian features induced by the transverse flow.

7 Opaque sources

The emission function (33) is only one of an infinity of possible source parametrizations. Its form permits, however, easy implementation of most of the important features of the sources created in heavy ion collisions. Still, there is one important physical situation which cannot be parametrized by the formula (33): if the source emits particles not from the entire volume, but only from a thin surface layer. This is how the sun radiates photons, and this is also an often suggested picture for the slow hadronization of long-lived QGP blobs through a deflagration-type strong first order transition.

The significance of such a phenomenon for HBT interferometry was realized by Heiselberg and Vischer [37] who pointed out that an effective emission region which is part of a thin surface layer has a much smaller extension in the "outward" or x-direction than in the "sideward" or y-direction. In other words, such "opaque sources" have $\langle \tilde{x}^2 - \tilde{y}^2 \rangle < 0$. Depending on the degree of opacity (the thickness of the surface layer relative to the source radius) this difference can be large and negative. The authors pointed out [37] that this leads to the possibility of a smaller "outward" than "sideward" HBT radius parameter in the Pratt-Bertsch parametrization, even at $K_\perp = 0$. Recently B. Tomášik showed [38] that in the YKP parametrization opacity effects would show up even more spectacularly by a "lifetime parameter" R_0^2 which would diverge to $-\infty$ in the limit $K_\perp \to 0$ resp. $\beta_\perp \to 0$ (see Eq. (32)).

The source (33) can be made opaque [37,38] by multiplying it by the factor $\exp\left(-\kappa l_{\text{eff}}/\lambda\right))$ where λ is the mean free path and κl_{eff} is the effective travelling distance of the emitted particle through matter in the source.[38]

Fig. 4c shows the "temporal" YKP radius parameter R_0^2 as a function of $M_\perp$ for sources with different degrees of opacity. For different opacities R/λ, the transverse source parameters R and η_f were readjusted [38] to give the same measured [39] single particle slope. The crucial features of opacity are clearly visible: the negative contribution $\sim \langle \tilde{x}^2 - \tilde{y}^2 \rangle$ in (32) drives R_0^2 to negative values at small $K_\perp$, and this happens the sooner the shorter the mean free path λ, i.e. the thinner the surface layer is.

Fig. 4c implies that thin surfaces with $\lambda < R$ are essentially excluded. Pion freeze-out occurs not in the form of surface emission, but happens in bulk. 160 A GeV Pb+Pb collisions are thus quite similar to the Early Universe: an early stage of *complete opaqueness* (here lasting for about 8 fm/c) is followed by a rather *sudden* ($\Delta\tau$ ~1-2 fm/c) transition to complete transparency. In both cases this transition is due to the expansion and cooling of the system, causing a rapid increase of the mean free path of the particles. In the Early Universe photon decoupling is triggered by the recombination of electrons and ions into neutral atoms; in heavy ion collisions at SPS energies pion decoupling is caused mostly by the rapid cooling and dilution of the baryon density, since baryonic resonances with their strong coupling to the pion channel provide the "glue" needed for keeping the system close to local thermal equilibrium.

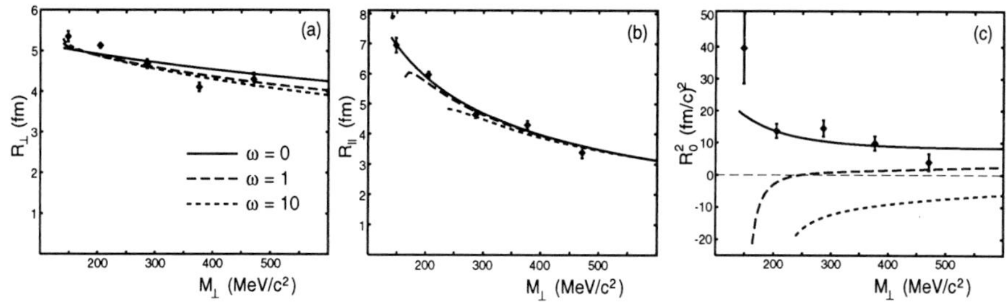

Figure 4: The YKP radius parameters $R_\perp, R_\parallel$, and R_0^2 as functions of $M_\perp$ for pion pairs from 158 A GeV/c Pb+Pb collision [39] at slightly forward pair rapidity $Y_{CM} = 1.25$. Solid, dashed and dotted curves correspond to $R/\lambda = 0, 1, 10$, respectively. (The corresponding pairs (η_f, R) are (0.4,5.74 fm), (0.345,4.83 fm), and (0.215,3.35 fm), respectively.[38]) Other source parameters: T=120 MeV, τ_0=8 fm/c, $\Delta\tau$=2 fm/c, $\Delta\eta$=1.3.

8 Analysis of Pb+Pb data

Fig. 4 shows a numerical fit [38,41] of the YKP radius parameters, using the expressions (30)-(32) with our model source (33), to data collected by the NA49 collaboration in 158 A GeV/c Pb+Pb collisions.[39] The fit includes resonance decay contributions to the single particle spectra, but not to the 2-particle correlations. Based on sample calculations the latter are, however, expected to be inside the systematic error of the data.

The width of the pion rapidity distribution is reproduced with $\Delta\eta$=1.3. τ_0 and $\Delta\eta$ are then fixed [38] by the magnitude of $R_\parallel$ and R_0. The magnitude of $R_\perp(K_\perp = 0)$ fixes the radius R once T and η_f are known. The latter are obtained from the $M_\perp$-dependence of $R_\perp$, albeit not independently: essentially only the combination $\eta_f\sqrt{M_\perp/T}$ (the velocity gradient divided by the thermal smearing factor) can be extracted.[10,42] This is similar to the single particle spectra whose $M_\perp$-slopes determine only an effective blushifted temperature,[5] $T_{\rm eff} = T\sqrt{\frac{1+\bar{v}_f}{1-\bar{v}_f}}$. The correlations between T and η_f are, however, exactly opposite in the two cases: for a fixed spectral slope T must be decreased if η_f increases while a fixed $M_\perp$-slope of $R_\perp$ requires decreasing values of η_f if T is reduced.[42] *The combination of single-particle spectra and two-particle correlation thus allows for a separate determination of T and η_f.*

The fit in Fig. 4 corresponds to an average transverse flow velocity $\bar{v}_f$=0.44, combined with a freeze-out temperature of about 120 MeV. Similar values were advocated by Kämpfer [43] based on a simultaneous analysis of single particle spectra of various different hadron species from Pb+Pb collisions.

Let us discuss in more detail the numbers resulting from this fit. First, the transverse size parameter $R \approx 6$ fm is surprisingly large. At mid-rapidity even larger values ($R \simeq 7$ fm) are required.[39,42] Resonance contributions are not expected to reduce it by more than 0.5 fm.[19] The transverse flow correction to $R_\perp$ is appreciable, resulting in a visible transverse homogeneity length of only about 5 fm at small $K_\perp$, but even this number is large. R=6-7 fm corresponds to an r.m.s. radius $r_{\rm rms} = \sqrt{\langle \tilde{x}^2 + \tilde{y}^2 \rangle} \approx$ 8.5-10 fm of the pion source, to be compared with an r.m.s. radius $r^{\rm Pb}_{\rm rms} = 1.2 \times A^{1/3} * \sqrt{2/5}$ fm $=$ 4.5 fm for the density distribution of the original Pb nucleus projected on the transverse plane. This implies a transverse expansion of the reaction zone by a linear factor $\simeq 2$. That we also find a large transverse flow velocity renders the picture consistent. The longitudinal size of the collision region at the point where the pressure in the system began to drive the transverse expansion can be estimated as follows: for the source to expand in, say, the y-direction from $\sqrt{\langle y^2 \rangle}_{\rm initial} = 1.2\, A^{1/3}/\sqrt{5}$ fm $=$ 3.2 fm to $\sqrt{\langle y^2 \rangle}_{\rm final} = R \approx 6\,$fm with an average transverse flow velocity of at most $\bar{v} = 0.44\, c$ (the freeze-out value determined from the fit shown in Fig. 4) requires a time of at least $(6-3.2)/0.44$ fm/$c \simeq 6.5$ fm/c. Due to the selfsimilarity of the longitudinal expansion the longitudinal dimension of the source grows linearly with τ. If the total expansion time until freeze-out is given by the fit parameter $\tau_0 = 8$ fm/c, the source expanded in the $\simeq 6.5$ fm/c during which there was transverse expansion by a factor 8/(8-6.5) $=$ 8/1.5 $\simeq 5$ in the longitudinal direction. We conclude that the fireball volume must have expanded by a factor $5 * 2^2 \approx 20$ between the onset of transverse expansion and freeze-out! This is the clearest evidence for strong collective dynamical behaviour in ultra-relativistic heavy-ion collisions so far.

The local comoving energy density at freeze-out can be estimated from the fitted values for T and η_f. The thermal energy density of a hadron resonance gas at $T = 120$ MeV and moderate baryon chemical potential is of the order of 100 MeV/fm^3. The large average transverse flow velocity of $\langle v_f \rangle \approx 0.44$ implies that about 25% flow energy must be added in the lab frame. This results in an estimate of about 0.1 GeV/fm^3 $\times$ 1.25 $\times$ 20 $\approx$ **2.5 GeV/fm^3** for the energy density of the reaction zone at the onset of transverse expansion. This is well above the critical energy density $\epsilon_{\rm cr} \leq 0.9$ GeV/fm^3 predicted by lattice QCD for deconfined quark-gluon matter.[44] Whether this energy density was fully thermalized is, of course, a different question. It must, however, have been accompanied by transverse pressure (i.e. some degree of equilibration of momenta must have occurred already before this point), because otherwise transverse expansion could not have been initiated.

9 Conclusions

I hope to have shown that

- two-particle correlation functions from heavy-ion collisions provide valuable information both on the geometry **and** the dynamical state of the reaction zone at freeze-out;
- a comprehensive and simultaneous analysis of single-particle spectra and two-particle correlations, with the help of models which provide a realistic parametrization of the emission function, allows for an essentially complete reconstruction of the final state of the reaction zone, which can serve as a reliable basis for theoretical back-extrapolations towards the interesting hot and dense early stages of the collision;
- simple and conservative estimates, based on the crucial new information from HBT measurements on the large transverse size of the source at freeze-out and using only energy conservation, lead to the conclusion that in Pb+Pb collisions at CERN, before the onset of transverse expansion, the energy density exceeded comfortably the critical value for the formation of a color deconfined state of quarks and gluons. There is, however, no evidence for long time delays due to hadronization of the QGP, and pions freeze out in bulk rather than from the surface of the collision fireball. This is in line with lattice results which predict at most a *weakly* first order confinement transition, and with other evidence [45] for rapid hadronization.

I would like to thank the organizers of CRIS98 for their hospitality and support. The results reported here were obtained in collaboration with D. Anchishkin, S. Chapman, P. Scotto, C. Slotta, B. Tomášik, U. Wiedemann, Y.-F. Wu, and Q.H. Zhang. Our work was supported by grants from DAAD, DFG, NSFC, BMBF, GSI and the Alexander von Humboldt Foundation.

References

1. R. Hanbury Brown and R.Q. Twiss, *Phil. Mag.* **45**, 633 (1954); and *Nature* **177**, 27 (1956); **178**, 1046 and 1447 (1956).
2. G. Goldhaber, S. Goldhaber, W. Lee, and A. Pais, *Phys. Rev.* **120**, 300 (1960).
3. J. Schmidt-Sörensen, this volume, and references therein.
4. U. Heinz and Q.H. Zhang, *Phys. Rev.* C **56**, 426 (1997); H. Heiselberg and A. Vischer, nucl-th/9707036.
5. K.S. Lee and U. Heinz, *Z. Phys.* C **43**, 425 (1989); K.S. Lee, U. Heinz and E. Schnedermann, *Z. Phys.* C **48**, 525 (1990); E. Schnedermann and U. Heinz, *Phys. Rev. Lett.* **69**, 2908 (1992); E. Schnedermann, J. Sollfrank and U. Heinz, *Phys. Rev.* C **48**, 2462 (1993); E. Schnedermann and U. Heinz, *Phys. Rev.* C **50**, 1675 (1994); U. Heinz, in *Hot Hadronic Matter: Theory and Experiment*, (J. Letessier *et al.*, eds.), NATO ASI Series **B346** (Plenum, New York, 1995), p. 413.
6. J. Stachel *et al.*, [E814 Collaboration], *Nucl. Phys.* A **566**, 183c (1994); P. Braun-Munzinger *et al.*, *Phys. Lett.* B **344**, 43 (1995).
7. See proceedings of *Quark Matter 96*, *Nucl. Phys.* A **610** (1996), and of *Quark Matter 97*, *Nucl. Phys.* A (1998), in press.
8. M. Herrmann and G.F. Bertsch, *Phys. Rev.* C **51**, 328 (1995).
9. S. Chapman, P. Scotto and U. Heinz, *Phys. Rev. Lett.* **74**, 4400 (1995); and *Heavy Ion Physics* **1**, 1 (1995).
10. S. Chapman, J.R. Nix and U. Heinz, *Phys. Rev.* C **52**, 2694 (1995).
11. S. Pratt, *Phys. Rev. Lett.* **53**, 1219 (1984); *Phys. Rev.* D **33**, 1314 (1986).
12. A.N. Makhlin and Yu.M. Sinyukov, *Z. Phys.* C **39**, 69 (1988).
13. S.V. Akkelin and Y.M. Sinyukov, *Phys. Lett.* B **356**, 525 (1995).
14. T. Csörgő and B. Lörstad, *Phys. Rev.* C **54**, 1396 (1996).
15. B. Tomášik and U. Heinz, *Eur. Phys. J.* C (23. Feb. 1998) (online publication) [nucl-th/9707001].
16. U.A. Wiedemann, P. Scotto and U. Heinz, *Phys. Rev.* C **53**, 918 (1996).
17. U. Heinz *et al.*, *Phys. Lett.* B **382**, 181 (1996); Wu Y.-F. *et al.*, *Eur. Phys. J.* C **1**, 599 (1998).
18. B.R. Schlei *et al.*, *Phys. Lett.* B **293**, 275 (1992); J. Bolz *et al.*, *ibid.* **300**, 404 (1993); *Phys. Rev.* D **47**, 3860 (1993).
19. U.A. Wiedemann and U. Heinz, *Phys. Rev.* C **56**, R610 (1997); and *Phys. Rev.* C **56**, 3265.
20. U. Heinz, in *Correlations and Clustering Phenomena in Subatomic Physics*, (M.N. Harakeh *et al.*, eds.), NATO ASI Series **B359** (Plenum,

New York, 1997), p. 137.
21. M. Gyulassy, S.K. Kauffmann, and L.W. Wilson, *Phys. Rev.* C **20**, 2267 (1979).
22. E. Shuryak, *Phys. Lett.* B **44**, 387 (1973); *Sov. J. Nucl. Phys.* **18**, 667 (1974).
23. S. Chapman and U. Heinz, *Phys. Lett.* B **340**, 250 (1994).
24. D. Anchishkin, U. Heinz, and P. Renk, *Phys. Rev.* C **57**, 1428 (1998).
25. Q.H. Zhang, U.A. Wiedemann, C. Slotta, and U. Heinz, *Phys. Lett.* B **407**, 33 (1997).
26. U.A. Wiedemann *et al.*, *Phys. Rev.* C **56**, R614 (1997).
27. T. Csörgő and J. Zimányi, *Phys. Rev. Lett.* **80**, 916 (1998); J. Zimányi and T. Csörgő, hep-ph/9705432.
28. U.A. Wiedemann, nucl-th/9801009
29. D. Pelte and H. Merlitz, this volume, and references therein [nucl-th/9806049].
30. H.W. Barz, *Phys. Rev.* C **53**, 2536 (1996); S. Pratt and M.B. Tsang, *Phys. Rev.* C **36**, 2390 (1987).
31. D. Anchishkin and U. Heinz, in preparation.
32. U. Heinz and U.A. Wiedemann, in preparation.
33. D.A. Brown and P. Danielewicz, *Phys. Lett.* B **398**, 252 (1997); and *Phys. Rev.* C **57**, 2474 (1998).
34. U. Heinz, in: Proceedings of the 5th International Workshop on *Relativistic Aspects of Nuclear Physics*, (T. Kodama *et al.*, eds.), World Scientific, Singapore, 1998, in press [nucl-th/9710065].
35. J. Ellis, K. Geiger, U. Heinz and U.A. Wiedemann, in preparation.
36. I.V. Andreev, M. Plümer, and R.M. Weiner, *Int. J. Mod. Phys.* A **8**, 4577 (1993).
37. H. Heiselberg and A.P. Vischer, *Eur. Phys. J.* C **2**, 593 (1998).
38. B. Tomášik and U. Heinz, nucl-th/9805016, subm. to *Eur. Phys. J.* C .
39. H. Appelshäuser *et al.* (NA49 Collaboration), *Eur. Phys. J.* C **2**, 661 (1998); P.G. Jones *et al.* (NA49 Collaboration), *Nucl. Phys.* A **610**, 188c (1996); R. Ganz, this volume.
40. T. Alber *et al.* (NA35 Collaboration), *Z. Phys.* C **66**, 77 (1995).
41. U.A. Wiedemann, B. Tomášik, and U. Heinz, in *Quark Matter 97*, *Nucl. Phys.* A (1998), in press [nucl-th/9801017].
42. S. Schönfelder, PhD thesis, MPI für Physik, München (1996).
43. B. Kämpfer, nucl-th/9612336, and private communication.
44. E. Laermann, *Nucl. Phys.* A **610**, 1c (1996).
45. J. Letessier, A. Tounsi, U. Heinz, J. Sollfrank, and J. Rafelski, *Phys. Rev.* C **51**, 3408 (1995).

MEASURING BLACK HOLES

G. G. C. PALUMBO
Università degli Studi di Bologna
Dipartimento di Astronomia
Via Zamboni 33, 40126 Bologna, Italy
E-mail: ggcpalumbo@astbo3.bo.astro.it

The current paradigm unifies all Active Galactic Nuclei, the central engine being a massive black hole fed by an accretion disk. Spectroscopy of iron lines observed in the X-rays provide, so far, the most sensitive probe to test the black hole environment.

> The black hole of nature are the most perfect macroscopic objects there are in the universe: the only elements in their construction are our concepts of space and time. And since the general theory of relativity provides only a single unique family of solutions for their descriptions, they are the simplest objects as well.
>
> S. Chandrasekhar
> (The mathematical theory of black holes, Oxford University Press Inc., N.Y. 1983)

1 Black Holes

1.1 What are they ?

In physical terms a "black hole" is a region where gravity is so strong that nothing, not even light, can escape. In a spacetime which is asymptotically flat a mathematically precise definition of a black hole can be given. The boundary of a black hole is called its event horizon. Any mass M of matter has a characteristic radius

$$R_S = 2\mathrm{MG}/\mathrm{c}^2$$

known as the Schwarzschild radius, which defines a volume of space containing M and from which no light can escape. Regardless of which kind of matter a black hole is made of it is indistinguishible from any other black hole but for its mass. A black hole is just a lump of gravity. As examples: the Schwarzschild radius of the Earth is about 1 cm, of the Sun 1.5 Km and of a 10^9 solar masses

BH (this is the mass one expects to be in the core of quasars) 100 Astronomical Units (1 A.U. = 1.5×10^{13} cm). Black holes are objects with extraordinary properties. Classical General Relativity Theory tells us that black holes must contain a spacetime singularity, in other words the complete gravitational collapse of a star, or any other body, must always result in a black hole rather than a "Naked Singularity". In spite of the little direct evidence of this statement it is generally believed to hold, this makes black holes physically relevant objects which describe the endpoint of gravitational collapse.

The most relevant results obtained from studies carried on in the late 1960's and early 1970's are: stationary (i.e. time-independent) black holes are uniquely determined by three parameters: mass, angular momentum and charge; energy can be extracted from rotating or charged stationary black holes; mathematical analogies between black hole mechanics and laws of thermodynamics have been found, the black hole mass plays the role of energy of a thermodynamic system, its surface gravity the role of temperature and the area of its event horizon the role of entropy.

Because of their nature black holes cannot be created and studied in a laboratory and one is forced to look for them in the vastness of the universe hoping to find some, perhaps remnant of some star or star cluster.

1.2 Where are they ?

Some celestial bodies, in the course of cosmic evolution, are expected to reach the right conditions to collapse into extremely compact spheres. There is overwhelming evidence, for example, that Supernovae explosions leave behind very dense Neutron Stars, most of which are detected at many wavelenghts and are known as pulsars. X-Ray bright binaries exchange matter from a less evolved star to the much more compact companion, the observational evidence towards this companion being a black hole of stellar origin is growing. The most convincing evidence, so far, that black holes exist in nature, however, comes from the study of Active Galactic Nuclei (AGNs). In what follows I will attempt to review the basic evidence so far collected, showing that AGNs, the most powerful engines in the universe, do indeed host a black hole. In doing this I will present the basic principles behind the measurements from which black hole parameters can be estimated. This is an interesting Physics exercice in itsef because it gives one the opportunity to conduct experiments and probe extreme gravity regimes. I will try to keep in mind that my audience is mostly made of non-astrophysicists.

2 Active Galactic Nuclei: the basics

Two dates can be considered milestones in the recent history of extragalactic astrophysics: in 1943 active galaxies were first noted by Carl Seyfert and in 1963 the 273rd radio source of the third Cambridge catalogue, 3C273, was identified with a distant star-like object, these extremely powerful objects were promptly called Quasi Stellar Radio Sources or quasars (objects with weak radio emission but similar properties are called Quasi Stellar Objects, QSO). These two classes of astronomical objects have one thing in common: their light output is so high, compared to other "normal" galaxies or quasars, that it seems unreasonable to account for it with the integral output of many luminous stars concentrated in their cores. Seyfert galaxies and QSOs therefore form the bulk of the AGN family [1].
Seyfert galaxies exibit broad and nerrow emission lines in their optical spectra and luminosities in the range 10^{41-43} erg s^{-1} while quasars, whose spectra also are rich in high ionization emission lines so strongly redshifted that QSOs are to be considered the most distant, and therefore youngest, objects in the observable universe. Their energy output is 10^{47-49} erg s^{-1}, which must be compared to a normal galaxy like our own Milky Way luminosity output of 10^{39} erg s^{-1}. Such amount of energy combined with flux variations observed in most wavebands across the electromagnetic spectrum as short as days and sometime hours have led to the conclusion that the core of AGNs ought to be confined in volumes of the size of our solar system across (10^{15} cm) [2]. In such volumes no reasonable number of stars can be contained and no acceptable mechanism can be envisaged for these stars to produce the requested energy output for the expected lifetime of a galaxy or quasar (10^{8-9} yrs).
Supernovae explosions have been advocated to explode almost dayly and indeed they might provide the requested energy, but no hope to have such processes going for any reasonable length of time, let alone the galaxy lifetime. Furthermore nuclear power has far too low efficiency ($\eta \sim 0.007$) to be considered the basic mechanism for converting stellar fuel into radiative energy. On the other hand when matter falling on a hole reaches the event horizion it can be converted into radiative energy with high efficiency. Assuming, as it seems quite natural in the highly dense stellar environment of a galactic core, that a star gets gravitationally trapped by the Black Hole pull it can be easily shown the star will get disrupted and the matter of which it is formed quickly distributed in a disk spiralling towards the hole and accreting on it [3].
The ingredients necessary to extract energy from accreting matter are large scale magnetic fields supported by electrical current flowing through the surrounding gas. The hole can be seen as an electrical conductor with resistance

of about 100 Ohms. In other words the hole acts as unipolar inductor driving a global electric current between poles and the equator. The large (10^{20} V, 10^{18} A) field should create electron-positron pairs and it is estimated that about 30% of the mass can be converted into energy. This is in principle, in practice a somewhat smaller efficiency ($< 15\%$) is expected which is however considered adequate to power the non-thermal emission of AGNs and the relativistic jets observed in many radio loud objects [3].

In order to use the above mentioned energy source for the central AGN engine a unified model has been proposed, based on several observations which all seem to indicate that AGNs are the same but appear different because of different observer's viewing angles. The AGN zoo comprises two main classes: radio loud (RL) and radio quiet (RQ) objects. The limits between the two are somewhat arbitrarely set in terms of radio power output. A useful criterion is the radio to optical ratio of the fluxes measured at 5 GHz (6 cm) and 4400 A respectivelly. For radio loud objects this ratio is in the range 10-1000 while for radio quiet 0.1-10.

As it frequently happens in astronomy classifications are qualitative criteria which inevitably introduce ambiguity for transition objects. AGNs are no exception, Seyfert galaxies were soon devided in type 1 and 2, then shortly after 1.2,..1.5,...1.9 types appeared. Radio loud AGNs are quasars and Blazars (Optically Violently Variable objects (OVV), BL Lac objects); radio quiet AGNs are Seyfert 1 and 2 galaxies, QSOs, LINERS and NELGs which are all galaxies with different emission line characteristics. This differences are not hard to understand since galaxies are clusters of some 10^{11} stars which evolved for some 10^{8-9} years and may therefore differ in some aspects such as overall gas distribution, elemental composition, interstellar density and morphology etc. However what concerns us here is mainly the bases of the AGN phenomenon well aware that fine tuning may be necessary from object to object.

The radio emitting objects are well explained by the unified model which pictures a BH surrounded by an accretion disk embedded in a thick molecular torus. Since these structures are elongated geometry plays a dominant role. Seyfert 2 are believed to be similar to Seyfert 1 but seen equatorially. X-rays are obscured and reflected by the thick equatorial disk. Seyfert X-ray spectra are explained as reflection of non-thermal power law from the surface of the disk. Compton recoil and photoelectric absorption are the responsible mechanisms for the spectral curvatures observed. RL objects are less known, although intrinsically brighter, because are further away and therefore appear dim.

The unified model, so far, has not failed although data from some objects can only fit it with difficulty. Recent observations of ionization cones in Seyfert galaxies and infrared emission from tori all work towards unification if one assumes that UV nuclear luminosity is re-radiated in the infrared[4].

3 Probing the Black Hole environment

Radio maps and optical spectroscopy are impressive methods to probe the environment of a AGN core but are giving information which is still some 100,000 gravitational radii from the BH itself. Matter of which the accretion disk orbiting the BH is made of, because of Kepler's law, spirals in with velocities which depend on the radius at which it finds itself from the hole. This velocity gradient generates viscosity and the spiral motion is therefore slow. This is the source of the X-ray emission from quasars and Seyferts and it generates within about 20 gravitational radii from the hole. Rapid X-ray variability, common in Seyfert galaxies, has been of paramount importance in determining the size of AGN cores and establishing their compactness. The emitting region of the nucleus of an active galaxy is as small as 100 light years yet the observed luminosity is some 10^{43} erg s^{-1}. In the 1990's Fe emission lines have been discovered in X-ray spectra of AGNs[5]. This turned out to be an extremely powerful diagnostic tool to probe the inner regions of the accretion disk where matter flows into a massive BH. The iron line observed is fluorescence from Fe atoms which retain their K- and L-shell electrons although at several million degrees.

Iron fluorescence has been shown to originate from reflection of X-rays generated by the corona above the disk. Some photons are back-scattered in our direction and some are absorbed by K-shell electrons of Fe which de-exites by emitting fluorescence 6.4 keV photons, the Fe line. To understand what happens one has to consider that, close to the BH, matter is spiraling very fast compared to the outer regions of the disk. In a purely Newtonian gravity field the line is simmetrically broadened by Doppler effect. Relativity plays an important role here, two effects are predicted: beaming and transverse Doppler effect. Beaming implies that the approching side of the disk appears brighter than the receding one. Since gravity is so strong close to the hole, gravitational redshift also has to be considered. As a result the broadest part of the line, which are coming from the innermost parts of the disk, are shifted to lower frequencies with respect to the narrow parts of the line originating from the outer parts of the accretion disk. The resulting emission line is very broad, with two peaks and skewed.

The Japanese-US X-ray satellite ASCA has been able to detect Fe lines of the expected shape in the active nucleus of the galaxy known as MCG–6-30-15 thanks to its sensitive and low background CCD detectors with excellent spec-

tral resolution (160 eV) around 6.4 keV, the expected Fe line energy [6]. Data from several other galaxies have confirmed both the broad shape double horn as well as the skewness of the lines. The line width implies velocities of 50,000 km s^{-1}, there is a sharp drop in the spectrum at 6.5 keV and a peak around 6.2-6.4 keV. The long tail extends down to 5 keV. This complex feature can be well fitted with a relativistic disk model and a Schwarzschild non rotating central black hole. To best explain the observed data one has to assume an inclination of about 30° for the disk respect to the observer's line of sight. This implies that the observed radiation is coming from about 20 gravitational radii from the hole, the closest to a Black Hole one has ever been able to probe an accretion disk. Other galaxies show lines with different shapes and different disk inclinations are implied. The Fe line width is of 380 eV, rather larger than expected if solar abundance for heavy elements is assumed for Seyfert 1 galaxies.
Data about the longest observations of the Fe line show extreme variability in both line and continuum. An overall variability of a factor of 2-3 and rapid variations of 1000 s or longer have been observed [7]. The variations have been attributed to coronal flashes supposed to move around with the disk. The cause of such flashes is, of course, as yet undetermined. It could be reconnection of magnetic field lines amplified by the differential rotation. The envisaged mechanism is similar to the one implied by X-ray observations of the Sun surface.
Another point is about similarities between Black Holes in AGNs and candidate Black Holes in binary stars which indeed show a power law X-ray continuum spectrum surprisingly similar to the one observed in AGNs but with no Fe line at all. The argument is the central Black Hole mass and disk. It is possible that in the stellar mass black hole case disks are far more compact than in the AGN case. If this is true they should be much hotter and therefore the Fe should be highly ionized and so should Oxygen. Large absorption above the Fe K edge is expected, in contrast with little or no absorption below it. Data however are still rather preliminary to confirm or disprove these arguments.

4 Conclusions

With X–ray spectroscopy it is possible to probe the immediate environment of a black hole, much closer than with radio and optical wavelenghts. Circumstantial evidence, mostly from one galaxy: MCG–6-30-15 shows gravitational effects on the detected iron line which suggests the presence of a rapidly spinning black hole.
Several X–ray missions are scheduled in the next few years (Axaf, XMM, ASTRO-E ...) which are about one order of magnitude more sensitive than

previous ones; from their data we will know masses and spin parameters for many black holes hidden in both radio-loud and radio-quiet objects. Properties of Kerr metrics will be explored. Strong gravity will be subject to observation.

References

1. B. Peterson in *An Introduction to Active Galactic Nuclei* (Cambridge University Press, 1997).
2. M.-H. Ulrich, L. Maraschi and C.M. Urry, *Ann. Rev. Astron. & Astroph.* 35, 445 (1997)
3. R. Blandford in *Relativistic Astrophysics and Particle Cosmology* 688, 311 (Annals of the New York Academy of Sciences, 1993)
4. C.M. Urry and P. Padovani, *Publ. Astr. Soc. Pacific* 107, 803 (1995).
5. M. Matsuoka *et al*, *Astroph. Journal* 361, 400 (1990)
6. Y. Tanaka *et al*, *Nature* 375, 659 (1995).
7. K. Iwasawa *et al*, *MNRAS* 282, 1038 (1996)

HBT correlation in 158 A·GeV Pb+Pb collisions

R. Ganz for the NA49 collaboration
Max-Planck-Institut für Physik München, Föhringer Ring 6, D-80805 München, Germany
E-mail: ganz@mppmu.mpg.de

H. Appelshäuser[7,#], J. Bächler[5], S.J. Bailey[17], D. Barna[4], L.S. Barnby[3], J. Bartke[6], R.A. Barton[3], H. Białkowska[15], A. Billmeier[10], C.O. Blyth[3], R. Bock[7], C. Bormann[10], F.P. Brady[8], R. Brockmann[7,†], R. Brun[5], P. Bunčić[5,10], H.L. Caines[3], D. Cebra[8], G.E. Cooper[2], J.G. Cramer[17], M. Cristinziani[13], P. Csato[4], J. Dunn[8], V. Eckardt[14], F. Eckhardt[13], M.I. Ferguson[5], H.G. Fischer[5], D. Flierl[10], Z. Fodor[4], P. Foka[10], P. Freund[14], V. Friese[13], M. Fuchs[10], F. Gabler[10], J. Gal[4], R. Ganz[14], M. Gaździcki[10], E. Gładysz[6], J. Grebieszkow[16], J. Günther[10], J.W. Harris[18], S. Hegyi[4], T. Henkel[13], L.A. Hill[3], I. Huang[2,8], H. Hümmler[10,+], G. Igo[12], D. Irmscher[2,7], P. Jacobs[2], P.G. Jones[3], K. Kadija[19,14], V.I. Kolesnikov[9], M. Kowalski[6], B. Lasiuk[12,18], P. Lévai[4] A.I. Malakhov[9], S. Margetis[11], C. Markert[7], G.L. Melkumov[9], A. Mock[14], J. Molnár[4], J.M. Nelson[3], M. Oldenburg[10], G. Odyniec[2], G. Palla[4], A.D. Panagiotou[1], A. Petridis[1], A. Piper[13], R.J. Porter[2], A.M. Poskanzer[2], S. Poziombka[10], D.J. Prindle[17], F. Pühlhofer[13], W. Rauch[14], J.G. Reid[17], R. Renfordt[10], W. Retyk[16], H.G. Ritter[2], D. Röhrich[10], C. Roland[7], G. Roland[10], H. Rudolph[2,10], A. Rybicki[6], A. Sandoval[7], H. Sann[7], A.Yu. Semenov[9], E. Schäfer[14], D. Schmischke[10], N. Schmitz[14], S. Schönfelder[14], P. Seyboth[14], J. Seyerlein[14], F. Sikler[4], E. Skrzypczak[16], G.T.A. Squier[3], R. Stock[10], H. Ströbele[10], Chr. Struck[13], I. Szentpetery[4], J. Sziklai[4], M. Toy[2,12], T.A. Trainor[17], S. Trentalange[12], T. Ullrich[18], M. Vassiliou[1], G. Veres[4], G. Vesztergombi[4], D. Vranić[5,19], F. Wang[2], D.D. Weerasundara[17], S. Wenig[5], C. Whitten[12], T. Wienold[2,#], L. Wood[8], T.A. Yates[3], J. Zimanyi[4], X.-Z. Zhu[17], R. Zybert[3]

[1] *Department of Physics, University of Athens, Athens, Greece.*
[2] *Lawrence Berkeley National Laboratory, University of California, Berkeley, USA.*
[3] *Birmingham University, Birmingham, England.*
[4] *KFKI Research Institute for Particle and Nuclear Physics, Budapest, Hungary.*
[5] *CERN, Geneva, Switzerland.*
[6] *Institute of Nuclear Physics, Cracow, Poland.*
[7] *Gesellschaft für Schwerionenforschung (GSI), Darmstadt, Germany.*
[8] *University of California at Davis, Davis, USA.*
[9] *Joint Institute for Nuclear Research, Dubna, Russia.*
[10] *Fachbereich Physik der Universität, Frankfurt, Germany.*
[11] *Kent State University, Kent, OH, USA.*
[12] *University of California at Los Angeles, Los Angeles, USA.*
[13] *Fachbereich Physik der Universität, Marburg, Germany.*
[14] *Max-Planck-Institut für Physik, Munich, Germany.*

[15] *Institute for Nuclear Studies, Warsaw, Poland.*
[16] *Institute for Experimental Physics, University of Warsaw, Warsaw, Poland.*
[17] *Nuclear Physics Laboratory, University of Washington, Seattle, WA, USA.*
[18] *Yale University, New Haven, CT, USA.*
[19] *Rudjer Boskovic Institute, Zagreb, Croatia.*
[†] *deceased.*
[#] *present address: Physikalisches Institut, Universitaet Heidelberg, Germany.*
[+] *present address: Max-Planck-Institut für Physik, Munich, Germany.*

The large acceptance TPCs of the NA49 spectrometer allow for a systematic multidimensional study of two-particle correlations in different part of phase space. Results from Bertsch-Pratt and Yano-Koonin-Podgoretskii parametrizations are presented differentially in transverse pair momentum and pair rapidity. These studies give an insight into the dynamical space-time evolution of relativistic Pb+Pb collisions, which is dominated by longitudinal expansion.

1 Introduction

Recent high statistics experiments have demonstrated that, in heavy ion collisions at relativistic beam energies, a single characteristic radius from intensity interferometry does not exist. First of all one deals with systems, in which the emission of particles is distributed over various ranges in three dimensional space as well as in time. Secondly, due to the dynamical behaviour of the source (eg. collective expansion) correlations arise between the momenta of particles and the point of emission in space-time, which results in a dependence of HBT radii on kinematical quantities. In the case of a "boost invariant" picture, first suggested by Bjorken[1], the hot and dense initial stage of the collision expands longitudinally until freeze-out such, that a distinct longitudinal velocity (v_z) profile is established: $v_z = z/t_f$. Here, z is the distance of particle emission at freeze-out from the center of the collision and t_f the freeze-out time. With respect to HBT correlations – appearing at low momentum differences of particle pairs – such a velocity profile disconnects different parts of the source along the beam axis. Only additional thermal motion with average velocities of $< v_{therm} > \approx \sqrt{T/m_\perp}$ can – over certain distances – compensate for the longitudinal velocity gradient. This means that, in case of relativistic heavy ion collisions, HBT radii do not measure the geometry of the source, but a "length of homogeneity" [2] $dz = R_z = t_f\sqrt{T/m_\perp}$, which may vary inside the source and hence depend on the phase space in which the pairs are observed.

Experimentally, the question of space-time size of the source is addressed by extracting HBT radii for all space-time components of the momentum difference vector $q^\mu = p_1^\mu - p_2^\mu; \mu = 0..3$ of pairs of identical particles 1 and 2. Unfortunately, the "on-mass-shell"- constraint for the observed (real) parti-

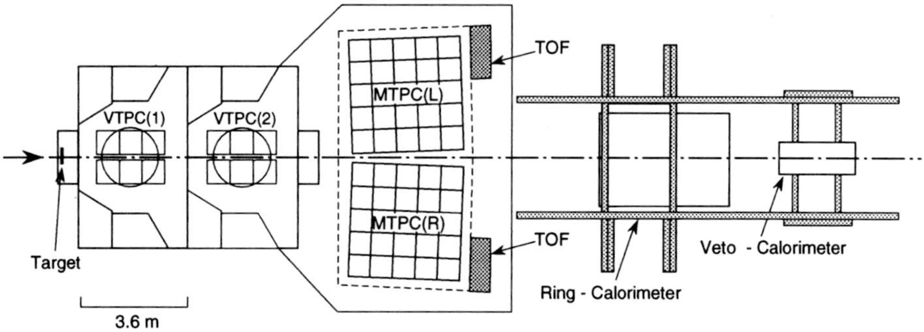

Figure 1: The NA49 spectrometer

cles reduces the degrees of freedom to three. For pairs this constraint can be written as

$$q_0 k_0 = \vec{q}\vec{k} \tag{1}$$

with the average pair momentum $k^\mu = (p_1^\mu + p_2^\mu)/2$. Later it will be shown how this condition is utilized to eliminate one of the four space-time components of q^μ to deduce parametrizations of the three dimensional correlation function $C_2(q)$ according to Bertsch-Pratt(BP[3]) or to Yano-Koonin-Podgoretskii (YKP[4]).

The second point – the dynamical space momentum correlation – is investigated experimentally by carrying out full three dimensional HBT analyses for intervals in the average transverse pair momentum $(k_\perp)$ and average pair rapidity $(y = (y_1 + y_2)/2)$, separately. All together this can be expressed by (q components introduced in section 3):

$$C_2(q^\mu) \longrightarrow C_2(q^\mu, k^\mu) = C_2\left(\begin{bmatrix} BP : q_{side}, q_{out}, q_{long} \\ YKP : q_\perp, q_0, q_\parallel \end{bmatrix}, [k_\perp, y]\right) \tag{2}$$

Such a detailed analysis requires of course a large data sample over a large fraction of phase space. Facilitated by the large number of particles produced in central Pb+Pb collisions at 158 A·GeV and the large acceptance of the NA49 spectrometer at the CERN SPS accelerator such a study has become feasible and is presented here.

2 NA49 spectrometer and data analysis

The NA49 spectrometer[5] (figure 1) is located in the North area of the CERN SPS. The results presented here are derived from the run period in 1995 with a 158 A·GeV Pb-beam. In every spill (5 s) about 10^5 Pb^{82+} ions impinge on a

224 mg/cm^2 Pb-target. 1% of these projectiles interact with the nuclei of the target foil. The most central 5% (impact parameter $b = 0 - 3.5$ fm) of these collision are selected by the NA49 trigger utilizing a *Veto- Calorimeter*, which measures the energy carried by the beam spectator fragments. In such central collision about 1200 charged particles are produced, roughly 800 of them are detected in at least one of the four *TPCs* (Time Projection Chambers) of NA49. At 2.0 m and 5.8 m downstream of the target two smaller chambers (*VTPC1, VTPC2*; volume=$2.0 \times 0.7 \times 2.5$ m^3) are placed inside magnetic dipole fields. In the "Standard Configuration" the fields amount for 1.5 Telsa (*VTPC1*) and 1.1 Tesla (*VTPC2*), respectively, whereas in case of the "Low field configuration", fields of 0.3 Tesla and 1.5 Tesla have been used, shifting the acceptance for pions towards central rapidities. Further downstream – outside the magnetic field – the acceptance is extended to forward rapidities and higher momenta by two large volume TPCs' (*MTPC*; volume=$3.8 \times 1.2 \times 3.8$ m^3).

All four *TPCs* are equipped with a charge sensitive read-out, highly segmented along the beam (z-axis) and perpendicular to it in horizontal direction (x-axis). For each of these 180.000 pads the charge is determined in 512 consecutive time slices (100 ns), to determine the vertical (y-axis) position of particle tracks via the drift velocity ($v_d \approx$1.4 cm/μs inside *VTPC* and $v_d \approx$2.2 cm/μs inside *MTPC*). This scheme allows for 3-dimensional track reconstruction. From the curvature of reconstructed tracks inside the *VTPC* the momentum of particles is determined to a precision of $\delta p/p^2 \approx 0.3\%$ in addition to their charge. Moreover, all chambers measure the energy (dE/dx) deposited by the particle in the detector gas, which – in a future stage of analysis – will be used to identify the particles. For the HBT analysis presented here, pairs of hadrons of the same charge (h^-h^- or h^+h^+) without further identification have been considered. These are dominated by pairs of identical pions ($\pi^-\pi^-$ or $\pi^+\pi^+$). It has been shown, that a contamination from other particles has negligible influence on the determination of HBT radii, with the exception of the chaoticy parameter λ, which will not be discussed in this article.

Independent analyses have been carried out for the *VTPC2* and the *MTPCs*. The *VTPC2* analysis[6] is based on 40.000 central Pb+Pb collisions, half of them taken in the "Low field configuration" and the other half in "Standard field configuration". The analysis of the *MTPC* data[7] includes the same 40.000 events and adds another 50.000 events in "Standard field configuration". In both analyses pairs of tracks with distances less than 2 cm inside the TPC have been excluded to eliminate the influence of two particle reconstruction inefficiency for very close tracks. The same requirement has been imposed on the distribution of uncorrelated pairs, which is generated by combining tracks from different events. This distribution is used as reference (denominator) in the determination of the correlation function. Both data sets have been corrected for Coulomb final state interaction based on measured correlations of opposite-sign charged particles. In case of the *VTPC2*–data this correction is carried out by the Gamov factor $G(q_{inv})$; $q_{inv} = \sqrt{-(p_1 - p2)^2}$ modified by a dumping term to account for the finite size of the source [8]:

$$C_{2,corr}^{--} = C_{2,meas.}^{--} \times \Big((G(q_{inv}) - 1)e^{-q_{inv}/q_{eff}} + 1\Big) \tag{3}$$

The parameter q_{eff} is determined from a fit of the 1-dimensional correlation function $C_2^{+-}(q_{inv})$.

The *MTPC*–data are corrected by the correlation function C_2^{+-} of the opposite charged particles, evaluated in the same three dimensionial relative momentum space as the like-sign pairs; eg. for BP-paramertization (see below):

$$C_{2,corr}^{--} = C_{2,meas.}^{--} \times C^{+-}(q_{side}, q_{out}, q_{long}) \tag{4}$$

No further corrections have been applied to the data. The systematic error in the determination of the HBT radii is estimated to about 7%. The correlations function have been obtained in different reference frames. The most intuitive choice of a reference frame for all pair momenta might be the **C**enter- of **M**ass **S**ystem (CMS) of the colliding ions, but in case of variations of the longitudinal velocity across the source due to a longitudinal expansion, the **L**ongitudinal **C**o-**M**oving **S**ystem (LCMS) might be better suited. It is defined on a pair-by-pair basis such that the longitudinal pair momentum $k_{\parallel}$ vanishes. The **F**ixed **L**ongitudinal **C**o-**M**oving **S**ystem (FLCMS), in which the observer frame for every interval in pair rapidity is fixed at the center of that interval, might be seen as a compromise between both. In case of narrow widths of rapidity intervals LCMS and FLCMS are equivalent and are therefore treated as one in the following discussion, even though the *MTPC* data [7] have been evaluated in the CMS and FLCMS frames, whereas the *VTPC2* data [6] use the CMS and LCMS frames for BP projections and CMS and FLCMS in case of YKP.

3 Q- parametrizations

As shown in references [9,10], the correlation function for a source distribution, expanded to second order at the (in general k-dependent) space time points of maximum emission ($\bar{X}$; saddle point) can be written as:

$$C_2 = 1 + e^{-q^\mu q^\nu <\hat{x}_\mu \hat{x}_\nu>} \tag{5}$$

with $\hat{x}_\mu := x_\mu - \bar{X}_\mu$; $<>$ denotes an averaging over the source distribution. Even though this "model independent" expression (5) can – because of the

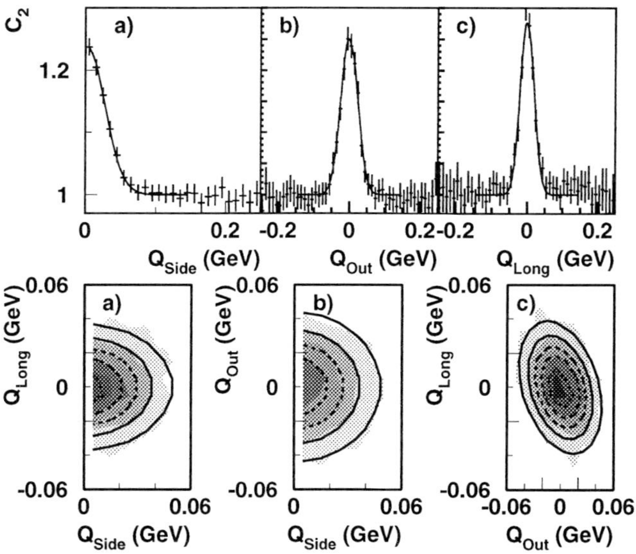

Figure 2: 1- and 2-dimensional projections of the h^-h^- correlation function measured in the NA49 *MTPC* using BP parameters for pair rapidities $4 < y < 5$ and transverse momenta $k_\perp < 100$ MeV/c (FLCMS). The projections on components q_j have been carried out by restricting the other components $q_i; i \neq j$ to $q_i < 30$ MeV/c.

constraint in eq. (1) – not be used to describe measured correlation functions, it is a generalization of both the Bertsch Pratt [3] as well as the Yano Koonin Podgoretskii parametrization. It therefore gains its importance as a tool when interpreting the radii extracted in the framework of both formulations; it furthermore provides a proof of consistency, when comparing results from both parametriztions.

3.1 BP parametrization

Eliminating the temporal component of (5) by condition (1) in the form $q_0 = \vec{\beta}\vec{q}$ ($\vec{\beta} = \vec{k}/k_0$), results in a correlation function parametrized according to reference [3]:

$$C_2 = 1 + \lambda e^{-q_{side}^2 R_{side}^2 - q_{out}^2 R_{out}^2 - q_{long}^2 R_{long}^2 - 2q_{out}q_{long}R_{out-long}^2} \tag{6}$$

Here q_{long} is the component of q^μ in beam direction, whereas q_{side} and q_{out} are those perpendicular to it, with $q_{out} || \vec{k}$ and $q_{side} \perp \vec{k}$.

Due to symmetries of the sources considered here, only the "out–long" cross term remains, all others ("out–side" and "side–long") vanish. This is consistent with the experimental observation presented in figure 3 for h^-h^--pairs at rapidities $4 < y < 5$ and transverse pair momentum $k_\perp < 100$ MeV/c. The contours lines in the 2-dimensional projection deviate from a circular shape (no cross term correlation) only in case of the "out–long" projection. The 1-dimensional projections demonstrate the good agreement between the function (6) and the data for small q, where the correlation signal is clearly visible, as well as for larger q, where the data points are consitent with $C_2 = 1$ within the error-bars. At this point it should be emphasized, that all HBT radii presented here are derived by a simultaneous fit of all three components of q in C_2. The projections are generated for better visibility only.

When interpreting the BP radii it is advantageous to explicitly write down the relation between equation 6 and ansatz 5:

$$R^2_{side}(\vec{k}) = < \hat{x}_y^2 > \tag{7}$$

$$R^2_{out}(\vec{k}) = < (\hat{x}_x - \beta_\perp \hat{t})^2 > \tag{8}$$

$$R^2_{long}(\vec{k}) = < (\hat{x}_z - \beta_\| \hat{t})^2 > \tag{9}$$

$$R^2_{out,long}(\vec{k}) = < (\hat{x}_x - \beta_\perp \hat{t})(\hat{x}_z - \beta_\| \hat{t}) > \tag{10}$$

With the exception of R_{side}, BP-radii mix spacial ($\hat{x}_{xyz}$)and temporal components ($\hat{t}$) of the source and an interpretation becomes therefore reference frame and model dependent. In case of a longitudinally expanding source the "length of homogeneity" observed in the center of mass frame, appears Lorentz contracted in different intervals of pair rapidity. Such a behaviour is supported by the rapidity dependence of the measured radius R_{long}, as shown for different intervals in $k_\perp$ in figure 3. The data points can be described by $R_{long} = t_f/\cosh(\mathrm{y})\sqrt{\mathrm{T}/\mathrm{m}_\perp}$, which is the Lorentz frame dependent expression of section 1. Assuming a temperature of $T = 150$ MeV, a freeze-out time of $t_f \approx 9 - 7$ fm/c can be derived for the different $k_\perp$-intervals. By comparing the measured "*side*"- and "*out*"-radii to equation (8)-(7) the duration time of freeze-out yields $\approx 2-4$ fm/c. Both radii ("*side*" and "*out*") appear to be constant over $y < 2.5$ within errors. In addition to the physical interpretation it is important to note the good overall agreement between the different analyses and between pairs of opposite total charge, i.e. h^-h^- compared to h^+h^+.

3.2 YKP parametrization

Instead of eliminating the temporal component of (5) in the BP formalism, condition (1) might also be used via the relation $q_x = q_0/\beta_\perp - q_\|\beta_\|/\beta_\perp$. With the picture of a boost invariant source in mind with different parts of the source moving at different longitudinal velocities, one explicitly introduces a

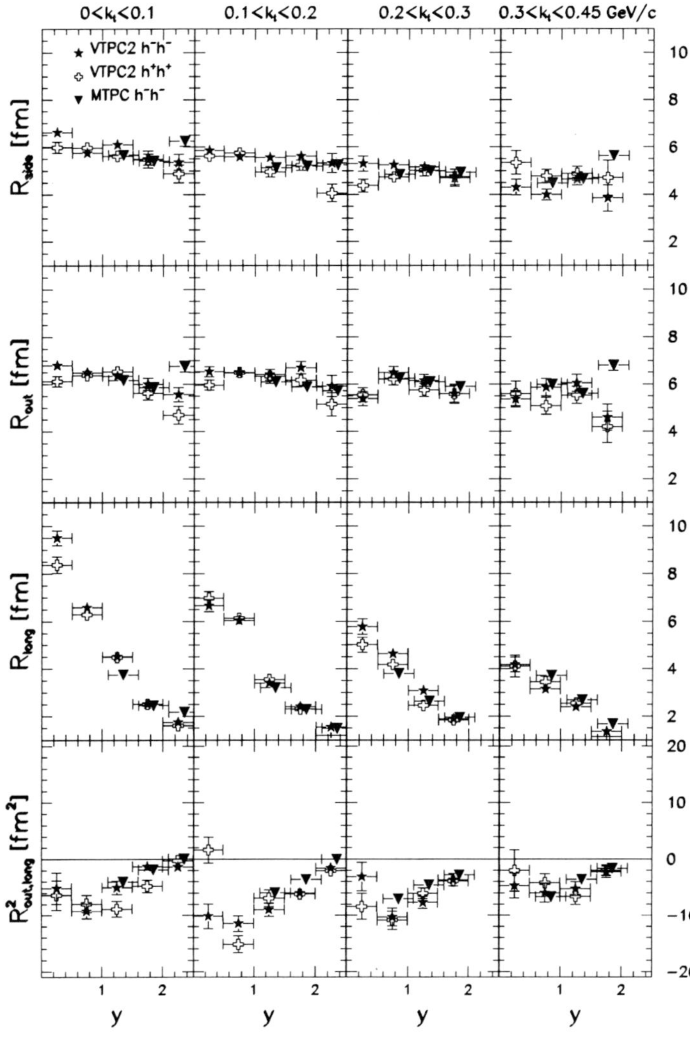

Figure 3: The dependence of BP radii on the pair rapidity for h^-h^--pairs ($\star$) and h^+h^+-pairs ($+$) in the *VTPC2* and h^-h^--pairs ($\bigtriangledown$) in the *MTPC* in four intervals of transverse momentum k_t (CMS). The rapidity scale is shifted to the center of mass system of the ions. The horizontal error bars correspond to the width of the intervals chosen in the analysis and the vertical to the statistical errors only.

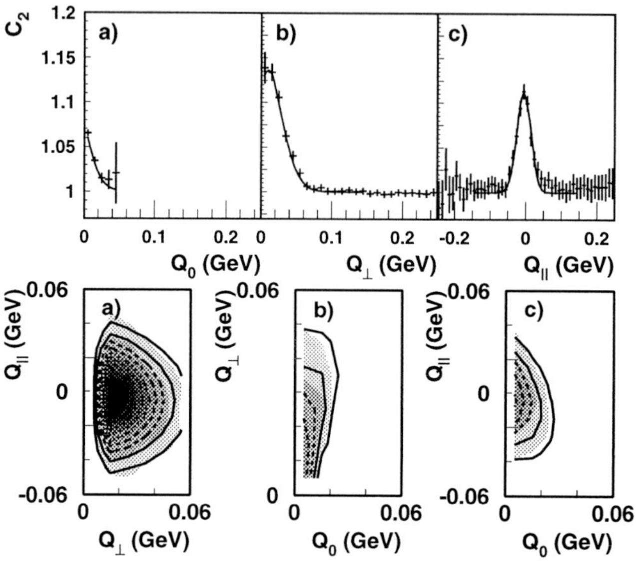

Figure 4: Projections of the same data sample as in figure 2, but this time in the YKP parametrization (FLCMS). The projection onto component(s) q_j have been carried out by restricting the other component(s) $q_i; i \neq j$ to $q_i < 70$ MeV/c.

longitudinal velocity parameter (β_{YKP}) into the correlation function of YKP type [4]:

$$C_2 = 1 + \lambda e^{-q_\perp^2 R_\perp^2 - \gamma_{YKP}^2 (q_\parallel - \beta_{YKP} q_0)^2 R_\parallel^2 - \gamma_{YKP}^2 (q_0 - \beta_{YKP} q_\parallel)^2 R_0^2} \tag{11}$$

with $\gamma_{YKP} = 1/\sqrt{1-\beta_{YKP}^2}$, $q_\perp = \sqrt{q_x^2 + q_y^2}$ for the transverse momentum difference, q_0 for the energy difference and $q_\parallel$ for the longitudinal component. In this case the interpretation of extracted radii becomes more evident, since space and time components are decoupled (the validity of the approximation is discussed in reference [11]):

$$R_\perp^2(\vec{k}) = < \hat{x}_y^2 > \tag{12}$$

$$R_0^2(\vec{k}) \approx < \hat{t}^2 > \tag{13}$$

$$R_\parallel^2(\vec{k}) \approx < \hat{x}_z^2 > \tag{14}$$

Figure 4 shows one and 2-dimensional projections of the measured correlation function in YKP coordinates evaluated in the LCMS frame for the same $k_\perp$ and y-interval as in figure 2. Again, the data are described well by the chosen gaussian ansatz. A problem of this type of parametrization manifests itself

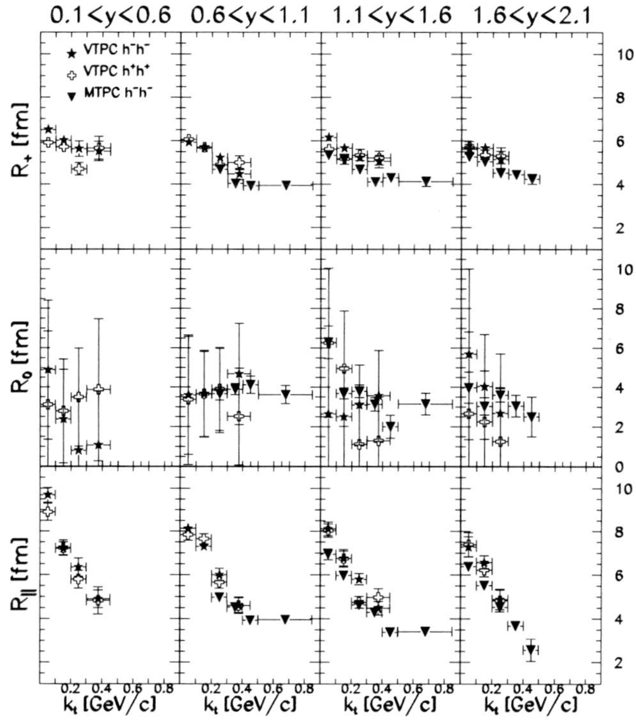

Figure 5: The dependence of YKP-HBT radii on the transverse pair momentum $k_\perp$ in different intervals of pair rapidity (LCMS) (otherwise same conventions as in figure 5).

in the projection of q_0. For a given interval in $q_\perp$ and $q_\parallel$ only a limited region of q_0 is kinematically available. This is the reason for the large uncertainties of R_0 in figure 5, which sumarizes the k_t-dependence for the YKP-radii in different intervals of rapidity. The clear decrease of $R_\parallel$ vs. k_t again points to strong space-momentum correlations in the source. Moreover, even in the transverse direction $R_\perp$ decreases for larger k_t-values, which can be interpreted by transverse flow [12]. The estimate of the duration of emission given in section 3.1 is confirmed by the extracted values of R_0.

In a parametrization of YKP type one can gain further insight into the dynamics of the source by utilizing the YKP velocity β_{YKP}. The YKP rapidity $y_{YKP} = \frac{1}{2}\ln\frac{1+\beta_{YKP}}{1-\beta_{YKP}} + y$ derived from the measured β_{YKP} is compared to the pair rapidity in figure 6. The data show the characteristics of a source, which expands in longitudinal direction. Even though deviations from an ideal boost invariant picture (line in figure 6), become apparent at forward rapidities, the

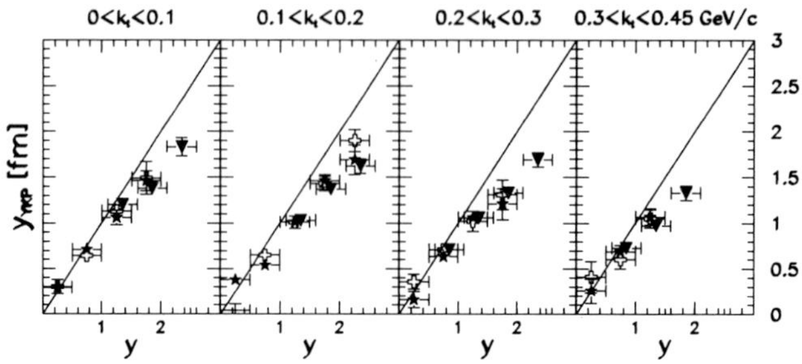

Figure 6: Dependence of the Yano Koonin rapidity vs. the pair rapidity for four intervals in $k_{\perp}$ (LCMS).

consitstency which such a model is good.

4 Conclusion and Outlook

Results from different analyses of multidimensional Bertsch-Pratt and Yano-Koonin-Podgoreskii parametrizations of two particle correlation functions in 158 A·GeV Pb+Pb collisions have been presented differentially in pair rapidity y and transverse pair momentum k_t. The results confirm the importantance of extracting HBT-radii seperately in different parts of phase space to disentangle the dynamical correlations of the source. The pion source appears to expand in a close to boost invariant way, as seen in the rapidity dependence of the Yano-Koonin velocity as well as of the R_{long} radius in the CMS system. Moreover, a finite duration time of emission of $2 - 4$ fm/c and a decreasing tranverse radius at large k_t are observed. For a better understanding of the behaviour of the radii a comparison to a similar analysis of proton-proton and proton-lead collisions is currently in progress. Moreover, the centrality dependence and beam energy dependence might constrain interpretations even further and will be investigated in upcoming analysis and further data taking.

Acknowledgments

Acknowledgements: This work was supported by the Director, Office of Energy Research, Division of Nuclear Physics of the Office of High Energy and Nuclear Physics of the US Department of Energy under Contract DE-ACO3-76SFOOO98, the US National Science Foundation, the Bundesministerium fur Bildung und Forschung, Germany, the Alexander von Humboldt Foundation, the UK Engineering and Physical Sciences Research Council, the Polish State Committee for Scientific Research (2 P03B 01912), the Hungarian Scientific Research Foundation under contracts T14920 and T23790, the EC Marie Curie

Foundation, and the Polish-German Foundation.

References

1. J.D. Bjorken, *Phys. Rev.* D **27**, 140 (1983).
2. Yu. M. Sinyukov *et el.*, *Nucl. Phys.* A **498**, 151 (1989) and *Nucl. Phys.* A **566**, 589 (1994).
3. G. F. Bertsch, *Nucl. Phys.* A **498**, 173 (1989); S. Pratt, *Phys. Rev.* D **33**, 1314 (1986).
4. F. B. Yano and S.E. Koonin, *Phys. Lett.* B **78**, 556 (1978); M.I. Podgoretskii, *Sov. J. Nucl. Phys.* **37**, 272 (1983) S. Chapman, P. Scotto and U. Heinz, *Phys. Rev. Lett.* **74**, 4400 (1995).
5. P. Seyboth *et al.*, *Proc. of XXV Int. Symposium on Multiparticle Dynamics* Stara Lesna Slovakia, 170 (1995).
6. H. Appelshäuser, PhD thesis, Univ. Frankfurt a.M. (1997).
7. S. Schönfelder, PhD thesis, MPI München MPI-PhE/97-09 (1997).
8. K. Kadija *et el.*, *Nucl. Phys.* A **610**, 248 (1996).
9. S. Chapman *et al.*, *Phys. Rev.* C **52**, 2694 (1995).
10. U. Heinz *et al.*, *Phys. Lett.* B **382**, 181 (1996).
11. Y.F. Wu *et al.*, *Euro. Phys. J.* C **1**, 599 (1998).
12. H. Appelshäuser *et al.*, *Euro. Phys. J.* C **2**, 611 (1998).

Two particle correlations from NA44

J.P. Sullivan for the NA44 collaboration
I.G.Bearden[1], H. Bøggild[1], J. Boissevain[2], J. Dodd[3], B. Erazmus[4], S. Esumi[5,c], C.W. Fabjan[6], D. Ferenc[7], D.E. Fields[2,d], A. Franz[6,e], J. Gaardhøje[1], M. Hamelin[10], A.G. Hansen[1], O. Hansen[1], D. Hardtke[8,h], H. van Hecke[2], E.B. Holzer[6], T.J. Humanic[8], P. Hummel[6], B.V. Jacak[11], R. Jayanti[8], K.Kaimi[5], M. Kaneta[5], T. Kohama[5], M. Kopytine[11], M. Leltchouk[3], A. Ljubicic, Jr.[7,e], B. Lörstad[9], N. Maeda[5,f], R.Malina[6], A. Medvedev[3], M. Murray[10],H.Ohnishi[5], G. Paic[8], S.U. Pandey[8,g], F. Piuz[6], J. Pluta[4], V. Polychronakos[12], M. Potekhin[3], G. Poulard[6], D.Reichhold[8], A. Sakaguchi[5,b], J. Simon-Gillo[2], J. Schmidt-Sørensen[9], W. Sondheim[2], M.Spegel[6], T. Sugitate[5], J.P. Sullivan[2], Y. Sumi[5], W.J. Willis[3], K.L. Wolf[10], N. Xu[2,h], and D.S.Zachary[8]

[1] *Niels Bohr Institute, DK-2100, Copenhagen, Denmark.*
[2] *Los Alamos National Laboratory, Los Alamos, NM 87545, USA.*
[3] *Columbia University, New York, NY 10027, USA.*
[4] *Nuclear Physics Laboratory of Nantes, 44072 Nantes, France.*
[5] *Hiroshima University, Higashi-Hiroshima 739, Japan.*
[6] *CERN, CH-1211 Geneva 23, Switzerland.*
[7] *Rudjer Boskovic Institute, Zagreb, Croatia.*
[8] *Ohio State University, Columbus, OH 43210, USA.*
[9] *University of Lund, S-22362 Lund, Sweden.*
[10] *Texas A&M University, College Station, TX 77843, USA.*
[11] *State University of New York, Stony Brook, NY 11794, USA.*
[12] *Brookhaven National Laboratory, Upton, NY 11973, USA.*
[b] *now at Osaka University, Toyonaka, Osaka 560-0043, Japan.*
[c] *now at Heidelberg University, D-69120 Heidelberg, Germany.*
[d] *now at University of New Mexico, Albuquerque, NM 87131, USA.*
[e] *now at Brookhaven National Laboratory, Upton, NY 11973, USA.*
[f] *now at Florida State University, Tallahassee, FL 32306, USA.*
[g] *now at Wayne State University, Detroit, MI 48202, USA.*
[h] *now at Lawrence Berkeley National Laboratory, Berkeley, CA 94720, USA.*

Three dimensional source size parameters measured via two-particle interferometry in experiment NA44 for 200 GeV/nucleon S+S and S+Pb and 158 GeV/nucleon Pb+Pb are presented. The emission volumes are larger than the beam nuclei. The size parameters decrease with increasing transverse mass and increase as the size of the projectile increases. Comparison to the RQMD model is used to relate the fit parameters to the actual emission volume.

1 Introduction

The NA44 spectrometer was optimized for the study of single particle spectra and two-particle particle spectra near mid-rapidity for transverse momenta below $\approx$ 1 GeV/c. This paper will concentrate on the source size parameters determined from fits to two-particle correlation functions for pions and kaons in the mid-rapidity region. If a first-order phase transition from a quark-gluon plasma is present the duration of particle emission can be comparable to the spatial extent of the source [1,2]. The duration of particle emission may be measurable through a multi-dimensional analysis of the two-particle correlation function, although the expansion dynamics of the particle emitting source and final state interactions complicate the interpretation [3]. The transverse momentum dependence of the correlation function gives insight into the dynamics of the system as well as the resonance decay contributions to the particle sample [4].

2 Results

The NA44 Focusing Spectrometer [5,6,7] uses two dipole magnets and three quadrupoles, covering a momentum range of $\pm 20\%$ around its central momentum setting. For the data shown here, the central momentum settings are 4 GeV/c (lab) for pions and 6 GeV/c for kaons. The two spectrometer angle settings (44 mr and 131 mr) used are referred to as low p_T ($< p_T > \approx 170$ MeV/c) and high p_T ($< p_T > \approx 480$ MeV/c), respectively. For the S beam data, the tracking and time-of-flight uses three scintillator hodoscopes whose time resolution is $\approx$ 100 ps, with a Cherenkov beam counter [8] for the time-of-flight start ($\sigma \approx 35$ ps). Particle identification is based on time-of-flight and two gas Cherenkov counters. The first Cherenkov counter (C1) fires on pions and electrons, but not for heavier particles. The second Cherenkov counter (C2) fires for electrons, but not for heavier particles. These two Cherenkov counters give one signal per event. For the Pb+Pb data, the first of the three scintillator hodoscopes was replaced with a pad chamber. Two strip chambers were added between the second and third hodoscopes. Also, a threshold imaging Cherenkov counter (TIC) [9] was added to the system. The TIC gave a Cherenkov signal for individual tracks. The TIC gives a signal for pions and electrons, but not for heavier particles.

The NA44 data have been analyzed in terms of three components [1,2] of the two-particle momentum difference ($\vec{Q} = \vec{p_1} - \vec{p_2}$). The data are analyzed in the frame in which the z-component of the pair momentum ($p_{z1} + p_{z2}$) is zero (the longitudinally co-moving system, or "LCMS"). The momentum difference is resolved into a component (Q_L) parallel to the beam direction and a component perpendicular to the beam direction. The perpendicular component is further

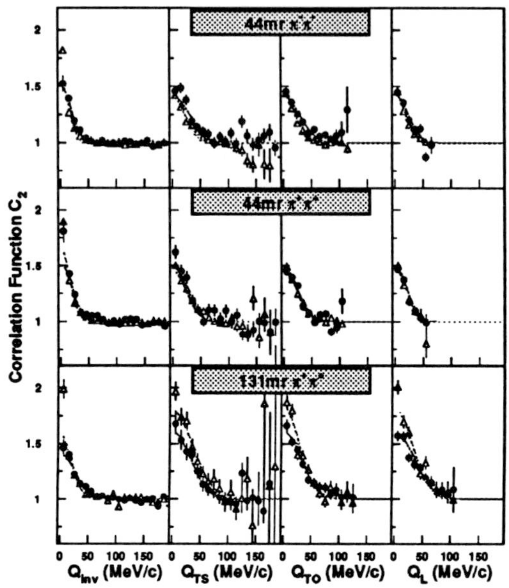

Figure 1: The one-dimensional correlation functions and the projections of the three-dimensional correlation functions for the 44 mr $\pi^-\pi^-$, 44 mr $\pi^+\pi^+$ and the 131 mr $\pi^+\pi^+$ data. Also included are the projections of the fitted Gaussian parameterizations. The projections are over the lowest 20 MeV/c in the other momentum difference directions. The solid circles are the data from the horizontal setting and the solid triangles are the data from the vertical setting. The data shown here use the Coulomb wave function integration correction.

resolved into Q_{TO} parallel to the sum of the pair momentum and Q_{TS}, which is perpendicular to the sum and to the beam. Three corresponding source size parameters (R_L, R_{TO}, R_{TS}) are determined by fitting Eq. 1:

$$C(Q_{TO}, Q_{TS}, Q_L) = A(1 + \lambda_{3D} e^{-Q_{TO}^2 R_{TO}^2 - Q_{TS}^2 R_{TS}^2 - Q_L^2 R_L^2}) \quad (1)$$

to measured 3D correlation functions from two different spectrometer settings. The "horizontal" focus spectrometer setting optimizes the acceptance for R_{TO}, while the "vertical" focus spectrometer setting is for R_{TS}. For the S beam data, the resolution in Q_{TO} and Q_L is ≈15 MeV/c and in Q_{TS} is ≈30 MeV/c. For Pb+Pb, the resolution is ≈15 MeV/c for all three components. One dimensional correlations functions are also shown. These data are fit to

$$C(Q_{INV}) = A(1 + \lambda_{INV} e^{-Q_{INV}^2 R_{INV}^2}) \quad (2)$$

where $Q_{INV} = \sqrt{\vec{Q}^2 - Q_0^2}$.

The one-dimensional fits and projections of the three-dimensional fits onto the three axes are shown together with the Coulomb wave corrected Pb + Pb data in Fig. 1. For the three-dimensional projections, the data from the horizontal and vertical settings are both shown. The top row shows the correlation function and fit for the low p_T $\pi^-\pi^-$ data, the middle row shows the low p_T $\pi^+\pi^+$ data, and the bottom row shows the high p_T $\pi^+\pi^+$ data.

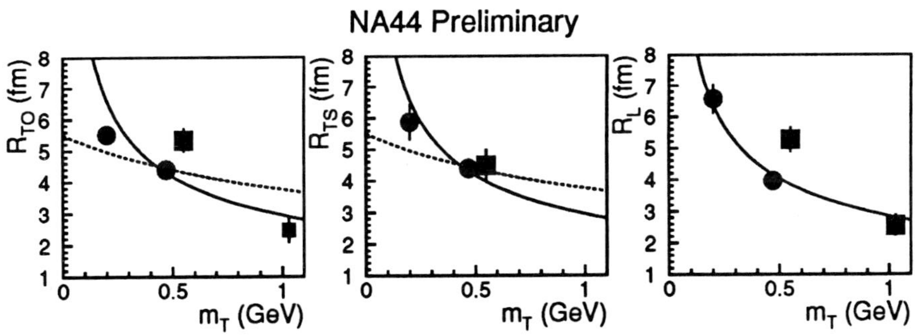

Figure 2: Radius parameters for $\pi^+\pi^+$ (circles) and K^+K^+ (squares) in Pb+Pb collisions at 158 GeV/nucleon as functions of m_T. The data points are compared with a simple model function $1/\sqrt{(m_T)}$ (solid curve) and the model function in Eq. 3 and 4 (dashed curve). For the later case, T_{FO}=140 MeV and $\langle v_T \rangle$=0.4C are used.

3 Discussion

The radius parameters from fitting Eqs. 1 and 2 to the measured correlation functions depend on the transverse mass ($m_T = \sqrt{m^2 + p_T^2}$ where m and p_T are the mass and transverse momentum of the particle). The lowest order estimates of the m_T dependence of the radius parameters from a theory with an analytic treatment of the particle emission function are as follows[10,11,12,13]:

$$R_L = \tau_{FO}\sqrt{\frac{T_{FO}}{m_T}} \propto \frac{1}{\sqrt{m_T}} \tag{3}$$

$$R_{TO} = R_{TS} = \frac{R_G}{\sqrt{1 + \frac{m_T}{T_{FO}}\langle v_T \rangle^2}} \tag{4}$$

where τ_{FO}, T_{FO}, and $\langle v_T \rangle$ are the freezeout time, the freezeout temperature, and the average transverse flow velocity, respectively. In Eq. 3, the effect of finite geometrical size is ignored since the effect is minor compared with the strong space-momentum correlation in the longitudinal direction, and R_L has the well known $1/\sqrt{m_T}$ dependence[6,14]. In Eq. 4, the effect of finite geometrical size is included but the time contribution to R_{TO}, which is believed to be small[3,5,6,7,15], is omitted. In the limit of rapid transverse flow $\langle v_T \rangle \approx 1$ or in the limit of large transverse mass ($m_T/T_{FO} >> 1$), R_{TO} and R_{TS} also have a $1/\sqrt{m_T}$ dependence.

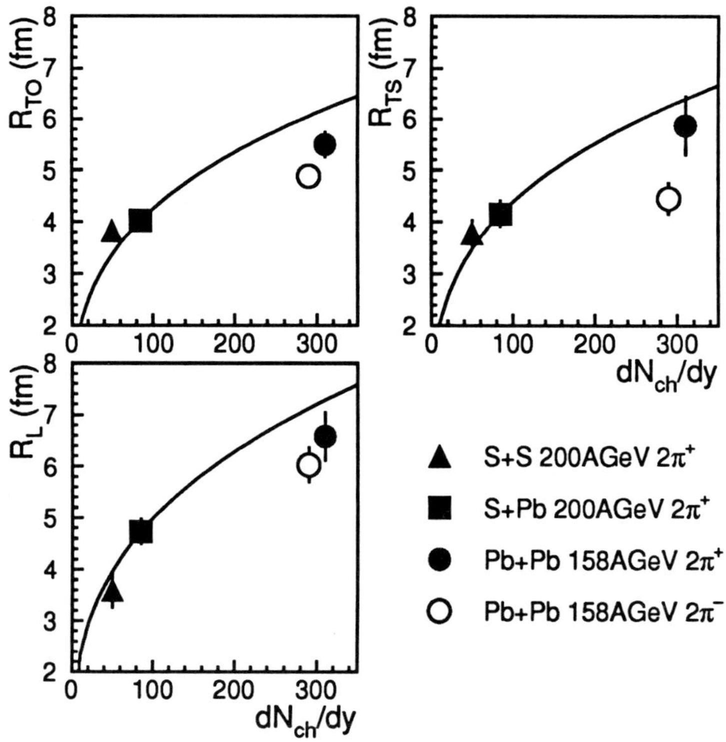

Figure 3: dN/dy dependence of Pb + Pb size parameters for $\pi^+\pi^+$, $\pi^-\pi^-$, and K^+K^+.

Fig. 2 shows the m_T dependence of the NA44 radius parameters for Pb+Pb at 158 GeV/nucleon. The parameters from the $\pi^+\pi^+$ correlations and a preliminary result for K^+K^+ correlations are plotted. The m_T dependence of the radius parameters is compared with the simple model function $1/\sqrt{m_T}$ (solid curve). The m_T dependence of the R_L radius parameter is well described by the simple model function. This is consistent with the picture of a rapid longitudinal expansion. For the other two radius parameters (R_{TO} and R_{TS}), the simple model function gives a slightly stronger m_T dependence than the experimental data. The dashed curves in Fig. 2 show results using Eq. 4 for the transverse radius parameters. The curves used T_{FO}=140 MeV and $\langle v_T \rangle$ =0.4c [16]. This model function gives an m_T dependence which is slightly weaker than the data.

One of the simplest models [17] used to understand the multiplicity dependence of the radius parameters assumes freezeout at constant density [15]. This model predicts the multiplicity dependence of the radius parameters as follows:

$$R_{TS}, R_{TO}, R_L \propto (dn/dy)^{1/3} \tag{5}$$

Fig. 3 shows the multiplicity dependence of the NA44 radius parameters for two-pion correlations in semi-central ($\sigma/\sigma_{total} \approx 0.15$) S+S and S+Pb collisions at 200 GeV/nucleon and Pb+Pb collisions at 158 GeV/nucleon [6,7]. The solid curves in Fig. 3 are the prediction of Eq. 5, normalized to the S+Pb data points. The multiplicity dependence of the semi-central events is close to the model prediction, but slightly weaker than the prediction. Another analysis [7] of minimum bias data of the S+S, S+Ag, and S+Pb collisions at 200 GeV/nucleon also indicates a weaker multiplicity dependence than the simple model prediction. If flow increases with multiplicity, the multiplicity could increase more slowly than $(dn/dy)^{1/3}$.

Fig. 4 compares size parameters from NA44 Pb+Pb data to size parameters calculated [18] by using the RQMD event generator [19,20] to generate the single particle emission probability distribution at the point of each particle's last interaction. The two-particle correlation function is calculated from the single particle distribution using a Wigner function formalism [21,22]. The RQMD results are fit with (filled circles) and without (open squares) including the point at the lowest q value in the fit. The fits to the NA44 data are only shown with the first point included, but the NA44 results are insensitive to this difference. The fits to the RQMD calculations are more sensitive to the inclusion for the first point because the error bar on the lowest q point is much smaller in RQMD than in the data. The NA44 spectrometer and the cuts used in the analysis result in very few counts in the first q bin – with the resulting insensitivity of the fits to this point. When the lowest q bin is included in the RQMD fits, the RQMD fit parameter are significantly larger than the NA44 fit parameters. When the lowest q bin is omitted in the RQMD fits, the agreement is reasonably good, but the RQMD parameters still tend to lie slightly above the data.

As a useful tool in understanding the relationship between the size pa-

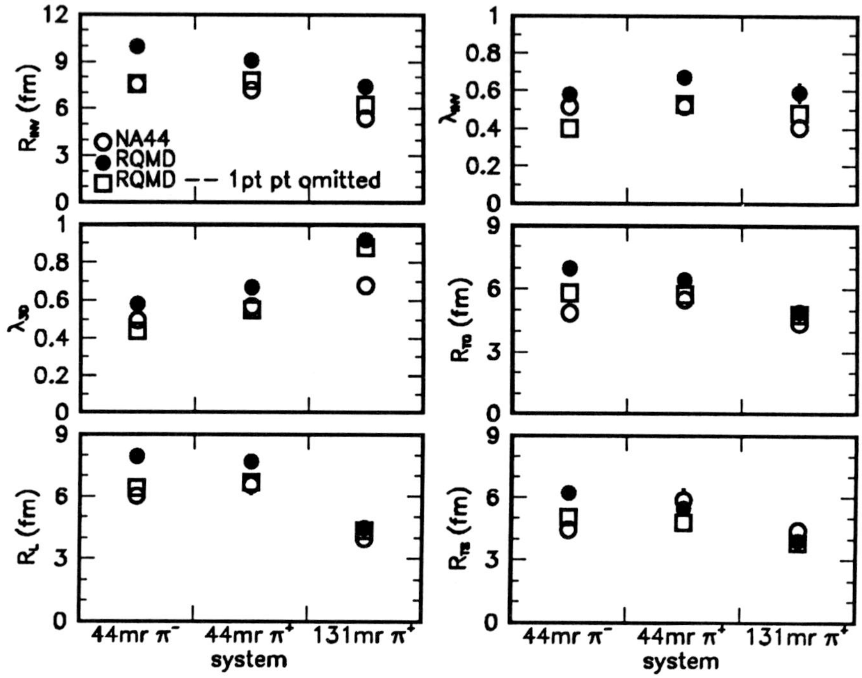

Figure 4: Correlation function fit parameters for Pb+Pb collisions. NA44 results are compared to RQMD results. The open circles show NA44 results. The filled circles show RQMD results with all data points included in the fits. The open squares show RQMD results with the point at the lowest q value omitted from the fit. Data for $\pi^+\pi^+$ (left) and $\pi^-\pi^-$ (middle) at 44 mr and for $\pi^+\pi^+$ at 131 mr (right) are shown.

rameters from fits to a correlation function and the size of the source which produced the particles, fig. 5 shows the freezeout position and time distributions of pions from RQMD. In these plots, x is defined as the Q_{TO} direction and y is along Q_{TS}. The beam direction is along the z axis. These plots are for positive pions and the horizontal focus setting of the spectrometer. The top part of Fig. 5 shows the position and time distributions of pions which contribute to the RQMD correlation function for the NA44 low p_T setting and the bottom shows the corresponding distributions for the high p_T setting. Each individual plot in Fig. 5 shows a histogram (solid line) which represents the distribution for all π^+ produced in an RQMD event – without an acceptance cut. These histograms are the same on the top (low p_T) and bottom (high p_T) halves of Fig. 5. The hatched histograms in each plot show the freeze-

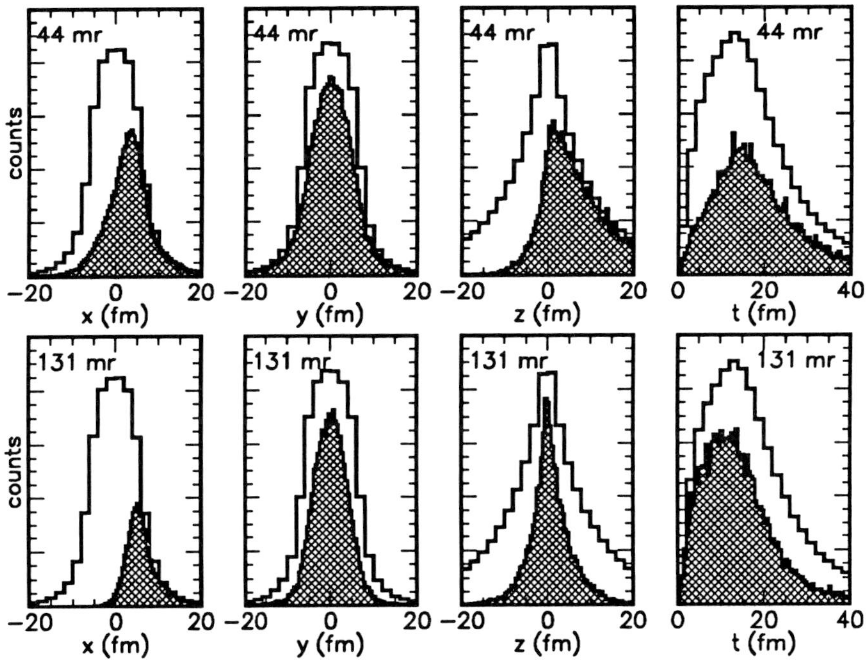

Figure 5: RQMD Freezeout distributions for pions. The (solid line) histograms are for all pions from RQMD, and the hatched histograms are for pions in the NA44 44mr horizontal (upper panels) and 131mr horizontal (lower panels) acceptances. The x axis is in the direction of Q_{TO}, the y axis is in the Q_{TS} direction, and z is the beam axis. The center of mass coordinate system is used.

out distributions for pions which are in the NA44 low p_T (top) and high p_T (bottom) acceptances – these are the pions which were used to construct the RQMD correlation functions. In these plots, the relative normalizations of the plots with and without the acceptance cuts are arbitrary – only the shapes (and centroids) of the distributions should be compared.

A number of interesting observations can be made from Fig. 5. First, the freezeout distributions of pions which contribute to the correlation functions are narrower than the complete freezeout distributions in all cases shown. Ideally, the size parameters from fitting the correlation functions should reflect the widths of the freezeout distributions for pions within the acceptance. The size parameters should therefore be smaller than the full size of the source. From Fig. 5 we can also see that all of the distributions become narrower as

p_T is increased – which is consistent with the experimental observation (and the RQMD result) in which the radius parameters get smaller with increasing p_T. Fig. 5 also shows that the x position distribution (where x is in the direction of Q_{TO}) for particles in the acceptance is centered at positive x and that the center of the distribution moves to large x values as p_T is increased. The HBT method only "sees" the side of the source closest to it. It is also interesting that the widths of the distributions of particles in the two transverse directions (x and y) are not the same for particles in the acceptance. Formulas which attempt to calculate the duration of pion emission from the expression [1,2,23] $c\Delta\tau = \sqrt{R_{TO}^2 - R_{TS}^2}/\beta$ are based on the assumption that the "true" size of the source in two transverse directions is the same. The durations of pion emission extracted from the above expression, when applied to the correlation function fit parameters from RQMD, are inconsistent (too small) with the widths of the lifetime distributions in Fig. 5.

4 Conclusions

The NA44 S+Pb size parameters showed a $1/\sqrt{m_T}$ dependence [6]. The Pb + Pb data shown here have a more complex m_T dependence – R_L seems consistent with $1/\sqrt{m_T}$, but the other two radius parameters are not. The Pb+Pb size parameters risely more slowly than $(dN/dy)^{1/3}$ – the dependence naively expected if freezeout is at a constant density. Perhaps this is caused by an increase in the flow velocity with multiplicty. The radius parameters are larger than the projectile size in all cases. The (rms) size parameters which should be compared to the radius parameters are 1.7 fm for the S beam and 3.2 fm for the Pb beam. Finally, the RQMD calculations are close to the data, but they tend to overestimate the Pb+Pb radius parameters (by a little) and the λ parameter (by more).

Acknowledgments

The NA44 Collaboration wishes to thank the staff of the CERN PS-SPS accelerator complex for their excellent work. We thank the technical staff at CERN and the collaborating institutes for their valuable contributions. We are also grateful for the support given by the Science Research Council of Denmark; the Japanese Society for the Promotion of Science, and the Ministry of Education, Science and Culture, Japan; the Science Research Council of Sweden; the US Department of Energy; and the National Science Foundation. We also thank Heinz Sorge for giving us the RQMD code.

References

1. S. Pratt, Phys. Rev. **D33**, 1314 (1986).
2. G. Bertsch and G.E. Brown, Phys. Rev. **C40**, 1830 (1989).
3. D. E. Fields, *et al.*, Phys. Rev. **C52**, 986 (1995).
4. B.R. Schlei and N. Xu, Phys. Rev. **C54**, R2155 (1996).
5. H. Beker, *et al.* (NA44 Collaboration) Z. Phys. **C64**, 209 (1994); H. Bøggild, *et al.* (NA44 Collaboration) Phys. Lett. **B349**, 386 (1995);
6. H Beker, *et al.* (NA44 Collaboration), Phys. Rev. Lett. **74**, 3340 (1995).
7. K. Kaimi, *et al.* (NA44 Collaboration) Z. Phys. **C75**, 619 (1997).
8. N.Maeda, *et al.*, Nucl. Inst. Meth. **A346**, 132 (1994).
9. C.W.Fabjan, *et al.*, Nucl. Inst. Meth. **A367**, 240 (1995); A. Braem, *et al.*, CERN-PPE/97-120.
10. S. V. Akkelin and Yu. M. Sinyukov, Phys. Lett. **B356**, 525 (1995).
11. U. A. Wiedemann, P. Scotto, and U. Heinz, Phys. Rev. **C53**, 918 (1996).
12. T. Csörgo and B. Lörstad, Phys. Rev. **C54**, 1390 (1996).
13. A. Sakaguchi for the NA44 collaboration, Quark Matter '97 proceedings.
14. A. N. Makhlin and Yu. M. Sinyukov, Z. Phys. **C66**, 69 (1988).
15. Th. Alber, *et al.* (NA35 Collaboration), Z. Phys. **C66**, 77 (1995).
16. I. G. Bearden *et al.* (NA44 collaboration), Phys. Rev. Lett. **78**, 2080 (1997).
17. S. Nagamiya, Nucl. Phys. **A544**, 5c (1992).
18. J. P. Sullivan *et al.*, Phys. Rev. Lett. **70**, 3000 (1993).
19. H. Sorge, H. Stöcker and W. Greiner, Nucl Phys. **A498**, 567c (1989); H. Sorge *et al.* *Z. Phys.* C **47**, 629 (1990).
20. H. Sorge *et al.* *Phys. Lett.* B **289**, 6 (1992).
21. T. Csörgo *et al.*, *Phys. Lett.* B **241**, 301 (1990).
22. S. Pratt, T. Csörgo, J Zimányi, Phys.Rev C**42**, 2646 (1990).
23. S. Pratt and T. Csörgő, KFKI-1991-28A, 75.

Particle Correlation Studies at NA49: Moment Analysis of Two Particle Correlation Functions

B. Lasiuk for the NA49 collaboration[a]
Department of Physics
Yale University
E-mail: Brian.Lasiuk@Yale.edu

NA49 is a large acceptance spectrometer which measures over almost all phase space, the charged particles produced in 158 A GeV Pb+Pb collisions at the CERN-SPS. Single events containing up to 1200 measured charged particles have provided unprecedented statistical precision in a single event. Such precision has allowed the study of the shape of two-particle correlation functions. A new method to extract radii parameters from the correlation function without using a fitting procedure is evaluated and compared with the conventional Gaussian fitting method.

1 Introduction

Intensity interferometry is the only measurement that is able to provide an estimate of the spatial and temporal extension of the system created in a relativistic heavy ion collision. The volume is an important parameter required for the calculation of the energy density, a critical parameter which specifies the initial conditions and links experimental observables to theoretical calculations. The statistical precision of large acceptance/high-rate experiments utilizing the 158 A GeV Pb beams at the CERN-SPS has afforded the luxury of studying correlation functions in great detail. This paper concentrates on a very specific aspect of correlation functions that has recently drawn interest; that is the shape of the two-particle correlation function deviating from Gaussian profile.

The NA49 experiment is a large acceptance spectrometer which is designed to measure the final state charged particles over nearly complete phase space. It does so with 4 large volume Time Projection Chambers (TPCs), 2 of which are located in a dipole magnetic field which cover from near target rapidity to just forward of mid-rapidity, and 2 larger downstream chambers in a field free region which cover from mid to near beam rapidity. The experiment also provides capabilities for Particle Identification via specific ionization measurements with the TPCs and Time-Of-Flight (TOF) information. A figure of the experimental apparatus is shown by Ganz [1].

A single central Pb+Pb event produces the order of some 1200 charged particles within the acceptance of the NA49 detector which corresponds to the

[a] *Athens - Birmingham - Bratislava - Budapest - CERN - GSI - Dubna - IKF - Krakow - LBNL - Marburg - MPI - Seattle - UCD - UCLA - Warsaw - Yale - Zagreb*

order of $\sim 10^6$ particle pairs in a single event. This large statistical power has allowed the construction of correlation functions with insignificant statistical error from ensembles of events. It should be noted that although the statistical error is small, there still exists rather large systematic uncertainties in the construction and interpretation of the correlation functions themselves. Nonetheless as statistical power has increased, deviations from the assumed Gaussian profile have become more evident in the quality of the fits. Such behavior has generated much interest as some recent studies have indicated that non-Gaussian structures in correlation functions are a result of real physical effects. These include exotic emission patterns such as particles being emitted from a thin shell instead of a volume[2], or the result of long-lived resonances decaying outside the source smearing the correlation function by introducing long tails[3]. This speculation has remained largely qualitative because no concrete method to quantify this deviation has existed until this point in time—especially in the context of the current formalism which relates correlation functions to the phase space density or emission function of the source [4].

Equation 1 shows how the spatial-temporal extension of a source is accessible through a measurement of a two particle coincident spectrum in momentum space.

$$C(q) = \frac{Y(p,p')}{Y(p)Y(p')}_{meas} = 1 + \frac{\int_{source} d^4x d^4x' \; S(x,p) e^{iq\cdot x} S^*(x',p') e^{-iq\cdot x'}}{\int_{source} d^4x d^4x' \; S(x,p) S^*(x',p')} \tag{1}$$

Here C denotes the two particle correlation function measured as a function of the momentum difference q of the particles. $Y(p)$ denotes the yield of particles with momentum p and $S_i(x,p)$ denotes the emission function which is the probability that a particle of type i is emitted from a position x with momentum p. The two particle correlation function is a measure of the Fourier transform of the emission function. Because the width of the correlation function contains the spatial and temporal information, a convenient method of extracting the width is by fitting the curve to a Gaussian function as seen in equation 2:

$$C(q) = 1 + \lambda e^{-q^2 R^2} \tag{2}$$

where q is the momentum difference of the pair, λ is the correlation strength, and R is the size parameter in question. Although convenient this parameterization imposes a Gaussian profile on the emission function which introduces an implicit model dependence in the extraction of the the correlation width parameters which does not need to be made. Furthermore, such a parameterization by definition does not allow any "non-Gaussian" structure to be dealt with easily. As such indications of non-Gaussian behavior have been crudely estimated previously by the χ^2 or some other parameter which quantifies the fit quality [5]. This is a problem because such a parameter is as much a property of the precision of the data (i.e. size of the error bars) as the shape of the function. As such its sensitivity is questionable, especially if the shape deviates only slightly from Gaussian form.

Recently Wiedemann formally presented a method which reduces the model dependence of the extraction of the HBT correlation parameters [3]. Rather than utilizing a fitting procedure, the width of a correlation function is calculated directly via its 2nd order moment:

$$< q_i q_j > = \frac{\int d^3\mathbf{q}\, q_i q_j \,[C(q,K)-1]}{\int d^3\mathbf{q}\,[C(q,K)-1]} = D_{ij}^{-1}(K) \tag{3}$$

Here the correlation functions are numerically integrated over the different momentum components. For a one-dimensional correlation function $i=j$ and the result of the integral is the value $R/2$. In the case of multi-dimensional correlation functions, the results of the integral is a component of a tensor. In the Pratt-Bertsch representation [6], the tensor **D** takes the form:

$$D(\mathbf{K}) = 2\begin{pmatrix} R_o^2 & 0 & R_{ol}^2 \\ 0 & R_s^2 & 0 \\ R_{ol}^2 & 0 & R_l^2 \end{pmatrix} \tag{4}$$

where **K** denotes the pair momentum (k_T) dependence on the radii parameters, however the formalism is also sufficiently general to allow for the use of any other parameterizations such as the Yano-Koonin representation [3]. Note that any arbitrary cross-term(s) is allowed in the tensor **D**. In a similar manner the correlation strength parameter, λ can be calculated:

$$\lambda = \frac{\sqrt{\det \mathrm{D}}}{\pi^{\frac{3}{2}}} \int d^3\mathbf{q}\,[C(q,K)-1] \tag{5}$$

Although this expression appears to be independent of any model assumptions, a direct relationship between λ and the radius parameters is introduced as the correlation strength is proportional to the determinant of **D**. This is in contrast to a fitting procedure where the two parameters are linearly independent.[b] In the limit of a Gaussian correlator the radii and correlation strength as given by equations 4 and 5 respectively should return the same result as a Gaussian fitting method. This allows the parameters extracted from the integral method

[b] Note: there is always some degree of coupling between parameters in a fitting routine but they are functionally independent.

to be directly compared to those extracted from a Gaussian fitting method. This is not always the case when a new method of analysis is introduced.

Although use of the 2nd order moment provides a means with which to extract a model independent width from a correlation function, the real power of this method lies in its generalization. Higher order moments can be used to quantify the degree of deviation from a Gaussian profile in a systematic and controlled manner. Specifically a quantity derived from the 4th order moment is specified in equation 6:

$$G_i = \frac{< q_i^4 >}{3 < q_i^2 >< q_i^2 >} - 1 \tag{6}$$

It is normalized in such a way that it vanishes for a purely Gaussian correlator and takes on a negative or positive value depending if the function is flatter or more sharply peaked than a Gaussian distribution as illustrated in figure 1.

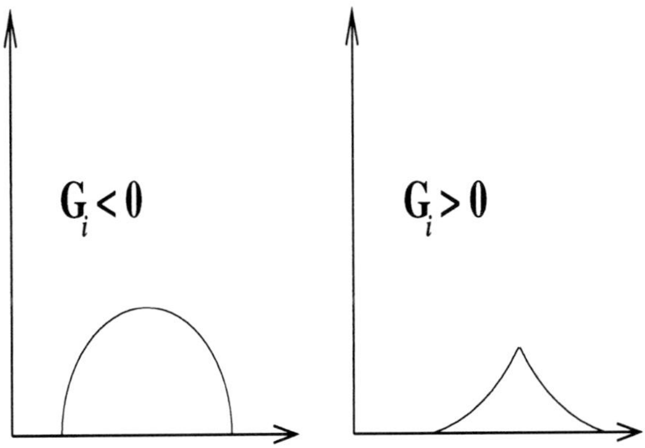

Figure 1: The value of G as specified in equation 6 is can take on positive or negative values depending on the profile of the correlation function.

2 Methodology

Perhaps the best way to illustrate the details of the new method is to directly compare the parameters extracted by a Gaussian fitting procedure and the integral approach on the same one dimensional correlation function. One dimensional projections of correlation functions are able to convey shape information most easily, so for simplicity and clarity, let us consider one-dimensional correlation functions in the Pratt-Bertcsh parameterization, which is specified by the Q_{Tout}, Q_{Tside}, and Q_{long} components. Results from a true multi-dimensional analysis will be reported later.

The construction of the correlation functions for this analysis was done with an ensemble of ~90k central Pb+Pb events that were taken with a 4% centrality trigger. The selection criteria was defined at three levels: the event level, the track level, and the pair level. Events with a multiplicity which fluctuated more than 3σ from the mean were removed from the event sample as a guard against pile-up. Tracks were selected on the basis of being within the acceptance of the VTPCs and passing through more than half the chamber. This criteria was imposed in order to reduce the possibility of double track counting. Primary tracks were selected by a strong constraint on the reconstructed event vertex position at the target. Tracks also had to be consistent with being a π according to the ionization information in the TPC. This had the effect of rejecting virtually all electrons while also reducing the contamination due to protons and kaons at higher momentum. At the particle pair level a two-track separation cut is made at 2 cm to ensure a constant track reconstruction efficiency as a function of track separation distance. A single correction was applied to the data at the particle pair level and this was to account for the effect of the Coulomb interaction. It is extracted from a correlation function constructed from unlike-sign pions [7]. A detailed discussion of the selection criteria and Coulomb corrections applied are outlined elsewhere [8].

Shown in figure 2 are the three one-dimensional projections of a correlation function constructed according to the Pratt-Bertsch parameterization. The projection in the other two components for each plot is ±30 MeV/c. As a consequence there is a finite dilution in the correlation. The curves shown are the best Gaussian fit to the data where a simple χ^2 minimization is used. The correlation parameters extracted from both methods are given in each panel; those extracted from the Gaussian fit method are denoted by "Fit" (left) while those deduced from the integral method specified by "$< q^2 >$" (right). The second order moment is calculated according to equation 3 by numerically integrating the correlation function on bin-by-bin with increasing Q starting at Q=0. A bin will contribute to the integral if the error is not larger than its contribution to the zeroth order moment; that is, if $[C(q,K)-1] > \delta C(q,K)$. When this condition is no longer satisfied, there is no further contribution to the integral. This point is indicated by an arrow in each plot and called the "cross-over" point.

There are several noteworthy differences between the parameters extracted from the different methods. First the integral method gives a radius that is smaller than that extracted by fitting the correlation function to a Gaussian by a factor of ~30%. The correlation strengths are also proportionally lower with

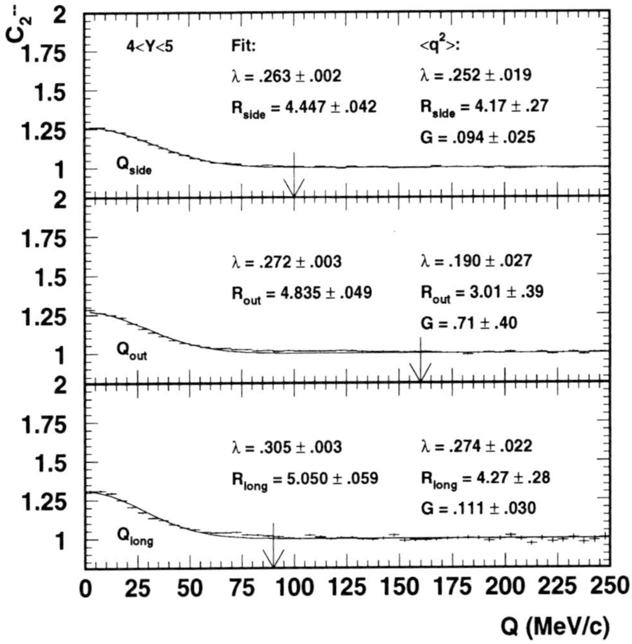

Figure 2: Projection of the three components of a multi-dimensional π^--π^- correlation function. The curves are the best Gaussian fit to the data. Due to the small error, deviations from a Gaussian profile are notable out to Q~160 MeV/c. This is still under investigation however, it is noteworthy that the Q_{out} component has the longest tail.

the use of the integral method. The errors are significantly larger from the integral method, and finally the magnitude of the G parameter can be visually correlated with the quality of the fit—it is largest when the cross-over point is large.

In regards to the significance of the correlation strength parameter being smaller as derived from the integral method, it is a consequence of the proportionality of the radii tensor as seen in equation 4; since the radii are smaller so too is λ. It is also not surprising that the errors are larger with the integral method because in the absence of the assumption of a functional profile of the distribution, the errors on each bin are independent and contributes linearly to the total error.

The discrepancy in the widths is probably the most significant finding and is due to contributions to the integral in the tail of the distribution which are much more heavily weighted than with the use of a fitting procedure. The criterion of the error on the bin being smaller than its contribution to the zeroth order moment implies that large Q bins will be used when statistical error is small. The value of the second order moment will increase with the number of bins that contribute to the integral. Since the value of $<q^2>$ is inversely proportional to the radius parameter, the radius will appear smaller, as observed. The Gaussian fit method is sensitive to a very localized region of the correlation function—typically between 25-85 MeV/c. Small Q bins (i.e. Q_{inv} <15 MeV/c) do not have a large impact because the errors in this region are usually substantially larger and bins at larger relative momentum have much smaller error and are very close to 1.

This insensitivity in the low Q region should not been viewed as serious because although the correlation strength is maximal in this region, it also carries the largest uncertainty as a result of the experimental difficulties in measuring closely separated tracks—either in momentum or configuration space, for both large acceptance and small acceptance experiments. In a large acceptance experiment like NA49, it is imperative that tracks not be split or fragmented in reconstruction as this can mimic low Q pairs and overestimate the yield in this region. On the other hand a small acceptance experiment is adversely affected by the need to generate a large event sample to be statistically competitive. A thick target is usually employed to increase the event rate, but this degrades the momentum resolution which distorts the low Q region of the spectrum.

Because the Gaussian fitting procedure is sensitive to such a small region, it gives very stable results if the region immediately around the inflection point of the correlation function (i.e. Q ~65 MeV/c) is included. Thus it is rather unimportant whether or not these low Q points are included when using a Gaussian fit to extract the correlation widths. This is not the case however with the integral method where the extracted width or 2nd moment appears to be a strong function of the position where the integral is truncated (i.e. the cross-over point). The integral method is sensitive over a more extended region both at small and large relative Q. Small relative momentum bins make a large contribution because the correlation function has its largest values in this region. Large relative momentum bins make a sizable contribution to the integral because they are weighted with Q^2 of the bin. In this respect the integral method is more democratic because it uses a wider range of the correlation function. However one must also consider that this method may be placing too much weight in the tails of the distribution. No matter, the errors must be small and well understood in order for one to make a judicious choice of a cross-over point because the parameters extracted are extremely sensitive to this position. Although there is a strong dependence of the radii as a function of the position of the cross-over point, it should be noted that many variations can be used on the specification of this point. For example a fixed interval can be isolated for study where a region from say $0<Q_{inv}$MeV/c<100 can be integrated. This allows the same region of all correlation functions to be treated equally and without bias. This is actually the procedure used in the analysis of correlation functions constructed from a single event [8].

3 Sensitivity and Limitations

Although there does appear to be indications of a non-Gaussian component, it is noteworthy that is is a small effect—at the per cent level in terms of deviation from Gaussian behavior. Furthermore no consideration of the systematic errors has been made. In order that this effect be established it is important to demonstrate that there is no introduction of artifacts in some other part of the analysis. Currently the Coulomb correction is the source of the large systematic uncertainty and it is very easy to introduce artifacts if one is not careful. The Coulomb correction of NA49 which uses the unlike particle 2-particle coincident spectra to parameterize the effect of the Coulomb interaction [7]. Although it is slightly controversial, it does not introduce large scale deviations into the correlation function as something more primitive like a Gamov factor does as seen in figure 3.

As mentioned previously the errors are larger with the integral method as well—an order of magnitude larger with the integral method as compared to the fit. This causes problems with multi-dimensional correlation functions. In order to get meaningful values for multi-dimensional correlation functions, it is necessary to project a correlation function into 2-dimensions and then require consistency between the 3 different projections. This is not ideal as there is a significant dilution in correlation strength whenever a projection is made. However the decrease in radii are still significant.

4 Physical Consequences

The most noticeable consequence of deriving the widths via moment analysis is that the sizes extracted are smaller than those extracted using a conventional fitting procedure. It is simply a result of increasing the weight and contribution of the tail in the distribution which increases the width of the correlation function as opposed to a fitting method is that the radii parameters are smaller; in fact by 15-30% which is shown in figure 4. As explained previously this is due to the large weighting at large relative momentum. This is a systematic effect and it is true not only for radii as a function of rapidity region but also for radii as a function of the pair transverse momentum or k_T [8]. The implications of such an effect are far reaching if the radii extracted using

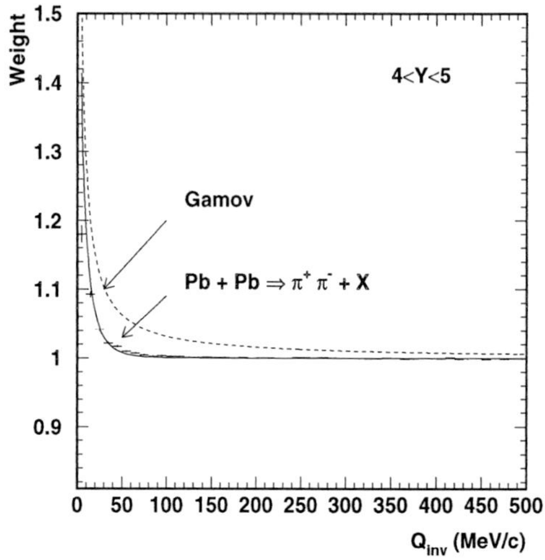

Figure 3: NA49 Coulomb correction extracted from an experimentally measured correlation function of π^+-π^- pairs. This is expected to parameterize the effect of the Coulomb effect. The solid line is the best fit to a one parameter exponentially damped Gamov factor and the dashed line is the Gamov weighting factor. The NA49 correction does not appear to introduce any non-Gaussian component or slope at large Q as the Gamov factor does.

the integral method are applied to the calculation of a source volume. Naively if one assumes an underestimate of 25% in the spatial extent of the source (or regions of homogeneity [9]) compare with those extracted from conventional analysis, the volumes of the sources are overestimated on the order of 75%! This has serious implications on the calculated energy densities, especially in the light of speculation that CERN is near or at the threshold of the transition to a deconfined state. However one must also bear in mind that this particular method may also over emphasize the contribution due to the tails of the distributions. Although the absolute magnitude of the effect is still not clear, it is true that tails will still reduce the size!

5 Conclusions and Outlook

The use of a second order moment is useful in extracting parameters from a correlation function without assuming any functional profile. This reduces model dependence. It is also true that the method is sensitive to the shape of the correlation function and can describe this behavior beyond qualitative discussions by the use of higher order moments. Although this method reduces

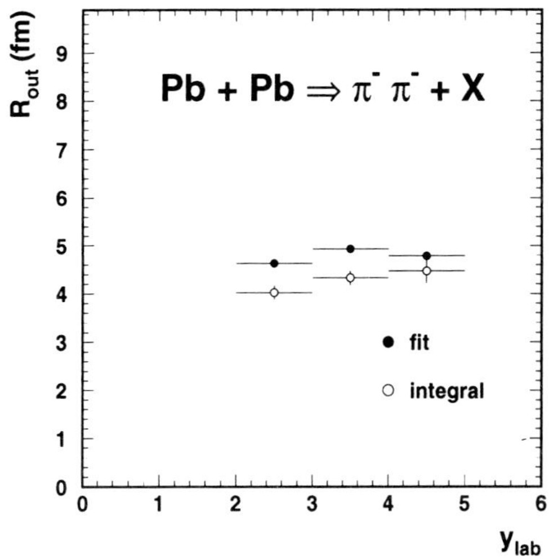

Figure 4: Although the same systematic trend of constant size versus rapidity region is followed for the radii extracted via the two methods, there is a systematic shift in the size. Although the absolute magnitude is dependent on the position of the cross-over point, the smaller sizes imply a smaller source size.

the model dependence in the extraction of correlation parameters, the interpretation of the parameters is no less model dependent. It is also possible that structures or regions of the correlation function can be isolated and studied in a manner much more convenient than a fitting algorithm allows by fixing the region to be integrated.

The question remains at what level is the measurement of smaller radii significant or an artifact of the tail of the distribution being weighted too heavily. If such an effect can be established unambiguously, it will have an effect on the physics in terms of overestimating source sizes. Still, this method has several drawbacks. This method requires much higher statistical precision than needed when using a fitting algorithm. This tends to make it difficult to report physically meaningful values with multi-dimensional correlation functions and simplifications and approximations are required. However using the moment analysis one is always assured of convergence which is not necessarily assured with fitting algorithms. The errors are also handled in a much more rigorous manner than with a fitting procedure.

Still to exploit the full capabilities of such a method further work is necessary. Now that there is a way to quantify the non-Gaussian component of a correlation function, this raises an obvious question—can this information be used to increase the information about the emission function and is the so-called "G-parameter" of use or is some other quantity derived from it more physically meaningful? Can resonance population be estimated, or further information regarding the emission structure be deduced via such a quantity or is there in fact a more appropriate way to extract this information? These are not easy questions to address because these are effects at the per cent level and systematic uncertainties must also be understood at this level if firm conclusions are to be drawn. As this is a small effect it is important to establish the effect accurately with the incorporation of appropriate systematic errors. In this regard the Coulomb correction plays a dominant role and limits the understanding.

Acknowledgements

This work is supported by the Director, Office of Energy Research, Division of Nuclear Physics of the Office of High Energy and Nuclear Physics of the US Department of Energy under contract DE-ACO3-76SFOOO98 and DE-FG02-91ER40609, the US National Science Foundation, the Bundesministerium fur Bildung und Forschung, Germany, the Alexander von Humboldt Foundation, the UK Engineering and Physical Sciences Research Counci, the Polish State Committee for Scientific Research (2 P03V01912), the Hungarian Scientific Research Foundateion under contracts T14920 adn T23790, the EC Marie Curie Foundation, and the Polish German Foundation.

References

1. see R. Ganz, these proceedings.
2. U. Heinz, seminar at Yale University, 16 Sept. 1997.
3. U. Wiedemann, Acta Phys. Slov **47** (1997) 15.
4. S. Chapman et al., Phys. Rev. Lett. **74** (1995) 4400.
5. L. Rosselet et al., Nucl. Phys. **A610** (1996) 256c.
6. G. F. Bertsch Nucl. Phys. **A498** (1989) 173.
7. T. Alber et al., (NA35 collaboration) Z. Phys **C73** (1997) 443.
8. B. Lasiuk, Ph.D. dissertation UCLA 1997, unpublished.
9. Yu. M. Sinyukov et al., Nucl. Phys. **A566** (1994) 589.

BOSE-EINSTEIN FINAL STATE SYMMETRIZATION FOR EVENT GENERATORS OF HEAVY ION COLLISIONS

U.A. Wiedemann[a], J. Ellis[b], U. Heinz[a] and K. Geiger[c]

[a] *Institut für Theoretische Physik, Universität Regensburg, D-93040 Regensburg, Germany, E-mail: Urs.Wiedemann@physik.uni-regensburg.de*

[b] *Theoretical Physics Division, CERN, CH-1211 Geneva 23, Switzerland*

[c] *Physics Department, BNL, Upton, N.Y. 11973, U.S.A.*

We discuss algorithms which allow to calculate identical two-particle correlations from numerical simulations of relativistic heavy ion collisions. A toy model is used to illustrate their properties.

1 Introduction

The current relativistic heavy ion program at CERN and BNL aims at investigating the equilibration properties of hadronic matter at extreme temperatures and densities where quarks and gluons are expected to be the physically relevant degrees of freedom for particle production processes. The theoretical discussion of these collision systems is complicated by their mesoscopic character. They are not sufficiently small to allow for an analytical description in terms of elementary processes. They are not sufficiently large to take a description in terms of macroscopic observables for granted. Even if simple thermodynamically or hydrodynamically inspired models account for the data, the task remains to understand the microscopic origin of their success, and to establish to what extent such an agreement is necessary or accidental.

Numerical simulations of relativistic heavy ion collisions are an adequate tool to attack these questions. They allow to propagate sufficiently many degrees of freedom from some initial condition, and they thus test the consequences of different assumptions about the microscopic in medium dynamics. Many event generators have been developed to this aim [1]. Irrespective of their particular model assumptions, they return for each simulated event a set of N_m phase space emission points $\{(\check{\mathbf{p}}_i, \check{\mathbf{r}}_i, \check{t}_i)\}_{i\in[1,N_m]}$ together with information, e.g., about the identity (PID) of the particle i, the parent resonance it originates from, the number of rescatterings it has undergone, etc.

The major aim of event generator studies is to reduce step by step the class(es) of dynamical scenarios consistent with experimental data, thus obtaining an increasingly better picture of the physics. To this end, one mainly compares the simulated particle ratios and one-particle momentum spectra to experiment. However, one can imagine different dynamical scenarios which all account for the same simulated momentum distributions $\{(\check{\mathbf{p}}_i)\}_{i\in[1,N_m]}$ but differ with respect to the simulated space-time structure $\{(\check{\mathbf{r}}_i,\check{t}_i)\}_{i\in[1,N_m]}$. Identical two-particle correlations $C(\mathbf{K},\mathbf{q})$, here written as a function of the measured pair momentum $K=\frac{1}{2}(p_1+p_2)$ and relative pair momentum $q=p_1-p_2$ are sensitive to the latter and can thus play an important complementary role in specifying the parameter space of event generators. Here, we discuss algorithms of the type

$$\left\{\{(\check{\mathbf{p}}_i,\check{\mathbf{r}}_i,\check{t}_i)\}_{i\in[1,N_m]}\right\}_{m\in[1,N_{\rm ev}]} \Longrightarrow C(\mathbf{K},\mathbf{q})\,. \tag{1}$$

For notational convenience, we restrict the following discussion to one particle species only, like-sign pions say. Also, we neglect final state Coulomb and strong interactions[3]. The set $\{(\check{\mathbf{p}}_i,\check{\mathbf{r}}_i,\check{t}_i)\}_{i\in[1,N_m]}$ denotes then the phase space emission points of the N_m like-sign pions generated in the m-th simulated event. The event generator simulates thus a classical phase space distribution

$$\rho_{\rm class}(\mathbf{p},\mathbf{r},t)=\frac{1}{N_{\rm ev}}\sum_{m=1}^{N_{\rm ev}}\sum_{i=1}^{N_m}\delta^{(3)}(\mathbf{r}-\check{\mathbf{r}}_i)\,\delta^{(3)}(\mathbf{p}-\check{\mathbf{p}}_i)\,\delta(t-\check{t}_i)\,. \tag{2}$$

2 From phase space densities to momentum correlations

The standard particle interferometric analysis of heavy ion collisions starts from the relation between the measured momentum correlations of identical particles and the quantum mechanical Wigner phase space density $S(x,K)$ of particle emitting sources in the collision region[2],

$$C(\mathbf{K},\mathbf{q})=\mathcal{N}\left(1+\frac{\left|\int d^4x\,S(x,K)\,e^{iq\cdot x}\right|^2}{\int d^4x\,S(x,p_1)\,\int d^4y\,S(y,p_2)}\right)\,. \tag{3}$$

This expression can be generalized to include final state interactions[3]. Depending on the definition of the two-particle correlator in terms of single- and two-particle cross sections[4,5] (respectively depending on its construction from experimental data), the normalization is either $\mathcal{N}=1$ or $\mathcal{N}<1$. The calculation of deviations of $\mathcal{N}$ from 1 typically involves multiparticle symmetrization effects[6,5], which we do not review here. Our discussion will thus be for $\mathcal{N}=1$.

In the light of Eq. (3), calculating the two-particle correlator $C(\mathbf{K},\mathbf{q})$ from an event generator output amounts to specifying the emission function $S(x,K)$ in terms of the generated phase space points $\{(\check{\mathbf{p}}_i,\check{\mathbf{r}}_i,\check{t}_i)\}_{i\in[1,N_m]}$, i.e., it amounts to an interpretation of $(\check{\mathbf{p}}_i,\check{\mathbf{r}}_i,\check{t}_i)$. In sections 2.1 and 2.2, we focus on two different interpretations referred to as *quantum mechanical* and *classical*, though they are both conceptually on an equal footing. Before doing that, we list here three requirements which an algorithm implementing (1) should fulfill:

1. Event generators use *probabilities* to describe the collision process. Bose-Einstein correlations are obtained from squaring production *amplitudes*. Hence, the generated momenta $\check{\mathbf{p}}_i$, $\check{\mathbf{p}}_j$ do not show the enhancement at low relative momenta characteristic for Bose-Einstein final state symmetrization. The prescriptions (1) should calculate this effect a posteriori.

2. The strength of Bose-Einstein correlations depends on the distance of the identical particles in phase space, not in momentum space. We thus require the prescriptions (1) to use the entire phase space information, and not only the generated momentum information. 'Weighting' or 'shifting' prescriptions which are based only on the latter may successfully match the measured momentum correlations but obviously do not allow to test the simulated space-time structure.

3. In general, Bose-Einstein statistics can affect particle multiplicity distributions during the particle production process. However, event generators are tuned to reproduce the average particle multiplicities in agreement with the data. We require hence that the prescriptions (1) conserve particle multiplicities. If the correlator is interpreted as a factor, relating measured two-particle differential cross sections to those simulated by the event generator, $d^6\sigma_{\rm meas}/d^3\mathbf{p}_1\, d^3\mathbf{p}_2 = C(\mathbf{K},\mathbf{q})\; d^6\sigma_{\rm sim}/d^3\mathbf{p}_1\, d^3\mathbf{p}_2$, then this implies[11,6] that $C(\mathbf{K},\mathbf{q}) < 1$ for some values of $\mathbf{q}$. This interpretation of $C(\mathbf{K},\mathbf{q})$ is however not necessary since the space-time analysis of correlation data can be based entirely on the momentum dependence of $C(\mathbf{K},\mathbf{q})$, disregarding its absolute normalization.

2.1 "Quantum" interpretation of event generator output

In the *quantum mechanical* interpretation[8,9,6,10], the set of phase space points $\{\{(\check{\mathbf{p}}_i, \check{\mathbf{r}}_i, \check{t}_i)\}_{i\in[1,N_m]}\}_{m\in[1,N_{\rm ev}]}$ is associated to the centers of Gaussian one-particle wavepackets according to[7,8,9]

$$(\check{\mathbf{p}}_i, \check{\mathbf{r}}_i, \check{t}_i) \longrightarrow f_i(\mathbf{x}, \check{t}_i) = \frac{1}{(\pi\sigma^2)^{3/4}} e^{-\frac{1}{2\sigma^2}(\mathbf{x}-\check{\mathbf{r}}_i)^2 + i\check{\mathbf{p}}_i\cdot\mathbf{x}}. \tag{4}$$

These wavepackets f_i are quantum mechanically best localized states, i.e., they saturate the Heisenberg uncertainty relation with $\Delta x_i = \sigma/\sqrt{2}$ and $\Delta p_i =$

$1/\sqrt{2}\sigma$.

Taking only two-particle symmetrized contributions into account, i.e., adopting the so-called pair approximation, all spectra can be written in terms of the functions $s_i(\mathbf{p})$, which coincide with the corresponding one-particle spectrum for a source emitting only one particle at the phase space point i. For the one- and two-particle correlator one finds [9,8]

$$E_p \frac{d^3N}{d^3p} = \frac{1}{N_{\rm ev}} \sum_{m=1}^{N_{\rm ev}} \nu_m(\mathbf{p}) = \frac{1}{N_{\rm ev}} \sum_{m=1}^{N_{\rm ev}} \sum_{i=1}^{N_m} s_i(\mathbf{p}) \,, \tag{5}$$

$$C(\mathbf{K},\mathbf{q}) = 1 + e^{-\sigma^2 \mathbf{q}^2/2} \frac{\sum_{m=1}^{N_{\rm ev}} \left[\left| \sum_{i=1}^{N_m} s_i(\mathbf{K})\, e^{iq^0 \check{t}_i - i\mathbf{q}\cdot\check{\mathbf{r}}_i} \right|^2 - \sum_{i=1}^{N_m} s_i^2(\mathbf{K}) \right]}{\sum_{m=1}^{N_{\rm ev}} \left[\nu_m(\mathbf{p}_a)\, \nu_m(\mathbf{p}_b) - \sum_{i=1}^{N_m} s_i(\mathbf{p}_a)\, s_i(\mathbf{p}_b) \right]} \tag{6}$$

$$s_i(\mathbf{p}) = \pi^{-3/2}\, \sigma^3\, e^{-\sigma^2(\mathbf{p}-\check{\mathbf{p}}_i)^2} \,. \tag{7}$$

Here, the terms subtracted in the numerator and denominator of $C(\mathbf{K},\mathbf{q})$ are finite multiplicity corrections which become negligible for large particle multiplicities [8]. This algorithm is consistent with an emission function $S(x,\mathbf{K})$

$$S(x,\mathbf{K}) = \int d\check{\mathbf{p}}_i\, d\check{\mathbf{r}}_i\, d\check{t}_i\, \rho_{\rm class}(\check{\mathbf{p}}_i, \check{\mathbf{r}}_i, \check{t}_i)\, s_0(\mathbf{x} - \check{\mathbf{r}}_i, t - \check{t}_i, \mathbf{K} - \check{\mathbf{p}}_i) \,, \tag{8}$$

$$s_0(x,\mathbf{K}) = \tfrac{1}{\pi^3}\, \delta(t)\, {\rm e}^{-\frac{1}{\sigma^2}\mathbf{x}^2 - \sigma^2 \mathbf{K}^2} \,, \tag{9}$$

which is given by a folding relation between a classical distribution $\rho_{\rm class}$ of wavepacket centers and the elementary source Wigner function $s_0(x,\mathbf{K})$. The latter has at emission an optimal quantum mechanical phase space localization around phase space points $(\check{\mathbf{p}}_i, \check{\mathbf{r}}_i, \check{t}_i)$ with spatial uncertainty σ and momentum uncertainty $1/\sigma$.

In the algorithm (5)-(7), the spectra are discrete functions of the input $(\check{\mathbf{p}}_i, \check{\mathbf{r}}_i, \check{t}_i)$ but they are continuous in the observable momenta $\mathbf{p}_1$, $\mathbf{p}_2$ and hence, no binning is necessary. Also, the number of terms to be calculated in (5) and (6) grows linear in N, not quadratic as in other prescriptions. However, both the one- and two-particle spectra depend in this algorithm on the wavepacket width σ. The role of this parameter will be discussed in the context of our toy model study in section 3.

2.2 "Classical" interpretation of event generator output

In the *classical* interpretation [9,10], the phase space points $(\check{\mathbf{p}}_i, \check{\mathbf{r}}_i, \check{t}_i)$ are seen as a discrete approximation of the on-shell Wigner phase space density $S(x,p)$,

$$S(x,\mathbf{p}) = \rho_{\rm class}(\mathbf{p},\mathbf{x},t) \,. \tag{10}$$

This defines both the two-particle correlator via (3) and the one-particle spectrum via the x-integration over (10). To simplify a numerical implementation of this calculation, it is useful to replace the δ functions of $\rho_{\rm class}$ in momentum space by rectangular 'bin functions'[9], or by properly normalized Gaussians[10] of width ϵ,

$$\delta^{(\epsilon)}_{\check{\mathbf{p}}_i,\mathbf{p}} = \frac{1}{(\pi\epsilon^2)^{3/2}} \exp\left(-(\check{\mathbf{p}}_i - \mathbf{p})^2/\epsilon^2\right) . \tag{11}$$

In the limit $\epsilon \to 0$, these Gaussian bin functions reduce to the properly normalized δ-functions $\delta^{(3)}(\check{\mathbf{p}}_i - \mathbf{p})$ and we regain Eq. (2). The one-particle spectrum and two-particle correlator read then[9,10]

$$E_p \frac{d^3N}{d^3p} = \int d^4x\, S(x,\mathbf{p}) = \frac{1}{N_{\rm ev}} \sum_{m=1}^{N_{\rm ev}} \sum_{i=1}^{N_m} \delta^{(\epsilon)}_{\check{\mathbf{p}}_i,\mathbf{p}} , \tag{12}$$

$$C(\mathbf{K},\mathbf{q}) = 1 + \frac{\sum_{m=1}^{N_{\rm ev}} \left[\left| \sum_{i=1}^{N_m} \delta^{(\epsilon)}_{\check{\mathbf{p}}_i,\mathbf{K}}\, e^{iq^0\check{t}_i - i\mathbf{q}\cdot\check{\mathbf{r}}_i} \right|^2 - \sum_{i=1}^{N_m} \left(\delta^{(\epsilon)}_{\check{\mathbf{p}}_i,\mathbf{K}} \right)^2 \right]}{\sum_{m=1}^{N_{\rm ev}} \left[\left(\sum_{i=1}^{N_m} \delta^{(\epsilon)}_{\check{\mathbf{p}}_i,\mathbf{p}_1} \right) \left(\sum_{j=1}^{N_m} \delta^{(\epsilon)}_{\check{\mathbf{p}}_j,\mathbf{p}_2} \right) - \sum_{i=1}^{N_m} \delta^{(\epsilon)}_{\check{\mathbf{p}}_i,\mathbf{p}_1} \delta^{(\epsilon)}_{\check{\mathbf{p}}_i,\mathbf{p}_2} \right]} . \tag{13}$$

The correlator (13) is the discretized version of the Fourier integrals in (3). The subtracted terms in the numerator and denominator remove discretization errors which would amount to pairs constructed from the same particles. The 'classical' algorithm has the same important advantages as the 'quantum' one, it involves only $O(N_m)$ rather than $O(N_m^2)$ numerical manipulations per event, and its observables are discrete functions of the input, but continuous functions of the measured momenta; hence, no binning is necessary. 'Classical' and 'quantum' algorithm then differ with respect to two points:

1. The classical algorithm has no analogue for the Gaussian prefactor $\exp(-\sigma^2\,\mathbf{q}^2/2)$ in (6) which is a genuine quantum effect stemming from the quantum mechanical localization properties of (4).

2. For the choice $\sigma = 1/\epsilon$, the bin functions $\delta^{(\epsilon)}_{\check{\mathbf{p}}_i,\mathbf{p}}$ is the classical counterpart of the Gaussian single-particle distributions $s_i(\mathbf{p})$. Finite event statistics puts a lower practical limit on ϵ, but the physical momentum spectra are recovered in the limit $\epsilon \to 0$, i.e., ϵ should be chosen as small as technically possible. In contrast, the quantum algorithm treats σ as finite physical localization of the single particle wavepackets. The limit $\sigma \to \infty$ (which corresponds to $\epsilon \to 0$) is not the physically relevant case. It amounts according to (8) to an emission function which is infinitely extended in space, and hence[8] $\lim_{\sigma\to\infty} C(\mathbf{K},\mathbf{q}) = 1 + \delta_{\mathbf{q},0}$.

3 The Zajc model

Our final aim is to apply the above algorithms to event generators of relativistic heavy ion collisions and high energy elementary processes. A first step in this direction will be reported in [10]. Here, we illustrate the properties of these algorithms for a toy model.

The strategy is to start from a simple expression for the classical distribution $\rho_{\rm class}$, from which the results of the algorithms of sections 2.1 and 2.2 can be calculated analytically. In a second step, we generate then in a Monte Carlo procedure sets of phase space points $\{(\check{\mathbf{p}}_i, \check{\mathbf{r}}_i, \check{t}_i)\}_{i\in[1,N_m]}$ according to this distribution, and we test the numerical results against our exact analytical expressions. This allows us to make in an analytically controlled setting statements about i) the σ-dependence of the quantum algorithm, ii) the ϵ-dependence of the classical algorithm (especially: how small ϵ has to be chosen to extract the limit $\epsilon \to 0$ numerically) and iii) the statistical requirements for the algorithms to work.

Here, we study these questions for the Zajc model which describes a position-momentum correlated phase space distribution of emission points [12]

$$\rho^{\rm zajc}_{\rm class}(\mathbf{p},\mathbf{r},t) = \mathcal{N}_s \exp\left\{ -\frac{1}{2(1-s^2)} \left(\frac{\mathbf{r}^2}{R_0^2} - 2s\frac{\mathbf{r}\cdot\mathbf{p}}{R_0 P_0} + \frac{\mathbf{p}^2}{P_0^2} \right) \right\} \delta(t)\,, \quad (14)$$

$$\mathcal{N}_s = E_p \, \frac{N}{(2\pi)^3\, R_0^3\, P_0^3\, (1-s^2)^{3/2}}\,. \quad (15)$$

This distribution is normalized such that the total event multiplicity calculated from the one-particle spectrum (12) is N. The parameter s smoothly interpolates between completely uncorrelated and completely position-momentum correlated sources. For $s \to 0$, the position-momentum correlation vanishes and we are left with the product of two Gaussians in position and momentum space. In the opposite limit

$$\lim_{s\to 1} \rho^{\rm zajc}_{\rm class}(\mathbf{p},\mathbf{r},t) \sim \delta^{(3)}\left(\frac{\mathbf{r}}{R_0} - \frac{\mathbf{p}}{P_0} \right) \delta(t)\,, \quad (16)$$

the position-momentum correlation is perfect. The corresponding phase space volume of the distribution is proportional to

$$V = 2^3\, R_0^3\, P_0^3\, (1-s^2)^{3/2}\,, \quad (17)$$

and goes to zero for $s \to 1$. This strong s-dependence allows to study the performance of our numerical algorithms for different phase space volumes. In the following subsections, we discuss the s-dependence of the one-particle spectrum and two-particle correlator, focussing in sections 3.1 and 3.2 on analytical results, and comparing these in section 3.3 to numerical calculations.

3.1 The Zajc model in the classical algorithm

Starting from an analytical emission function $S(x, \mathbf{p}) = \rho_{\text{class}}(\mathbf{p}, \mathbf{x}, t)$ according to (10), the one-particle spectrum and two-particle correlator read

$$E_p \frac{dN}{d^3p} = E_p \frac{N}{(2\pi)^{3/2} P_0^3} \exp\left(-\frac{\mathbf{p}^2}{2P_0^2} \right), \tag{18}$$

$$C(\mathbf{K}, \mathbf{q}) = 1 + \exp\Big(-\mathbf{q}^2 R_{\text{class}}^2(\epsilon) \Big), \tag{19}$$

$$R_{\text{class}}^2(\epsilon) = R_0^2 \left(1 - s^2 \frac{1}{1 + \epsilon^2/2P_0^2} - \frac{1}{(2P_0R_0)^2} \frac{1}{1 + \epsilon^2/2P_0^2} \right). \tag{20}$$

We recover the physical HBT radius from (20) in the limit $\epsilon \to 0$ or by inserting (10) directly into (3). The remarkable fact is that for sufficiently large s [12],

$$s > s_{\text{crit}} = \sqrt{1 - \frac{1}{(2R_0P_0)^2}}, \tag{21}$$

the HBT radius R_{class}^2 becomes negative and the two-particle correlator shows an unphysical rise of the correlation function with increasing $\mathbf{q}^2$. The change of sign in (19) seems to be related to the violation of the uncertainty relation by the emission function (14): In the limiting case (16), $\rho_{\text{class}}^{\text{zajc}}$ reduces to a perfect position-momentum correlated source which is not consistent with the Heisenberg uncertainty relation. Also, the phase space volume V in (17) drops below 1 for $s > s_{\text{crit}}$. The tentative interpretation of $V = 1$ as the volume of one elementary phase space cell further supports that the distribution $\rho_{\text{class}}^{\text{zajc}}$ is a quantum mechanically allowed emission function $S(x, \mathbf{p}) = \rho_{\text{class}}(\mathbf{p}, \mathbf{x}, t)$ for $s < s_{\text{crit}}$ only.

What do we learn from these analytical results? They show that an ansatz for a quantum mechanical Wigner phase space density $S(x, K)$ has to satisfy quantum mechanical localization properties. However, realistic event generators satisfy this requirement anyway. The practical importance of (this limiting case of) the Zajc model is that it provides an (extreme) testing ground for our numerical algorithms. Analytically, we conclude already from (20) that a bin width sufficient to investigate the limit $\epsilon \to 0$, has to be small on the scale of the width of the generated momentum distribution,

$$\epsilon \ll \sqrt{2}\, P_0 \,. \tag{22}$$

At least for the present model study, this requirement is independent of the strength of position-momentum correlations in the source.

3.2 The Zajc model in the quantum algorithm

We now interprete the classical probability distribution $\rho^{\rm zajc}_{\rm class}$ of (14) as a distribution of phase space points associated with the centers of Gaussian wavepackets according to (4). The one-particle spectrum and two-particle correlator are then readily calculated via the emission function (8),

$$E_p \frac{dN}{d^3p} = E_p \frac{N}{(2\pi)^{3/2}(P_0^2 + 1/2\sigma^2)^{3/2}} \exp\left(- \frac{\mathbf{p}^2}{2P_0^2 + 1/\sigma^2} \right) , \tag{23}$$

$$C(\mathbf{K}, \mathbf{q}) = 1 + \exp\left(- \mathbf{q}^2 R_{\rm qm}^2 \right) , \tag{24}$$

$$R_{\rm qm}^2 = \frac{\sigma^2}{1 + 2\sigma^2 P_0^2} \left(\sigma^2 P_0^2 + R_0^2/\sigma^2 + 2R_0^2 P_0^2 (1 - s^2) \right) . \tag{25}$$

In contrast to the classical algorithm, the radius parameter is now always positive, irrespective of the value of s. Even if the classical distribution $\rho^{\rm zajc}_{\rm class}(\check{\mathbf{p}}, \check{\mathbf{r}}, \check{t})$ is sharply localized in phase space, its folding with minimum uncertainty wave packets leads to a quantum mechanically allowed emission function $S(x, \mathbf{p})$ and to a correlator with falls off with increasing $\mathbf{q}^2$ as expected. However, the spread of the one-particle momentum spectrum (23) receives an additional contribution $1/\sigma^2$. Choosing σ too small increases this term beyond the phenomenologically reasonable values, choosing it too large widens the corresponding HBT radius parameters significantly. It was argued[8] that σ can be interpreted as quantum mechanical 'size' of the particle, $\sigma \approx 1$ fm. Given the heuristic nature of these arguments and the significant modifications this implies for the spectra (23) and (24), it is however fair to say that presently σ mainly plays the role of a regulator of unwanted violations of the quantum mechanical uncertainty principle while a deeper understanding of its origin in the particle production dynamics is still missing.

3.3 Numerical simulation in the Zajc model

We have studied the performance of our Bose-Einstein algorithms by generating with a random number generator sets $\{(\check{\mathbf{p}}_i, \check{\mathbf{r}}_i, \check{t}_i)\}_{i \in [1, N_m]}$ of phase space points according to the distribution $\rho^{\rm zajc}_{\rm class}$ and comparing the numerical results of our algorithms to the analytical expressions of section 3.1 and 3.2. Figure 1 shows such a comparison for the limiting case $s = 1$ of the Zajc model. We have chosen the model parameters $R_0 = 1$ fm and $P_0 = 50$ MeV, to explore the model for very small phase space volumes. Our numerical results are for a small 'event sample' of $N_{\rm ev} = 100$ with event multiplicity $N_m = 100$. The small differences between analytical and numerical results have their origin in

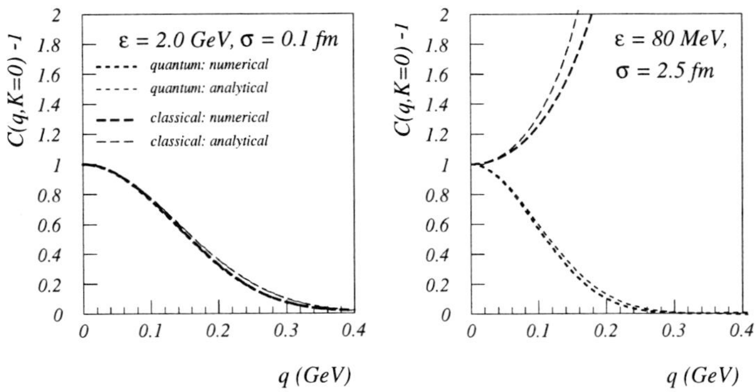

Figure 1: Numerical results for the Zajc model in the limit $s = 1$. Irrespective of the chosen width parameters σ and ϵ, the numerical algorithms reproduce the correct analytical results even for small event statistics and even for unphysically sharp phase space localizations. For further information, see the text.

statistical fluctuations and disappear for larger values of N_{ev} and/or N_m. The low statistical requirements of our algorithms come from the fact that they associate to the *discrete* set of generated momenta $\check{\mathbf{p}}_i$ *continuous* functions of the measured momenta $\mathbf{p}_1$, $\mathbf{p}_2$. This smoothens any statistical fluctuation significantly.

The relative momentum dependence shown in Fig. 1 is as expected from the analytical calculations. Fig. 1a shows that the values $\sigma = 0.1$ fm, $\epsilon = \frac{1}{\sigma} = 2000$ MeV approximate already very well the limiting case

$$\lim_{\sigma \to 0} R_{\text{qm}}^2 = R_0^2 = \lim_{\epsilon \to \infty} R_{\text{class}}^2(\epsilon)\,, \tag{26}$$

in which the prefactor of (6), $\exp\left(-\sigma^2 \mathbf{q}^2/2\right)$, approaches unity and the classical (13) and quantum (6) algorithm become identical. In Fig. 1b, we see that for $\sigma = 2.5$ fm, the HBT radius R_{qm} has picked up already a significant quantum contribution, $R_{\text{qm}}^2 > R_0^2$, which is reflected in a smaller width of the corresponding correlator. On the same scale, $\epsilon = \frac{1}{\sigma} = 80$ MeV, the bin width $\delta^{(\epsilon)}$ of the classical algorithm is already small enough to clearly resolve the unphysical increase, corresponding to the sharp phase space localization (16) in the classical interpretation. The slightly larger difference between numerical and analytical results indicates that the statistical requirements are more severe if one wants to resolve with our algorithms source characteristics which are localized in small phase space regions.

Acknowledgments

This is a report on work in progress. The presenting author, U.A.W., would like to acknowledge a DFG-Habilitationsstipendium. He also thanks the CRIS'98 Local Organizing Comittee for generous support which made his participation at this Conference possible.

References

1. Y. Pang, Talk at the 13th Intl. Conf. on Ultra-Relativistic Nucleus-Nucleus Collisions (Quark Matter'97), Nucl. Phys. **A** to be published.
2. U. Heinz, in: *Correlations and Clustering Phenomena in Subatomic Physics*, ed. by M.N. Harakeh, O. Scholten, and J.H. Koch, NATO ASI Series (Plenum, New York, 1996), p. 137.
3. D.V. Anchishkin, U. Heinz and P. Renk, Phys.Rev. C57 (1998) 1428.
4. D. Miśkowiec and S. Voloshin, nucl-ex/9704006.
5. Q.H. Zhang, P. Scotto and U. Heinz, nucl-th/9805046.
6. U.A. Wiedemann, nucl-th/9801009, Phys. Rev. **C** in press.
7. H. Merlitz and D. Pelte, Z. Phys. **A357**, 175 (1997).
8. U.A. Wiedemann, P. Foka, H. Kalechofsky, M. Martin, C. Slotta and Q.H. Zhang, Phys. Rev. C**56**, R614 (1997).
9. Q.H. Zhang, U.A. Wiedemann, C. Slotta and U. Heinz, Phys. Lett. **B**407, 33 (1997).
10. J. Ellis, K. Geiger, U. Heinz and U.A. Wiedemann, in preparation.
11. L. Lönnblad and T. Sjöstrand, Phys. Lett. **B351**, 293 (1995); Eur.Phys.J. C2 (1998) 165-180
12. W.A. Zajc, Particle Production in Highly Matter, ed. H.H. Gutbrod and J. Rafelski, NATO ASI Series B303 (Plenum Press, New York), 1993, p. 435.

PION INTERFEROMETRY USING WAVEPACKETS

D. PELTE, H. MERLITZ
Physikalisches Institut der Universität Heidelberg,
Philosophenweg 12, D-69120 Heidelberg, GERMANY
E-mail: pelte@pel5.mpi-hd.mpg.de

The conditions for the HBT interferometry from stars are compared to those encountered in heavy ion reactions. As a consequence the formalism using wavepackets is developed. The modifications relative to the standard formalism with plane waves are presented, and the effects of the residual Coulomb interactions between charged pions and with the charged source are calculated.

1 Introduction

Originally the interferometric method of Hanburry Brown, Twiss (HBT) was developed in astronomy and used to determine the angular radii of stars [1]. Shortly afterwards it was recognized that a similar procedure can also be applied to pions emitted in high energy reactions to measure the size of the pion emitting source [2]. However, it is clear that the different conditions prevailing in high-energy or heavy-ion reactions ask for certain modifications of the original formalism. The table 1 lists the relevant parameters encountered in the study of stars or hot nuclear matter. The correlation areas $\Delta\Omega$ are given by

Table 1: Parameters relevant to stars and nuclei.

	star (photon)	nucleus (pion)
source size R	10^9 m	10^{-14} m
momentum p	2 eV/c	$2 \cdot 10^8$ eV/c
correlation area $\Delta\Omega = \pi/(pR)^2$	$3 \cdot 10^{-32}$ sr	$3 \cdot 10^{-2}$ sr
de Broglie wavelength $\lambda = h/p$	$6 \cdot 10^{-7}$ m	$6 \cdot 10^{-15}$ m
coherence length Λ	$c\tau \approx 3$ m	$\lambda^2/\Delta\lambda \approx 10^{-14}$ m

the solid angle necessary to observe the speckle pattern in the emission of a large number of identical bosons as displayed in Fig.1 of ref. [3]. Their difference by 30. order of magnitudes is compensated by the equivalent difference in the distance between the star, respectively the nucleus and the interferometer. The large difference between the source size R and the coherence length Λ in case of a star ensures that the star can be treated as a chaotic radiator. In general one finds for stars $R \gg \Lambda \gg \lambda$ and the emitted photons may be described by plane waves. In case of the nucleus one finds $R \approx \Lambda \approx \lambda$, and wavepackets are the more appropriate description of the emitted pions.

2 Wavepacket Formalism

The width of a Gaussian wavepacket at time t=0 is given by σ_0 which from the principles of quantum mechanics is related to the pion coherence length: $\sigma_0 \approx \Lambda/4\pi \approx 1$ fm.At later times t the free, i.e. non-interacting wavepacket in coordinate space is [4]

$$\Psi(\mathbf{r},t) = \left(2\pi s^2\right)^{-\frac{3}{4}} exp\left\{\frac{i}{\hbar}\left(\mathbf{Pr} - Et\right) - \frac{\left(\mathbf{r} - \mathbf{R}\left(t\right)\right)^2}{4s\sigma_0}\right\}, \tag{1}$$

where $\mathbf{P}, E$ are the centre momentum , energy of the wavepacket, $\mathbf{R}(t)$ is its spatial coordinate which changes with time according to $\mathbf{R}(t) = \mathbf{R} + (\mathbf{P}/m)\cdot t$, and the dispersion of the width is given by

$$\sigma = |s| = \sigma_0\sqrt{1 + \frac{\hbar^2}{4m^2\sigma_0^2}t^2}. \tag{2}$$

Similarly the wavepacket may be described in momentum space, where it turns out to be stationary:

$$\Phi(\mathbf{p},t) = \left(2\pi\sigma_p^2\right)^{-\frac{3}{4}} exp\left\{\frac{i}{\hbar}\left(\mathbf{R}\left(\mathbf{p} - \mathbf{P}\right) + \frac{\mathbf{p}^2}{2m}t\right) - \frac{\left(\mathbf{p} - \mathbf{P}\right)^2}{4\sigma_p^2}\right\} \tag{3}$$

with $\sigma_p = \hbar/2\sigma_0$. It is obvious that the wavepacket representation of the pions fulfills Heisenbergs uncertainty relation

$$\sigma_p\sigma \geq \frac{\hbar}{2}. \tag{4}$$

In the following we will restrict our discussion to SIS energies, i.e. the like-charge pion multiplicities are in central collisions of very heavy systems $m_\pi \approx 20$, and multi pion effects can be neglected. We parametrize the pion source function by $g(R,P) = \rho(R)f(P)$ with

$$\rho(R) = \left(\pi R_s^2\right)^{-\frac{3}{2}} exp\left\{-\frac{R^2}{R_s^2}\right\} \tag{5}$$

$$f(P) = \left(2\pi mT\right)^{-\frac{3}{2}} exp\left\{-\frac{P^2}{2mT}\right\} \tag{6}$$

and with the temperature T. The single pion observables in spatial and momentum space are then easy to calculate and yield the identical forms as in eqs.(5,6) except that the relevant parameters have to be replaced by

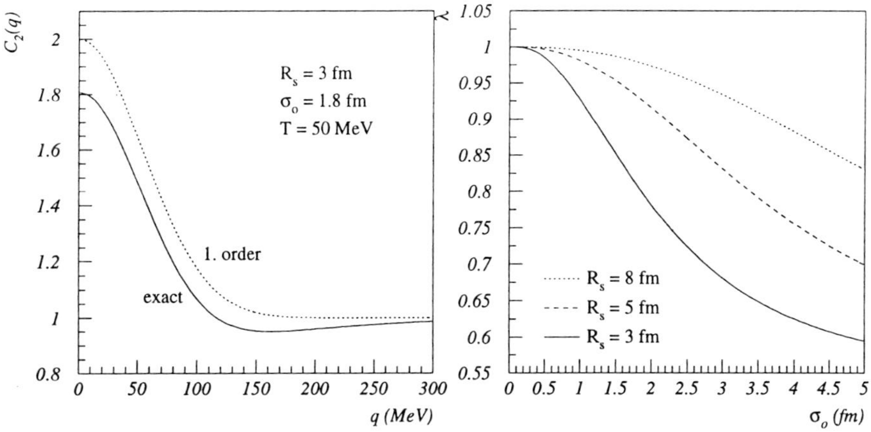

Figure 1: Left panel: The exact and 1. order correlation functions. Right panel: The dependence of the chaoticity λ on the localization σ_0.

$\tilde{R}_s^2 = R_s^2 + 2\sigma_0^2$ and $T_{eff} = T + T_{qm}$ where $T_{qm} = \frac{\hbar^2}{4m\sigma_0^2}$. This implies that for small pion multiplicities the localization σ_0 of the pions in the source does not change the single pion distributions, but the relevant parameters are changed: The effective source radius is modified by the pion localization, and the effective temperature is modified by a quantum contribution which takes into account the zero point energy of the pions.

Similarly the two pion correlation function $C_2(\mathbf{p}_1, \mathbf{p}_2) = \frac{\mathcal{P}_2(\mathbf{p}_1,\mathbf{p}_2)}{\mathcal{P}_1(\mathbf{p}_1)\mathcal{P}_1(\mathbf{p}_1)}$ may be calculated. This calculation is more involved than in case of the single pion observables, the exact result was published in[4]. As an approximate expression one obtains

$$C_2(\mathbf{p}_1, \mathbf{p}_2) = 1 + \lambda exp\left\{-\frac{R_{eff}^2}{2\hbar^2}\mathbf{q}^2\right\} \quad , \quad \mathbf{q} = \mathbf{p}_1 - \mathbf{p}_2 \tag{7}$$

with an effective radius parameter $R_{eff}^2 = R_s^2 + 2\sigma_0^2 \frac{T}{T_{eff}}$, so that $R_s < R_{eff} < \tilde{R}_s$. The chaoticity parameter λ is a function of the ratios $R_s^2/(2\sigma_0^2)$ and T/T_{eff} and can be expanded into the series

$$\lambda = 1 + \sum_{k=1}^{\infty}(-)^k \left(1 - k\frac{R_s^2}{2\sigma_0^2}\right)^{-\frac{3}{2}} \left(1 - k\frac{T}{T_{eff}}\right)^{-\frac{3}{2}} . \tag{8}$$

The Fig.1 displays on the left side the exact and first order ($\lambda = 1$) results for $C_2(q)$ using specific values for R_s, σ_0 and T, and on the right side the dependence of λ on σ_0 is shown for 3 different values of R_s and the temperature $T = 50$ MeV. From this dependence we infer that λ approaches 0 only when σ_0 becomes very large and T goes to zero.

The wavepacket formalism for pions introduces a new parameter σ_0 which was called localization because it determines the probability to localize the pion within a certain volume of the hot nuclear matter. Naively the value of σ_0 is therefore experimentally bounded by the two limits $R_{eff} = \sqrt{2}\sigma_0$ and $T_{eff} = T_{qm}$ which yields 0.8 fm $< \sigma_0 <$ 4 fm under SIS conditions. However it is more appropriate to relate σ_0 to the coherence length Λ of the pion, which in the case that the pion is emitted in the decay of the $\Delta(1232)$ resonance is coupled to the lifetime of that resonance in hot nuclear matter. Alternatively one may also consider the mean free path of pions in hot nuclear matter as the appropriate quantity. We have used this latter conjecture in our calculations and have assumed $\sigma_0 = 1.8$ fm. This implies a quantum contribution of $T_{qm} =$ 21 MeV to the effective temperature of pions. The wavepacket formalism was applied in other recent papers to the HBT interferometry with pions [4,5,6,7,8]. In cases where σ_0 was specified its value was set to $\sigma_0 \approx 0.7$ fm which yields $T_{qm} \approx 100$ MeV, a value which appears to be too high for the SIS energy range.

3 The Coulomb Residual Interactions

HBT interferometric studies in relativistic heavy-ion reactions are usually performed with charged pions. In this case the pions after emission cannot be described by free wavepackets but experience residual interactions from the charge of the source and from their own charges. As a result the centre motion of the wavepacket is modified and the wavepacket gets deformed. There are two ways to study these effects with respect to HBT interferometry, both ways are based on the time dependent Schrödinger equation: The multiconfigurational method [9], and the molecular dynamic method [4].

3.1 Coulomb Interaction with the Source

The second method was used to study the modifications of the two-pion correlation function when both pions interact with the source via the Coulomb force. Depending on the charge of the source the width of the correlation function is decreased for (π^-,π^-) but increased for (π^+,π^+) pairs, yielding a larger effective source radius from the former pairs, and a smaller radius from the latter pairs. The situation is depicted in the left panel of Fig.2 for a source with radius $R_s = 5.5$ fm and temperature $T = 50$ MeV. Surprisingly the increase respectively decrease are not of same size, but stronger for the (π^-,π^-) pairs. The reason for this asymmetry is found in the right panel of

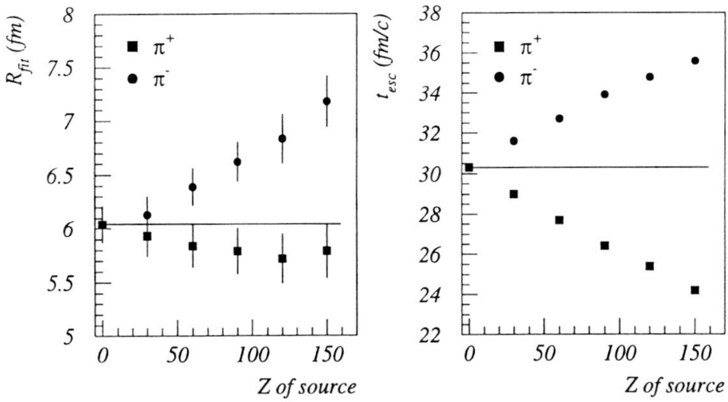

Figure 2: Left panel: The dependence of the effective source radii from (π^-,π^-) and (π^+,π^+) pairs on the charge Z of the source. Right panel: The escape times t_{esc} for π^- and π^+ as functions of Z.

Fig.2 where the average time is shown which a charged pion spends within the escape radius $R_{esc} = 20$ fm of the source. In case of the (π^+,π^+) pairs the correlation signal develops to a large fraction after the π^+ have left the interaction region, and it is therefore only little influenced by this interaction. In case of the (π^-,π^-) pairs the times the π^- spend inside the interaction region are longer and therefore this interaction disturbs the correlation signal more strongly. The asymmetry in the radius distortions is a direct proof that the appearance of the correlation does not occur instantaneously but is a process in time.

3.2 *Coulomb Interaction between Charged Pions*

The first and the second methods were used to study the effects of their mutual Coulomb interaction onto the two-pion correlation function. The results were similar: The Coulomb interaction between like-charge pions alters the correlation by such a small amount that it becomes unobservable under normal experimental conditions [9]. This conclusion is in apparent contradiction to experimental results from (π^+,π^-) pairs [10], we will therefore focus our discussion on this type of pairs.

The reason why the suppression of the correlation signal for very small values of q is not observed for like-charge pion pairs is connected to the zero point energy $E_0 = \frac{3}{2}T_{qm}$ of the pion which is almost two orders of magnitude

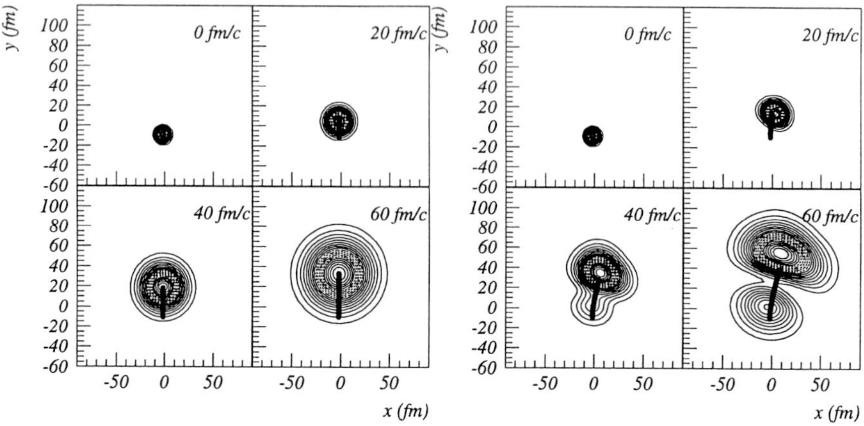

Figure 3: Left panel: The time development of the relative wavepacket of unlike-charge pairs with charge $Z = 1$. Right panel: The same for pairs with charge $Z = 20$.

larger than the Coulomb energy between two pions with localization $\sigma_0 = 1.8$ fm. This is particularly easy to see in the case of two pions with different charges where this Coulomb energy attains its maximum value at complete overlap and is, for particles with charge number Z, of size $E_{Coul} = -\frac{Z^2 e^2}{\sqrt{\pi}\sigma_0}$. One may define the parameter [11]

$$\xi = \frac{|E_{Coul}|}{E_0} = \frac{8e^2}{3\sqrt{\pi}\hbar^2} Z^2 m \sigma_0 \tag{9}$$

the value of which determines whether or not the Coulomb interaction becomes discernible. For $\sigma_0 = 1.8$ fm and $Z = 1$ one obtains $\xi = 1.4 \cdot 10^{-2}$, but $\xi = 5.6$ in case of a hypothetical charge number $Z = 20$. The time development of the pion wavepacket in **relative** coordinates is shown in Fig.3 for both cases, the contour lines give the shape of the wavepacket, the dark curves the movement of its centre. It is evident that in the first case the wavepacket movement and its deformation is not modified by the Coulomb interaction whereas it is strongly modified in the second case with respect to the centre as well as to the deformation. In order to observe the Coulomb interaction the ξ parameter has to be of order 1. This condition cannot only be met by an increase of the charge number Z but also by increasing the reduced mass of the particle pair or by increasing the localization σ_0. Finally one may also test the validity of the present procedure by approaching the classical limit $\hbar \to 0$. These dependences were studied in detail, as examples the Fig.4 displays in the left

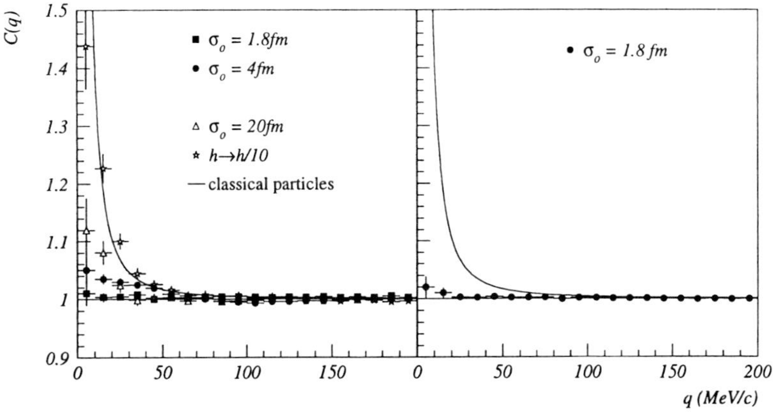

Figure 4: The correlation function of (π^+,π^-) pairs with different localizations σ_0. Also shown is the classical result (full curve) and its limit $\hbar \rightarrow \hbar/10$. The results in the left panel are from method 1, in the right panel from method 2 (see text).

panel the (π^+,π^-) correlation functions for three different values of σ_0 and for the case $\hbar \rightarrow \hbar/10$. In this latter case the calculated result almost agrees with the classical expectation shown by the continuous curve. The dependence on σ_0 is as predicted, as a cross check, shown in the right panel, we have also performed the calculation for $\sigma_0 = 1.8$ fm using the second method.

We conclude that the Coulomb interaction only modifies the correlation function at small values of q when the ξ parameter is of order one. Whereas this is not to be expected at SIS energies in case of pion pairs, it can occur for such pairs at much higher energies. At high energies pions are also emitted from the decay of long-living resonances which were produced by the nucleus - nucleus collision. Such pions have a large coherence length with the related extended localization. The strength of the Coulomb distortion depends on the relative abundance of pions from the interaction zone to those from the long-living resonances, only the former will contribute to the observation of the HBT signal. The strength of the HBT signal at $q \rightarrow 0$ also depends on the pion localization and on the source temperature, dependences which experimentally have not been explored up to now.

Acknowledgments

This work was supported by the Bundesministerium für Forschung und Technologie under contract No 06 HD 525 I and by the Gesellschaft für Schwerionenforschung mbH under contract No HD Pel K.

References

1. R. Hanburry Brown and R. Twiss, *Proc. Roy. Soc. (London)* A **243**, 291 (1957).
2. G. Goldhaber *et al*, PR **120**, 300 (1960).
3. H. Merlitz and D. Pelte, *Z. Phys.* A **351**, 187 (1995).
4. H. Merlitz and D. Pelte, *Z. Phys.* A **357**, 175 (1997).
5. U. A. Wiedemann *et al*, *Phys. Rev.* C **56**, 614 (1997).
6. U. A. Wiedemann, nucl-th/9801009.
7. J. Zimanyi and T. Csörgö, nucl-th/9705432.
8. T. Csörgö and J. Zimanyi, nucl-th/9705433.
9. H. Merlitz and D. Pelte, *Phys. Lett.* B **415**, 411 (1997).
10. J. Barrette *et al*, *Phys. Rev. Lett.* **78**, 2916 (1997).
11. H. Merlitz and D. Pelte, nucl-th/9805028.

COMBINED EFFECT OF NUCLEAR COULOMB FIELD AND RADIAL FLOW ON TWO-PION CORRELATIONS

H. W. BARZ
Institut für Kern- und Hadronenphysik,
Forschungszentrum Rossendorf,
Pf 510119, 01314 Dresden, Germany
E-mail:barz@fz-rossendorf.de

The common influence of the nuclear Coulomb field and the radial flow on two-pion correlations is investigated for a spherical expanding hot nuclear system. The radii extracted from the correlation function are determined in sideward and outward direction and compared with the real extension of the source. Corrections for the analysis of recent measurements at SIS and AGS energies are calculated.

1 Introduction

Measurements of correlations of pions as a function of their relative momenta are used to determine the source radius in heavy ion reactions at intermediate and relativistic energies. On the basis of the Hanbury-Brown and Twiss (HBT) effect the correlation function is related to the source radius via $R_0 = \hbar c/\Delta q$, where Δq is the width of the correlation function. However it is well known [1,2] that the correlation is a complicated function not only of the relative momentum but also of the total momentum of the pion pair. The correlation function is determined by the size of the region from which pions are emitted with roughly the same momenta. This has the consequence that for a collectively streaming matter this region is much smaller due to the strong correlation between the momenta and the emission points of the emitted particles.

On the other hand the Coulomb force of the source changes the momentum distribution in the source to the measured distribution at asymptotic distance. This latter effect was investigated in refs.[3,4,5]. Pions with small momenta are mostly influenced by the Coulomb forces which act quite differently on positively and negatively charged pions. Indeed quite different radii have been observed recently in collision of Au on Au at bombarding energies of 1 A· GeV and 11 A· GeV.

The aim of this investigation is to calculate the combined effect of the charge and the flow on the extracted radii. As a model we consider the emission of pions from a hot nucleonic system the expansion of which can described by radial hydrodynamical flow. For this source we construct a corresponding source function in Sect. 3. Once we have obtained the source function the standard technique is used to calculate the correlation function. The Coulomb distortion is included using distorted instead of plane waves for the pions.

2 Basic equations

Two-particle correlation of a system are defined by the two-particle density matrix in the final state. The basic assumptions for applying interferometry to nuclei is the chaoticity of the source i.e. the absence of initial correlations in the source between the two emitted pions. This allows one to express the two-particle density-matrix by the symmetrized product of one-particle density matrices $\rho(p,p')$, where p denotes the four-momentum $p = (\omega, \mathbf{p})$.

Assuming that we observe two pions with momenta $\mathbf{p}$ and $\mathbf{p}'$ we introduce the average pair momentum $\mathbf{K}$ and relative momentum $\mathbf{q}$

$$\mathbf{K} = \frac{1}{2}(\mathbf{p} + \mathbf{p}'), \quad \mathbf{q} = \mathbf{p} - \mathbf{p}'. \tag{1}$$

and correspondently $K = \frac{1}{2}(p + p')$ and $q = p - p'$. Following the standard derivation [1,6] one can write the probability to observe the two pions as being proportional to

$$C_2(\mathbf{K}, \mathbf{q}) = \left[1 + \frac{|\rho(K - \frac{q}{2}, K + \frac{q}{2})|^2}{\rho(K - \frac{q}{2}, K - \frac{q}{2})\,\rho(K + \frac{q}{2}, K + \frac{q}{2})}\right]. \tag{2}$$

This correlation function is the two-particle cross section normalized to the product of the single particle cross sections. In the derivation of Eq. (2) the final state interaction between the two pions is not taken into account. A thorough investigation of this issue can be found in refs. [3,4,7,8].

The one-particle density-matrix $\rho(p,p')$ in Eq. (2) contains the properties of the source. It is connected with the current-current correlator [1,9] which depends on the emission and absorption points x and x',

$$\rho(p, p') = \int d^4x d^4x' \psi_p^*(x) \psi_{p'}(x') \langle J^+(x') J(x) \rangle \ . \tag{3}$$

Via a Wigner transformation

$$\langle J^+(x') J(x) \rangle = \int \frac{d^4k}{(2\pi)^4} e^{-ik(x-x')} S(k, \frac{x + x'}{2}) \ . \tag{4}$$

one introduces the classical source term $S(k, x)$ which describes the creation of a pion at space-time x with four momentum k.

The problem is further reduced by assuming that the interactions do not depend on time. This assumption allows the use of stationary solutions for

$\psi_p(x) = exp(-i\omega t)\,\psi_{\mathbf{p}}(\mathbf{r})$. Inserting the definition(4) into Eq.(3) and integrating over one time-variable one obtains

$$\rho(p,p') = \int\int d\mathbf{r}\, d\mathbf{r}'\, \psi^*_{\mathbf{p}}(\mathbf{r})\psi_{\mathbf{p}'}(\mathbf{r}') \times$$
$$\int dt\, e^{i(\omega-\omega')t} \int \frac{d\mathbf{k}}{(2\pi)^3}\, e^{i\mathbf{k}(\mathbf{r}-\mathbf{r}')}\, S[(\frac{\omega+\omega'}{2},\mathbf{k}),(t,\frac{\mathbf{r}+\mathbf{r}'}{2})]. \quad (5)$$

In the case of a thermal momentum distribution of the source without correlation between space and momentum the integration over $\mathbf{k}$ introduces a correlation length in the order of the thermal wave length $\sqrt{2\pi/(mT)}$.

3 Hydrodynamical Picture

Here, we restrict our selves to a simple situation. We consider a hot spherical source from which pions are emitted. Since the source expands the problem is not stationary. However, we circumvent this difficulty by considering only pions which move with a sufficiently large velocity. Those pions are essentially outside the source and always feel a time independent Coulomb potential. Having in mind a collision of Au on Au nuclei at a few A·GeV generating a source of a temperature of about 100 MeV, the thermal velocity of the protons in the source is about 0.4 c embedded in a flow field of an average velocity of about 0.3 c. Thus, the model would be applicable for pion velocities larger than 0.7 c corresponding to pion momenta above 100 MeV/c if the pions were emitted from the surface region. Pions coming from the center will feel a reduced Coulomb force as it was observed in ref.[10].

For ultrarelativistic energies the system does not expand spherically but is preferentially stretched in longitudinal direction. Since our treatment neglects this dimension we do not consider the longitudinal correlation function. Due to the longitudinal expansion the Coulomb force decreases with time. As an approximate measure of the Coulomb action one can replace the central charge Z with twice the rapidity density $2(dN^+/dy - dN^-/dy)$ of the net charge as was shown in ref. [11].

In the hydrodynamical approach the momentum distribution is defined by a local temperature T and a velocity field given by the four-vector u^μ. We use parameters which are typical for heavy ion reactions: fixed temperature T, radial mean velocity $\langle\beta\rangle$ and size R_0. Thus, the source function reads

$$S(p,x) = e^{-\frac{1}{T}p^\mu u_\mu}\, e^{-\frac{\mathbf{r}^2}{2R_0^2}-\frac{t^2}{2\tau^2}}, \quad (6)$$

where the thermal distribution is assumed to be the Jüttner distribution in the fluid cell moving with the velocity u^μ. The radial flow velocity field is given by

$$u^\mu = (u^0, \mathbf{u}), \quad u^0 = \sqrt{1+\mathbf{u}^2}, \quad \mathbf{u} = \beta_0 \mathbf{r}/R_0, \tag{7}$$

where the four-velocity scales with the distance from the center. The parameter β_0 is related to the mean flow velocity $\langle\beta\rangle = \langle|\mathbf{u}|/u^0\rangle$ averaged over the density (6) which in the limit of small β_0 reads $\langle\beta\rangle = \sqrt{8/\pi}\beta_0$. Since β_0 characterizes a four-velocity it does not have an upper bound while in contrary $\langle\beta\rangle$ never exceeds the velocity of light. According to Eq.(6) the pions are radiated off during the emission time τ.

Now we insert the source function (6) into Eq.(5) and obtain

$$\begin{aligned}\rho(p,p') &= (2\pi)^{-5/2}\tau e^{-(\omega-\omega')^2\tau^2/2}\int d\mathbf{r}\, d\mathbf{r}'\, \psi^*_{\mathbf{p}}(\mathbf{r})\psi_{\mathbf{p}'}(\mathbf{r}') \\ &\times \int d\mathbf{k} e^{i\mathbf{k}(\mathbf{r}-\mathbf{r}')} e^{\mathbf{k}\mathbf{u}/T} e^{-(\mathbf{r}+\mathbf{r}')^2/(8R_0^2)-(\omega+\omega')\sqrt{1+\mathbf{u}^2}/(2T)}\,, \end{aligned}\tag{8}$$

where the velocity field $\mathbf{u}$ is proportional to $\frac{\mathbf{r}+\mathbf{r}'}{2}$. In the range of the convergence of Eq. (8) the integration over $\mathbf{k}$ leads to a δ function $\delta(\mathbf{r}-\mathbf{r}'-i\mathbf{u}/T)$ with the help of which we arrive at the final expression

$$\begin{aligned}\rho(p,p') = \sqrt{2\pi}\tau e^{-(\omega-\omega')^2\tau^2/2} &\int d\mathbf{r}\, [\psi_{\mathbf{p}}(\alpha\mathbf{r})]^*\psi_{\mathbf{p}'}(\alpha\mathbf{r}) \\ &\times e^{-\mathbf{r}^2/(2R_0^2)-(\omega+\omega')\sqrt{1+(\beta_0/R_0\mathbf{r})^2}/(2T)}\,. \end{aligned}\tag{9}$$

The factor $\alpha = 1 - i\beta_0/(2TR_0)$ rotates the integration path for the distorted wave functions $\psi_{\mathbf{p}}$ into the complex r plane.

4 Numerical Method

To incorporate the Coulomb field we solve the Klein-Gordon equation

$$[-\frac{\partial^2}{\partial\mathbf{r}^2} - (\sqrt{m^2+\mathbf{p}^2} - U)^2 + m^2]\psi_{\mathbf{p}}(\mathbf{r}) = 0, \tag{10}$$

where U means the Coulomb potential of the source with charge number Z. For the Gaussian source we take the potential $U = \pm Ze^2\,\Phi(r/\sqrt{2}R_0)/r$, with Φ being the error function. The solution is obtained numerically by using partial-wave expansion technique. We expand the Coulomb distorted wave into partial waves f_l

$$\psi^{(-)}_{\mathbf{p}}(\mathbf{r}) = \frac{4\pi}{pr}\sum_{l\,m} i^l e^{-i\sigma_l} f_l(r) Y^*_{lm}(\mathbf{p}) Y_{lm}(\mathbf{r}), \tag{11}$$

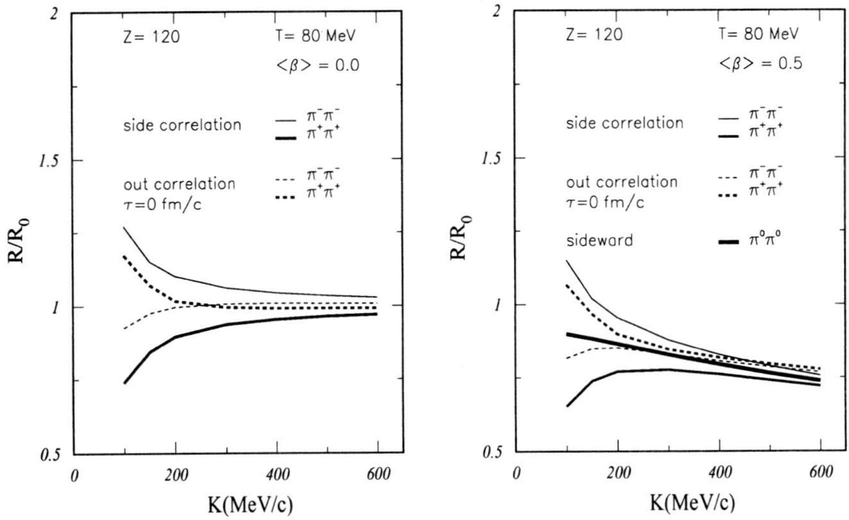

Figure 1: Ratios of sideward and outward HBT radii to the radius R_0 of a Gaussian shaped source without/with radial flow on the left/right hand side panel.

where the quantities Y_{lm} denote the spherical harmonics and the symbols σ_l are the Coulomb scattering phases. The boundary conditions for the partial waves f_l are chosen such that ψ behaves asymptotically like an outgoing wave in the direction of $\mathbf{p}$ with incoming spherical waves. This is indicated by the upper index (-). Once the radial function has been matched to the boundary condition it is analytically continued from $r = R_{max}$ to $r = \alpha R_{max}$. The value obtained at this point is used to normalize the function $\psi(\alpha r)$ obtained by integrating the radial equation along the path αr.

5 Results

The correlation function (2) has been calculated for various pair momenta as function of the relative momentum q. We investigate the correlation as a function of the momentum $\mathbf{q}_{out}$ pointing in the direction of $\mathbf{K}$ and momentum $\mathbf{q}_{side}$ being orthogonal to $\mathbf{K}$, respectively. In all cases considered the obtained correlation function has nearly a Gaussian shape which is easily fitted to the standard form

$$C_2 = 1 + \exp\left(-\mathbf{q}_{side}^2 R_{side}^2 - \mathbf{q}_{out}^2 R_{out}^2\right) \tag{12}$$

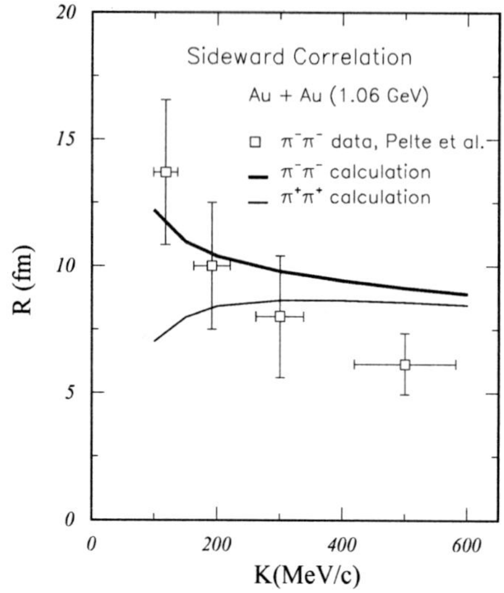

Figure 2: Measured sideward radii [15] as a function of the average pion momentum. in comparison with calculated HBT radii for a source radius of 10 fm.

providing us the HBT radii R_{out} and R_{side}.

As a typical situation we consider the collision of Au on Au nuclei at bombarding energies of 1 GeV per nucleon. In nearly central collisions a system of charge $Z = 120$ with a temperature of about $T = 80$ MeV for pions and a flow velocity of about $\langle\beta\rangle = 0.32$ is formed [12,13]. We use a hard-sphere radius of $R_h = 10$ fm which corresponds to the parameter $R_0 = R_h/\sqrt{5}$.

In Fig. 1 the fitted source radii are displayed as a function of the pair momentum $|\mathbf{K}|$ for two extreme cases assuming flow velocities $\langle\beta\rangle = 0$ and $\langle\beta\rangle = 0.5$, respectively. Compared to R_0 the observed radii R_{side} are increased for negatively charged pion and diminished for positively charged pions in comparison to the true radius. An opposite effect is seen for the radii R_{out}. These changes are significant for pions with momenta below 300 MeV/c. Pion momenta below 100 MeV/c have not been considered as the stationary approach is not justified. In addition we have inserted into the second panel of Fig. 1 the HBT radius extracted for π^0 mesons. Comparing the curves for the two flow velocities in Fig. 1 one recognizes that the corrections arising from the Coulomb field and the flow field add up nearly independently.

Differences between extracted HBT radii for positively and negatively charged pions have been measured at AGS [14] in the projectile rapidity region. A ratio of $R_{side}(\pi^-\pi^-)$ to $R_{side}(\pi^+\pi^+)$ of (5.6 ± 0.7) fm / (3.9 ± 0.8) fm $= 1.4 \pm 0.3$ has been found which is in agreement with our predictions. The ratio of the outward radii of (5.8 ± 0.5) fm / (6.5 ± 0.5) fm $= 0.9 \pm 0.15$ is smaller than unity although the relatively large error bars do not allow a definite comparison. These measurements also qualitatively agree with the ratio of 1.2 ± 0.4 of the sideward radii obtained by Pelte at al.[15].

A strong dependence of the side-correlation on the average pion momentum has been found in the reaction of Au on Au nuclei at a bombarding energy of 1.06 GeV per nucleon[15]. In Fig. 2 we compare this data with the HBT radii from a source with the true radius of $R_h = 10$ fm taking into account the combined effect of flow and Coulomb field. It is clearly seen that the measured sideward radius depends stronger on the pair momentum than one would expect from flow and Coulomb corrections. This means that indeed the fast pions come from an earlier more compressed stage of the matter while the low energy pions are emitted from a zone with a larger size.

In conclusion, we have found that the nuclear Coulomb field increases (decreases) the observed HBT radii extracted from sideward correlations for negatively (positively) charged pions. The influence is opposite and smaller for the outward correlation. This effect is largest for small momenta and is superimposed on the overall reduction caused by the radial flow.

Acknowledgments

Support by the German ministry BMBF under contract 06DR827 is acknowledged.

References

1. M. Gyulassy, S.K. Kaufmann, and L.W. Wilson, *Phys. Rev.* C **20**, 2267, (1979)
2. S. V. Akkelin and Yu. M. Sinyukov, *Phys. Lett.* B **356**, 525 (1995); U.A. Wiedemann, P. Scotto, and U. Heinz, *Phys. Rev.* C **53**, 918 (1996); J. Helgesen, T. Csörgő, M. Asakawa, and B. Lörstad, *Phys. Rev.* C **56**, 2626 (1997)
3. S. Pratt, *Phys. Rev.* D **33**, 72 (1986)
4. G. Baym and P. Braun-Munzinger, *Nucl. Phys.* A **610**, 286c (1996)
5. H. W. Barz, *Phys. Rev.* C **53**, 2536 (1996)
6. S. Pratt, *Phys. Rev. Lett.* **53**, 1219 (1984)

7. L. Martin, C.K. Gelbke, B. Erazmus, and R. Lednitzky, *Nucl. Phys.* A **604**, 69 (1996)
8. D. Anchishkin, U. Heinz, and P. Renk, *Phys. Rev.* C **57**, 1428, (1998)
9. P. Danieliewiz and P. Schuck, *Phys. Lett.* B **274**, 268 (1992)
10. A. Wagner et al., *Phys. Lett.* B **420**, 20 (1998)
11. H.W. Barz, J.P. Bondorf, J.J. Gaardhøje, and H. Heiselberg, *Phys. Rev.* C **57**, 2536, (1998)
12. D. Pelte et al., *Z. Phys.* A **357**, 215 (1997)
13. C. Müntz et al., *Z. Phys.* A **357**, 399 (1997)
14. J. Barette et al., *Phys. Rev. Lett.* **78**, 2916 (1997)
15. D. Pelte, FOPI collaboration, *Proc. of Int. Research Workshop on Heavy Ion Physics at Low, Intermediate and Relativistic Energies using 4π Detectors*, Poiana Brasov, October 1996, Romania, Edts. M. Petrovici et al., World Scientific Singapore, 1997, p. 271

Central Coulomb Effects on Pion Interferometry

D. Hardtke
Lawrence Berkeley National Laboratory
Berkeley, CA 94720
E-mail: DHHardtke@lbl.gov

Although much effort has been devoted to the understanding of the final-state Coulomb interactions between the outgoing particle pairs in nucleus-nucleus collisions, relatively little effort has been devoted to understanding the Coulomb interaction between the outgoing particle pair and the particle emitting source. Single particle measurements at the Bevelac, AGS, and SPS show evidence for strong central Coulomb effects on the single particle spectra, so there is no reason to assume that the central Coulomb effects on the interferometry measurements are negligible. Since the many-body Coulomb problem has not been solved, it is necessary to use approximations. Recent calculations for the Pb+Pb system at 158 GeV/c per nucleon will be presented. These calculations suggest that the central Coulomb potential leads to a non-negligible reduction in the measured pion correlation radii and lambda parameters. These calculations will be compared to recent NA44 data and be shown to agree qualitatively. Since this central Coulomb effect is generally ignored, the implications on the freeze-out scenario in high energy nucleus-nucleus collisions are explored.

1 Introduction

For many years, the large experimental uncertainty in pion correlation measurements made it useless to worry about theoretical uncertainties of order 10% or less. Recent experiments at the AGS and SPS have reduced the experimental uncertainties to the level where 10% theoretical uncertainties are no longer acceptable. One example is the recent effort has thus been devoted to the understanding of the final-state Coulomb interaction between the outgoing particle pair. It has long been known that the Gamow factor, which is the point source approximation for the two-body Coulomb interaction, overpredicts the Coulomb interaction between a pair of particles emitted from a source of finite size[1]. This did not cause much concern until recent experiments began to measure larger source sizes with increasing statistical accuracy. Suddenly the use of the Gamow factor to correct the data became an unacceptable approximation[2].

In contrast, very little effort has been to devoted to the understanding of the Coulomb interaction between the outgoing pion pair and the charged source. An exact understanding of this central Coulomb interaction would require a solution to the three-body Coulomb problem and this has yet to be solved analytically. Early attempts to gain an approximate understanding of the three-body Coulomb effects on the measured pion correlation functions showed that the effects appeared to be much smaller than the then current level of experimental uncertainty[3]. With the large reductions in the experimental uncertainties in the intervening years, and the application of the pion interferometry tool to an increasing number of systems, it is surprising that the assumption that the three-body Coulomb effects are unimportant has remained largely unchallenged.

The importance of the central Coulomb potential in the collision of heavy nuclei has been well demonstrated at Bevelac[4], AGS[5], and SPS[6] energies. In all cases, an enhancement in the π^-/π^+ ratio is observed at low transverse momentum (p_T). These data have been explained quite successfully in terms of the Coulomb interaction between the outgoing pion and the residual nuclear charge[7].

In this paper, recent NA44 data on the π^+/π^- ratio in 158 GeV/nucleon Pb+Pb collisions will be reviewed along with a comparison of the $\pi^+\pi^+$ and $\pi^-\pi^-$ correlation functions. A simple rescattering model will then be coupled with a classical model of the central Coulomb potential in order to explore the effect of the central Coulomb potential on the measured correlation functions. The possible impact of the central Coulomb potential on the interpretation of the correlation data will then be discussed.

2 NA44 Data

The NA44 data[6] on the π^+/π^- ratio in 158 GeV/nucleon Pb+Pb collisions is shown in Figure 1. The data show a suppression of the π^+/π^- ratio at low transverse mass ($m_T = \sqrt{p_T^2 + m^2}$). Using RQMD it was shown that this suppression cannot be entirely accounted for using only hadronic interactions[6]. The results of a simple classical Coulomb calculation[8] are also shown in Figure 1. This model uses energy and momentum conservation to calculate the expected mid-rapidity π^+/π^- ratio given a central Coulomb potential V_s:

$$\frac{N(\pi^+)}{N(\pi^-)} \propto \frac{\sqrt{p_t^2 - 2E(p_t)|V_s| + V_s^2}\,(E(p_t) - |V_s|)}{\sqrt{p_t^2 + 2E(p_t)|V_s| + V_s^2}\,(E(p_t) + |V_s|)}. \tag{1}$$

As shown in Figure 1, the data can be described reasonable well using this model and assuming a central Coulomb potential of 5 MeV.

NA44 has also measured $\pi^+\pi^+$ and $\pi^-\pi^-$ correlations in the same acceptance region as the single particle data[9]. The data are shown in Figure 2. These data have a mean p_T of 150 MeV/c. The correlation functions look very similar, and in fact there is no statistically significant difference between the two correlation functions[9]. The data thus suggest that the possible central

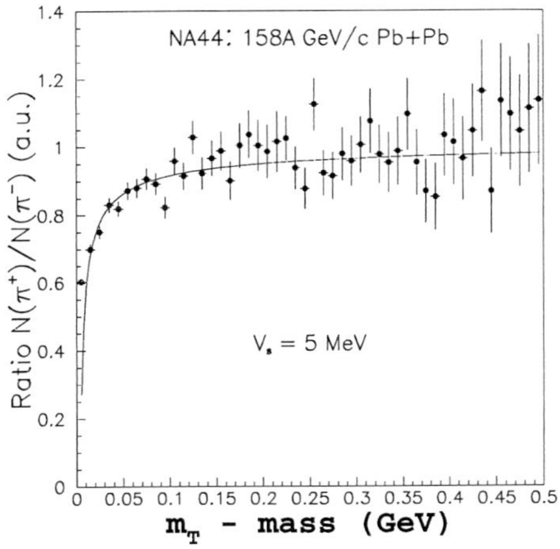

Figure 1: The π^+/π^- ratio as measured by NA44. Also shown is the prediction of a classical Coulomb model with a central Coulomb potential of 5 MeV.

Coulomb effects observed in the π^+/π^- ratio do not lead to a dramatic difference in the $\pi^+\pi^+$ and $\pi^-\pi^-$ correlation functions. As will be shown later, this similarity in the measured positive and negative pion correlation functions does not necessarily mean that central Coulomb effects are unimportant in these interferometry measurements.

3 Rescattering + Coulomb Model

In order to evaluate the effects of the central Coulomb potential on pion interferometry measurements in 158 GeV/nucleon Pb+Pb collisions, a simple rescattering + Coulomb model is used[10]. The rescattering model[11] begins with the measured particle multiplicites and a parameterized set of initial conditions. The particles then undergo billiard ball rescattering and a final phase-space distribution is produced. It should be noted that this model does not take into account isospin effects so any difference in the predictions for π^+ and π^- observables are created after the rescattering model. Figure 3 shows a comparison of NA44 low p_T $\pi^-\pi^-$ data and the rescattering model. The model agrees qualitatively with data, but the correlation functions predicted by the model are narrower than the data. In particular, the fitted longitudinal radius R_L from the model is roughly 8 fm, while the data give a longitudinal radius

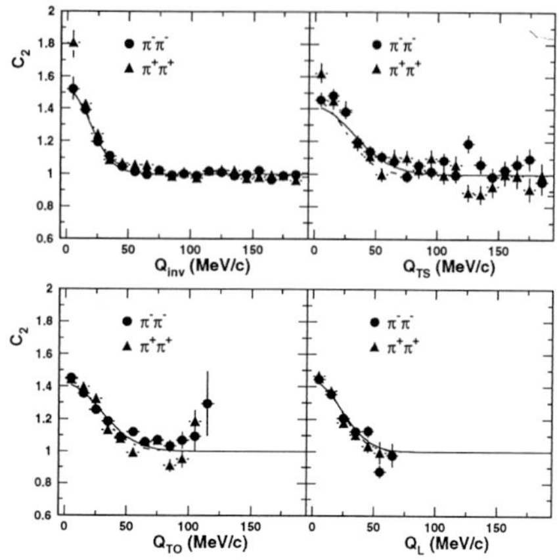

Figure 2: A comparison of the $\pi^+\pi^+$ and $\pi^-\pi^-$ correlation functions measured by NA44. Plotted are Q_{inv}, and the three projections of the standard Bertsch-Pratt parameterization

of only 6 fm. This discrepancy is not unique to this model. RQMD predicts a larger R_L than seen in NA44 data[9], and the BNC model predicts a much larger R_L than seen in NA49 data[12]. The general failure of these three hadronic models to describe quantitatively the correlation data for the Pb+Pb system at SPS energies may suggest new physics, but it is important to make sure that all final-state interactions have been properly taken into account before drawing such a conclusion.

It was shown previously[10] that the addition of a final-state central Coulomb interaction leads to a reduction in the predicted radius and λ parameters from the rescattering model. The introduction of the central Coulomb potential leads to both a net shift of the momentum difference and a momentum difference smearing effect. Because of the large source sizes in the Pb+Pb system at 158 GeV/nucleon, the momentum difference smearing leads to a net reduction in the radius and λ parameters for both the $\pi^+\pi^+$ and $\pi^-\pi^-$ systems. This is in contradiction with previous studies[8,13], but this is to be expected since neither took the momentum difference smearing due to the Coulomb potential into account. Figure 4 shows a comparison of the NA44 data to the rescattering model when the central Coulomb potential is added to the model calculation. The magnitude of the central Coulomb potential is extracted from the π^+/π^-

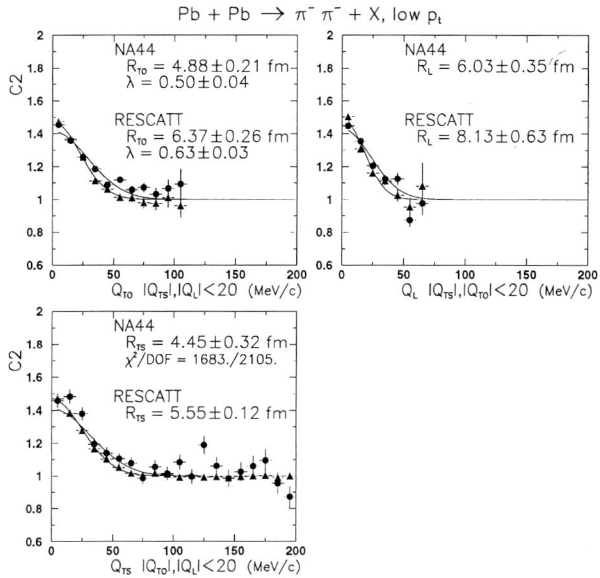

Figure 3: A comparison of the $\pi^-\pi^-$ correlation function measured by NA44 (circles) and the prediction of the rescattering model (triangles).

ratio using Equation 1.

The addition of the central Coulomb potential reduces the discrepancy between the rescattering model and the NA44 data. With no central Coulomb potential, the ratio $R_L(\text{model})/R_L(\text{data})$ is approximately 1.3. If the central Coulomb potential is added, this ratio is reduced to approximately 1.1. This effect should also reduce the discrepancy between the NA44 data and RQMD, and the discrepancy between NA49 data and the BNC model.

It is also important to note that the central Coulomb effect should have an impact on any estimates of the phase-space density using the one- and two-particle measurements. Bertsch showed that the phase-space density ($\langle f \rangle$) is related to the HBT radii and lambda parameters[14]:

$$\langle f \rangle \propto \frac{\lambda}{R_{TO} R_{TS} R_L}. \tag{2}$$

The results of the rescattering+Coulomb model indicate that the central Coulomb potential leads to and increase of the estimated phase-space density by 10-20%.

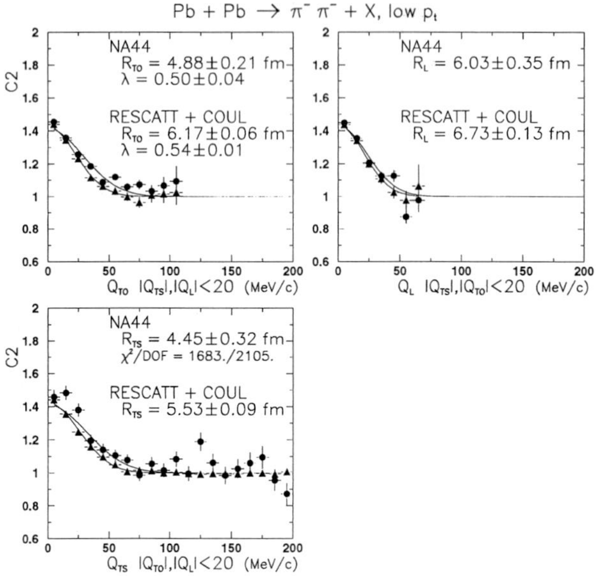

Figure 4: A comparison of the $\pi^-\pi^-$ correlation function measured by NA44 (circles) and the prediction of the rescattering + Coulomb model (triangles) assuming a central Coulomb potential of 5 MeV.

4 Conclusions

NA44 data on the π^+/π^- ratio in 158 GeV/nucleon Pb+Pb collisions show a clear suppression at low m_T. A simple classical Coulomb model is able to reproduce the data and gives an estimated effective Coulomb potential of 5 MeV. The NA44 $\pi^+\pi^+$ and $\pi^-\pi^-$ correlation measurements for the same system do not show any noticeable differences. A rescattering+Coulomb model suggests that for the large source sizes present in the Pb+Pb system, the central Coulomb potential leads to a reduction in the measured radius and lambda parameters due to momentum difference smearing for both $\pi^+\pi^+$ and $\pi^-\pi^-$ correlation measurements. Quantitative comparison to the NA44 data shows that the central Coulomb potential can largely explain the discrepancies between the rescattering model and the data. This statement can be extended to most cascade models since most tend to give predicted correlation functions similar to the rescattering model.

References

1. S. Pratt, *Phys. Rev.* D**33**, 72 (1986); M.G. Bowler, *Phys. Lett.* B**270**, 69, (1991).
2. J. Barrette *et al.*, *Phys. Rev. Lett.***78**, 2916 (1997); Th. Alber *et al.*, *Z. Phys.* C**73**, 443 (1997).
3. W.A. Zajc *et al.*, *Phys. Rev.* C**29**, 2173 (1984).
4. W. Benenson *et al.*, *Phys. Rev. Lett.***43**, 683 (1979); J.P. Sullivan *et al.*, *Phys. Rev.* C**25**, 1499 (1982).
5. L. Ahle *et al.*, *Nucl. Phys.* A**610**, 139c (1996).
6. H. Bøggild, *et al.*, *Phys. Lett.* B**372**, 339 (1996).
7. M. Gyulassy and S.K. Kaufmann, *Nucl. Phys.* A**362**, 503 (1979); H.W. Barz, J.P. Bondorf, J.J. Gaardhøje, H. Heiselberg, *Phys. Rev.* C**56**, 1553 (1997); T. Osada, S. Sano, M. Biyajima, G. Wilk, *Phys. Rev.* C**54**, R2167 (1996).
8. G. Baym, P. Braun-Munzinger, *Nucl. Phys.* A**610**, 286c (1996); P. Braun-Munzinger, Private Communication.
9. I.G. Bearden *et al.*, *Phys. Rev.* C, in print.
10. D. Hardtke and T.J. Humanic, *Phys. Rev.* C**57**, 3314 (1998).
11. T.J. Humanic, *Phys. Rev.* C**50**, 2525 (1994); T.J. Humanic, *Phys. Rev.* C**53**, 901 (1996); T.J. Humanic, *Phys. Rev.* C**57**, 866 (1998).
12. S. Pratt and J. Murray, *Phys. Rev.* C**57**, 1907 (1998).
13. H.W. Barz, *Phys. Rev.* C**53**, 2536 (1996).
14. G. Bertsch, *Phys. Rev. Lett.***72**, 2394 (1994).

Anti–Deuteron Production in 158 GeV/c Pb+Pb Collisions

Ian G. Bearden and Jens Jørgen Gaardhøje
for the NA44 collaboration
Niels Bohr Institute, Blegdamsvej 17
DK-2100 Copenhagen, Denmark.

Abstract

The invariant cross section of antideuterons produced in central Pb+Pb collisions at 158 AGeV/c has been measured by the NA44 experiment at CERN. The two–antibaryon coalescence parameter, sensitive to the freezeout volume, is deduced from antideuteron and antiproton spectra measured at the same rapidity. The transverse momentum and centrality dependence of the coalescence factor is presented. The data suggest a larger coalescence radius for antibaryon clusters than for the corresponding two baryon clusters.

In central collisions of heavy ions, e.g. Pb +Pb, at ultrarelativistic energies, a region of highly compressed and heated nuclear matter is formed. This region expands after formation, in the longitudinal and transverse directions, until the constituent particles cease to interact as a consequence of the decreasing density (freezeout). In order to understand the properties and the dynamical evolution of this fireball and to derive characteristic energy and matter densities, it is of interest to determine the spatial extent and the density profile of the expanding source at freezeout time.

Such information has so far mainly come from Hanbury-Brown and Twiss (HBT) interferometry in which the extent of the source can be estimated from the dependence of the two particle correlation function on the relative momentum of the two particles, as for example in π–π, K–K, and proton–proton correlations [1,2,3]. The interpretation of HBT data normally relies on assumptions on the density profile of the emitting source.

Independent information on source characteristics can be obtained from another two body correlation effect, namely the formation probability of small clusters of baryons or antibaryons. The simplest such cluster, the deuteron, consists of a neutron and a proton. Due to the small binding energy ($E_b = -2.225 MeV$) of the deuteron, compared to the typical kinetic energies of the particles in the fireball, it is evident that deuterons can only be formed when the probability for further collisions vanishes, i.e. close to freezeout. At the same time the formation of the bound two-nucleon state requires the presence of a third body, thus introducing a delicate dependence of the cluster formation and survival probability on the size and density profile of the source.

The cluster formation probability can be described in a simple coalescence approach in which composite particles are formed when their constituents overlap in momentum space [4,5]. The cross section for observing a cluster of A nucleons is related to the A'th power of the single particle cross section. In terms of the invariant cross sections for a cluster with A=2 :

$$(E_{\rm d}\frac{d^3N_{\rm d}}{dp_{\rm d^3}}) = B_2(E_{\rm p}\frac{d^3N_{\rm p}}{dp_{\rm p}^3})^2, \tag{1}$$

where $p_{\rm d} = 2p_{\rm p}$, and where it is assumed that the neutron and protons distributions are identical. The proportionality constant B_2 is called the coalescence factor and will in general depend on the properties of the source.

The coalescence factor can be directly related to the source size. Naively one would expect that the probability for cluster formation is smaller the more dilute the source, i.e. the larger it is. For a source in thermal and chemical equilibrium it can be shown [6] that:

$$B_2 = \frac{3}{4}R_{\rm np}(\frac{4}{3}\pi p_o^3)\frac{m_{\rm d}}{m_{\rm p}^2}, \tag{2}$$

where $R_{\rm np}$ is the ratio of neutrons to protons in the colliding system, and p_o^3 is the radius of a sphere in momentum space whithin which two particles coalesce.

Assuming a source with a Gaussian density profile the coalescence radius can be calculated [7]:

$$R_G^3 = \frac{3}{4}R_{\rm np}\pi^{\frac{3}{2}}(\hbar c)^3\frac{m_d}{m_p^2}(B_2)^{-1}. \tag{3}$$

The factor $R_{\rm np}$ in eqs. 2, 3 and accounts for the difference in the neutron and proton numbers of the colliding nuclei. In general, this factor depends on the beam energy and will tend towards unity as more baryons are produced in the collision.

It is of particular interest to study the production of antibaryon clusters and the associated coalescence radius. Antibaryons are all produced in the collision and it is expected that the antibaryon cluster production rate will depend on the production probability of antiparticles and on the anihilation probability in the dense medium.

In this contribution we report on the first systematic measurement of antiproton and antideuteron yields at non–zero transverse momentum from 158AGeV/c Pb+Pb central collisions at the CERN SPS accelerator. From the measured spectra we calculate the coalescence parameter B_2 for antideuterons and compare it to existing systematics.

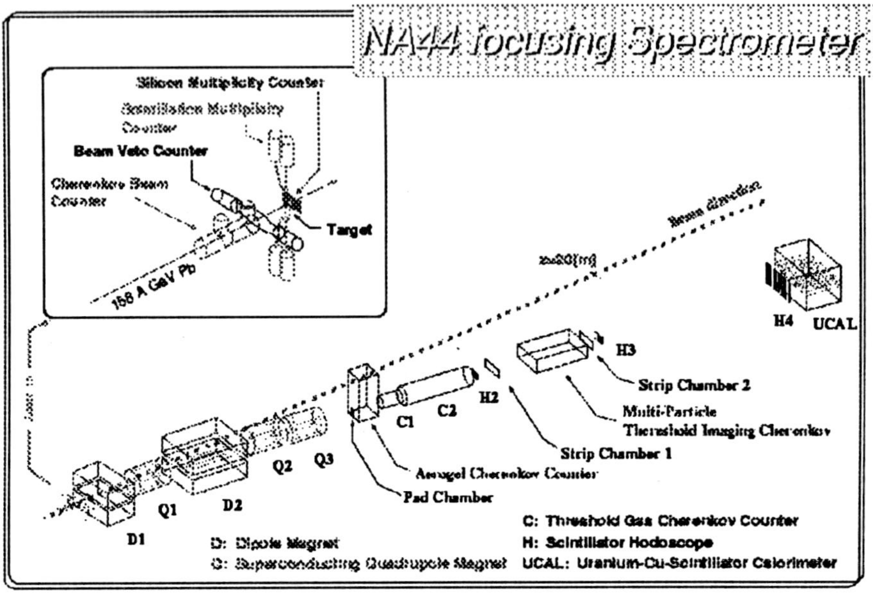

Figure 1: Schematic view of the NA44 spectrometer as used for the experiments discussed in the text. Of particular importance for the identification of rare antideuterons are the time of flight hodoscopes H3 and H4 and the UCAL calorimeter.

The measurements were done with the NA44 focussing spectrometer, which allows an accurate particle identification through the use of various tracking, time of flight and Cerenkov detectors (fig. 1). Details of the spectrometer, as employed in this study can be found in [1]. Anti-deuterons were measured at a momentum of $8 GeV/c \pm 20\%$ and compared to antiprotons measured at $4 GeV/c \pm 20\%$. With these settings both antideuterons and antiprotons are accepted in a narrow rapidity interval around y=2 (midrapidity is at 2.9), enabling a meaningful extraction of the antideuteron coalescence parameter. Fig. 2 shows the acceptance of the NA44 spectrometer for the antideuterons and antiprotons obtained in this study. Of particular importance for the present identification of rare particles was the use of the segmented time of flight hodoscopes H3 and H4 located 18 and 25 meters downstream from the target, respectively, and the segmented Uranium and scintillator calorimeter UCAL, positioned at 25.5 m from the target.

The first panel in figure 3 shows two dimensional distributions of the m^2 calculated from the time of flight in the H3 hodoscope and the particle momen-

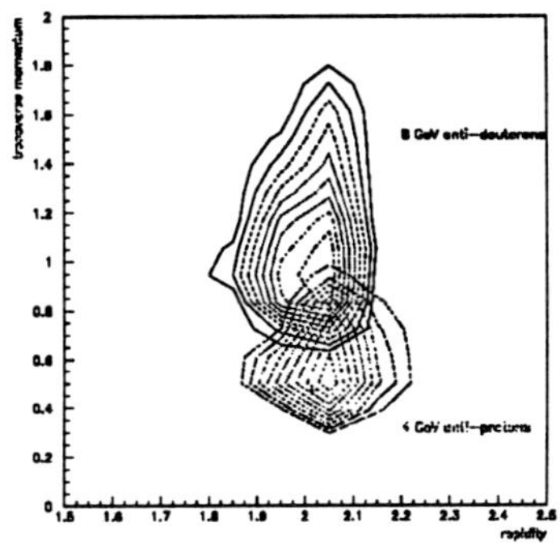

Figure 2: Acceptance, in the variables p_t and y of the NA44 spectrometer protons and deuterons and their respective antiparticles for the two nominal momentum settings, 4 and 8 GeV/c.

tum versus similar information obtained from the H4 hodoscope. The trigger employed accepted K^-, $\bar{p}$ and heavier negatively charged particles in the 4 GeV running, but only $\bar{p}$ and heavier in the 8GeV momentum setting . In panel (b) additional restrictions in the analysis on track identification quality and single target interaction have been applied. It is seen that this cleans up the background considerably in the region where the (rare) antideuterons are expected. Panel (c) shows the obtained m^2 distribution when further requiring that the total energy measured in UCAL be equal to the particle kinetic energy plus twice its rest mass. It may be seen that this requirement of annihilation of the particle in the UCAL material effectively selects the antiparticles and removes the background of misidentified particles. Antideuterons are explicitly recognized as the island around $m^2 = 3.54\mathrm{GeV}^2\mathrm{c}^{-4}$. Panel (d) shows the projected m^2 distribution in which a total 24 antideuterons are identified in the interval $m^2 = 3 - 4\mathrm{GeV}^2\mathrm{c}^{-4}$.

Antideuterons measured at the 8GeV/c setting fall in the transverse momentum range $\mathrm{p_T} = 0.6 \leq \mathrm{p_T} \leq 1.6\mathrm{GeVc}^{-1}$ and in the rapidity interval $y = 1.8 - 2.2$. Anti-protons in the 4GeV/c setting fall in the same rapidity interval and in the range $0.3 \leq \mathrm{p_T} \leq 0.9\mathrm{GeVc}^{-1}$. The invariant cross section Ed^3N/dp^3 for the anti-deuterons is presented as a function of transverse momentum in fig 4a. Acceptance effects and tracking inefficiencies have been corrected for using a Monte Carlo procedure. Due to the limited statistics of the data set, it is not possible to extract a meaningful slope parameter from the data. The anti-proton spectrum is presented in fig 4b as a function of $\mathrm{p}_t(\bar{\mathrm{d}})$, (i.e. $2\mathrm{p}_t(\bar{\mathrm{p}})$).

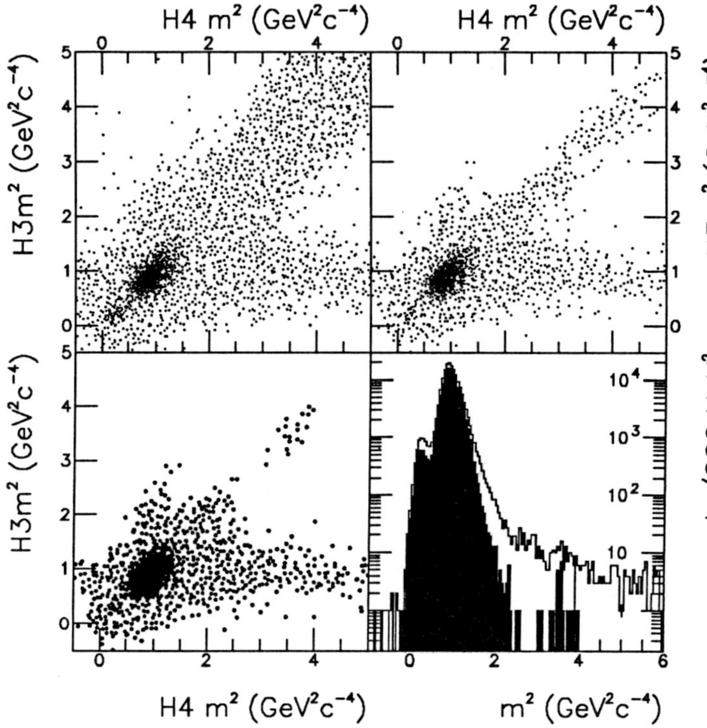

Figure 3: (a)Mass squared defined as $m^2 = (1/\beta^2 - 1)p^2$, where $\beta = L/t$, as determined in time of flight hodoscope H3 versus the similar quantity determined for hodoscope H4. (b) As panel (a) but with additional cuts on 'good tracks' and rejecting 'double interactions'. (c) As panel (b) but furthermore requiring that the energy measured in UCAL is equal to the kinetic energy of the particle plus the energy released by annihilation at rest in the calorimeter.

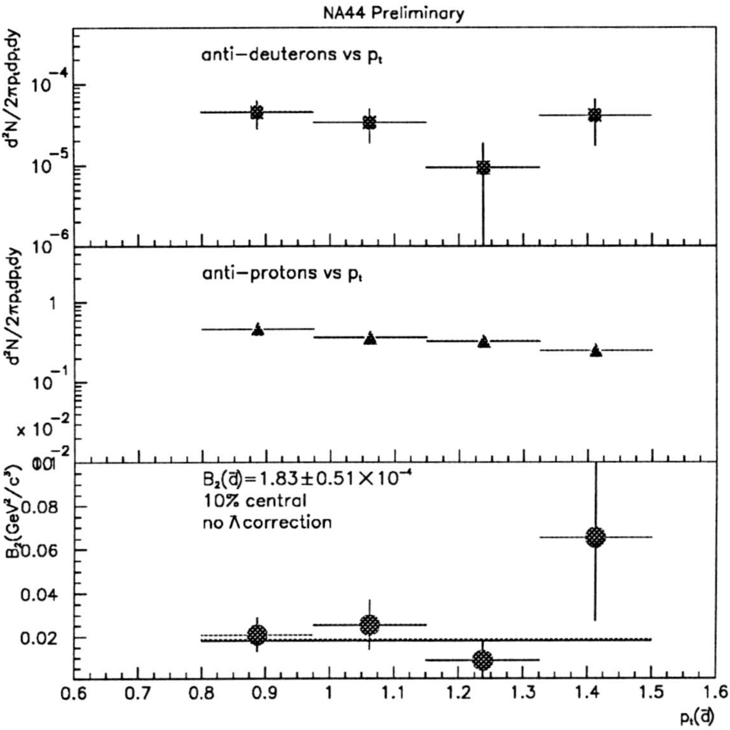

Figure 4: (a) Measured anti–deuteron yield as a function of transverse momentum, (b) anti–proton yield as a function of the transverse momentum of the anti–deuteron, (c) B_2 as a function of the anti–deuteron transverse momentum.

The coalescence parameter, as defined by eq. 1, is plotted as a function of $p_t(\bar{d})$ in figure 4c. It should be noted that the $\bar{p}$ sample used to obtain B_2 has not been corrected for the (unknown) contributions from the decay of $\bar{\Lambda}$ and $\bar{\Sigma}^0$. Strange baryons which decay to anti-protons and pions close to the target contribute to the measured antiproton yield and increase B_2. With the NA44 spectrometer acceptance, it is not possible to distinguish $\bar{p}$ originating from the decay of strange baryons from directly produced $\bar{p}$. It is possible to calculate the expected contribution to the anti–proton yield, and it should be noted that within the NA44 acceptance, this correction depends on transverse momentum with more $\bar{p}$ from weak decays at low p_t.

Figure 5 shows $\bar{d}$ (3a) and $\bar{p}$ (3b) cross sections for two bins of the collision

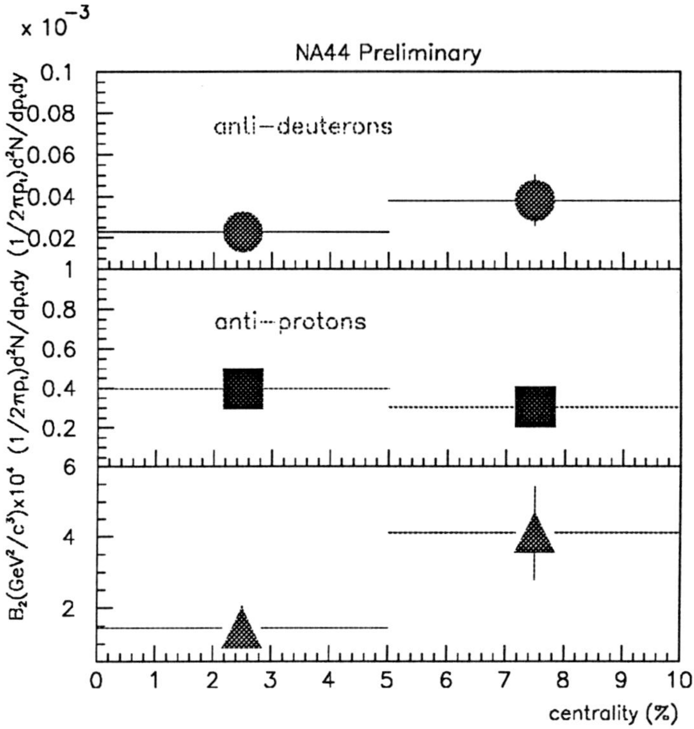

Figure 5: (a) Anti–deuteron yield vs centrality. The yield decreases for more central collisions, whereas the yield of anti–protons (b) increases. Thus the B_2 decreases with increasingly central collisions (c).

centrality, corresponding to $5 - 10\%$ and top 5% of the geometric interaction cross section, respectively. Again $\bar{p}$ cross sections are not corrected for contamination from weak decay products. Also shown is the associated B_2 for the two centrality bins which shows a marked decrease with increasing centrality suggesting a larger emitting source for the more central collisions, from the less central to more central bin. This decrease translates to a change of the freezeout radius of approx. 25%. A similar trend has previously been observed in 11.6 GeV Au +Au collisions at the AGS (ref. Abbott et al). For reactions corresponding to the top 10% of the reaction cross section a value factor $B_2 = 1.8 \pm 0.5 \cdot 10^{-4} GeV^2 c^{-3}$ is obtained. This can be translated to a coalescence radius of $R = (7.0 + 0.8 - 0.5) fm$ assuming that the source has a

Gaussian density profile and that antineutrons and antiprotons are produced in equal amounts.

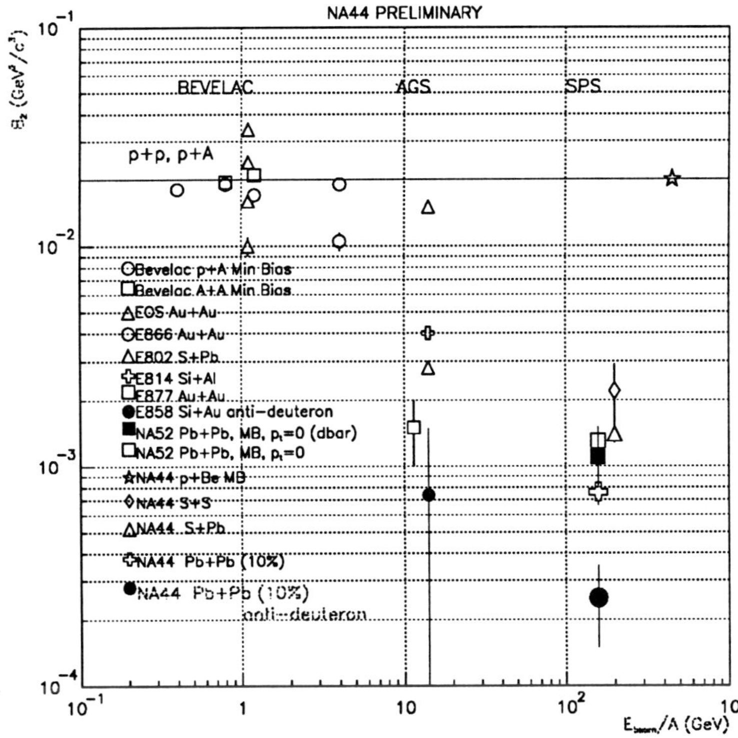

Figure 6: Comparison of B_2 for deuterons (open symbols) and anti–deuterons (filled symbols). The solid line represents the results of p+p and p+A collisions. The values do not include corrections for feed–down from strange baryon decays. The data are taken from references [10–21]

Finally, the result obtained here (filled circle) is compared to the available systematics for B_2 for both deuterons and anti-deuterons as a function of beam energy for a variety of colliding systems (fig. 6). A clear systematic decrease in B_2 with increasing beam energy is seen. This is consistent with the picture of significantly increasing expansion of the system (flow) as a function of beam energy that is obtained from the systematics of the slopes of single particle spectra [8]. This rapid expansion increases the size of the system at freezeout. More detailed inspection of fig. 6 shows that this trend also holds on the average when comparing similar colliding systems at similar collision centralities.

It should be noted that the result for antideuteron coalescence obtained here suggests that antideuteron freeze out occurs at about 3 times larger volume than for deuteron freeze out in similar Pb+Pb collisions (open circle). This corresponds to an increase in radius by about 50%. Nevertheless we stress again that this discrepancy may diminish considerably if there is a significant antiproton contribution arizing from the decay of strange particles. Likewise we note that the present result lies below the B_2 measured for antideuterons by the NA52 experiment for minimum bias collisions Pb+PB at SPS. It is tempting to attribute this difference solely to the differing centrality, however the comparison is complicated by the fact that the two data sets correspond to widely different regions in transverse momentum.

We have so far discussed the B_2 coalescence factor with reference to a static picture of the collision evolution. Recent theoretical work by Polleri [9] considers the expansion of the fireball and calculates the B_2 for different transverse expansion profiles and different density profiles (gaussian and box like) for the source. A pronounced difference between the p_t dependence of B_2 for a Gaussian density profile of the source and for a box-like (as for a Hubble expansion) is expected. Unfortunately, the limited number of antideuterons measured does not permit an unambiguous choice between these two scenarios, although the data suggest a constant B_2 with increasing p_t, as would be expected for a gaussian density profile.

In summary, we have presented the first measurement of antideuterons from Pb+Pb collisions at non zero p_t. This result, together with the measured yield of antiprotons in the same rapidity range has allowed us to extract a coalescence radius for the 10% most central collisions. In the absence of a significant $\bar{p}$ contribution from mainly anti–lambda decays this value corresponds to a Gaussian source with of radius 7 fm. This is consistent with transverse radii measured for the same reaction using two pion interferometry. Generally speaking, the study of multibaryon coalescence in relativistic heavy ion physics is a sensitive tool to study the global properties of the fireball at freezeout.

Acknowledgements.
The support of the Danish Natural Science Research Council is appreciated. We also wish to acknowledge useful discussions with our colleagues Ole Hansen and Allan Hansen.

1. I. G. Bearden et al., Phys Rev C58 (1998) 1656.
2. H. Beker et al., Z. Phys. C64 (1994) 209.
3. R. Ganz, these proceedings.
4. S. T. Butler and C. A. Pearson, Phys. Rev. Lett. 7 (1961) 69.
5. S. T. Butler and C. A. Pearson, Phys. Rev. 129 (1963) 836.

6. A. Z. Mekjian, Phys. Rev. C 17 (1978) 1051.
7. W. J. Llope et al., Phys. Rev. C 52 (1995) 2004.
8. I.G. Bearden et al., Phys. Rev. Lett. 78 (1997) 2080.
9. Alberto Polleri et al., Phys. Lett. B419 (1998)19.
10. S. Nagamiya, et al. Phys Rev. C 24 (1981) 971
11. Lemaire, Phys Lett. 37 (1976) 667
12. R.L. Auble et al., Phys Rev C 28 (1983) 1552.
13. S. Wang et al., Phys Rev Lett. 74 (1995) 2648.
14. J. Barrette et al., Phys Rev. C 50 (1994) 1077.
15. J. Barrette et al., Phys Rev. C 50 (1994) 3047.
16. J. Barrette et al., Phys Rev Lett. 70 (1993) 1763.
17. T. Abbot et al., Phys. Rev. C 50 (1994) 1024.
18. J. Simon–Gillo, Nucl. Phys. A590 (1995) 483c.
19. J. W. Cronin et al., Phys. Rev. D 11 (1975) 3105.
20. G. Appelquist et al., Phys Rev Lett (1996) 3907.
21. G. Appelquist et al.,Phys Lett. B376 (1996) 245.

Photon versus Hadron Interferometry*

R. M. Weiner†

Physics Department, University of Marburg,
Wieselacker 8, 35041 Marburg, Germany
and
Laboratoire de Physique Théorique et Hautes Énergies, Univ. Paris-Sud,
177 rue de Lourmel, 75015 Paris, France

Abstract

Photons and mesons are both bosons and therefore satisfy the same Bose-Einstein statistics. This leads to certain similarities in the corresponding Bose-Einstein correlations which underly photon and hadron intensity interferometry. However there are also important differences between the two effects and these will be analyzed in the following.

1 Introduction

Hanbury-Brown and Twiss (HBT) [1] developed in the mid fifties the method of photon intensity interferometry to be used as an alternative to the amplitude interferometry of Michelson. Initially this "alternative" was considered merely as a technical improvement of interest only for astronomy and it is therefore not surprising that Goldhaber, Goldhaber Lee and Pais (GGLP) [2] were not aware of the HBT experiment when they discovered in 1959-1960 that pairs of identical pions were bunched and interpreted this effect as due to Bose-Einstein correlations. This initial separation [1] between the two developments is in part due to the fact that the techniques used in the original HBT experiment and in the GGLP experiment were very different: the HBT interferometry in astronomy consists in measurements of distance correlations (actually correlations of time arrivals) in order to determine (angular) diameters of stars, while in GGLP experiments one measures momentum correlations in order to derive radii and lifetimes of sources of elementary particles.

On the other hand for (some) people working in optics it did not take much time to realize the quantum statistical significance of the HBT experiment and it turned out that the apparently small step in the history of interferometry due to HBT represented a huge step in the history of physics, leading to the creation of quantum optics with all its theoretical and practical developments. The implications and the importance of the HBT and GGLP effects for particle physics were appreciated only much later. It is therefore very timely that a conference like the present one where astronomers, particle and

*Invited talk at the CRIS98 meeting on HBT interferometry, Catania, June 1998

†E.Mail: weiner@mailer.uni-marburg.de

[1] As far as we can gather the link between the two experiments is mentioned for the first time in ref. [3].

nuclear physicists meet, is organized. At present the HBT/GGLP effect is an important tool in particle and nuclear physics, being the only direct experimental method known so far for the determination of space-time characterisitcs of particle sources. Moreover, the phenomenon of Bose-Einstein correlations (BEC) presents interesting and important theoretical problems in itself and it is thus understandable that in the last decenium of this century it has become an independent subject of research [2].

Although both the HBT effect in quantum optics and in astronomy use photons, quantum optics, being a microscopic discipline, is of course much more related to particle physics than to astronomy. Among other things, in quantum optics, too, one measures momenta, rather than distance correlations. On the other hand photon interferometry is not restricted only to astronomy and quantum optics, but finds applications also in particle and nuclear physics. As a matter of fact, photon interferometry in particles physics is from a certain point of view superior to hadron interferometry, because photons are weakly interacting particles, while hadrons interact strongly. This has two important consequences in photon BEC: (i) there is (up to higher order corrections) no final state interaction between photons, so that the BEC effect is "clean"; (ii) in a high energy reaction, hadrons are produced only at the end of the reaction (at freeze-out), while photons from the beginning, so that photons can provide unique information about the initial state. For the search of quark-gluon plasma this is essential, because if such a state of matter is formed, then this happens only in the early stages of the reaction. This is also important in lower energy heavy ion reactions where the dynamics of the reaction as well as its space-time geometry are studied in this way (cf. the talk by R. Barbera in these proceedings).

These advantages of photon interferometry have stimulated theoretical and experimental studies, despite the technical difficulties due to the small rates of photon production and the background due to π^0 decays.

Besides the difference in the coupling constant, photons and hadrons (for the sake of concreteness we shall refer in the following to pions) have also other distinguishing properties like spin, isospin, and mass which manifest themselves in the corresponding BEC and which sometimes are overlooked. This is the subject of this talk .

2 Comparison between photon and hadron BEC

Table 1 contains an enumeration of differences between photons and pions which appear relevant from the point of view of intensity interferometry. We will comment upon three topics in the following [3]: classical versus quantum fields, condensates, and the role of spin in photon BEC.

2.1 Classical versus quantum fields; coherence and chaos

As is well known BEC are sensitive to the amount of coherence of the source and this makes intensity interferometry a useful tool in the determination of coherence, both for photons and for hadrons. While classical fields are always coherent, quantum fields may be coherent or chaotic. Electromagnetic fields which are at the basis of optical phenomena are "classical", i.e. quantum phenomena do not play a role there (the Planck constant

[2] From 1990 meetings dedicated (almost) entirely to this subject were hold, beginning with CAMP [4]

[3] For more details cf. e.g. a forthcoming textbook on Bose-Einstein correlations by the author to be published by J. Wiley and Sons in 1999.

Table 1

Photons versus hadrons

Photons	**Properties**	**Hadrons(Pions)**
Trivial (electromagnetic)	Classical fields	Remarkable (Higgs,sigma meson)
Remarkable	Quantum fields	Trivial
Trivial	Chaos	Remarkable
Lasers	Condensates	Pion condensates
No	Final state interactions	Yes
$m = 0$	Mass	$m \neq 0$
Yes, if effective coupling is big enough (lasers)	Multiparticle production	Yes, if energy is big enough
$S = 1$	Spin	$S = 0$
$I = 0$	Isospin	$I = 1$
$1/3 \leq C_2 \leq 3$	**Correlations**	$2/3 \leq C_2 \leq 2$, (for charged pions) $1/3 \leq C_2 \leq 3$ (for neutral pions)
Astronomy, gravitational waves, quantum optics, atomic physics, chemistry, biochemistry	**Applications**	Particle and nuclear physics, search for quark-gluon plasma, determination of the mass of the W

h does not appear in the Maxwell equations). Therefore the discovery of photons i.e. of quanta of light was so important, as it lead to the creation of modern quantum physics. Particle physics developped much later and it was quantum from the very beginning. Therefore the fact that the associated particle fields are quantized is from this point of view trivial. On the other hand in the seventies it became clear that the symmetries observed in particle physics are spontaneously broken. This fact, which was brilliantly confirmed by the discovery of intermediate bosons, lead via the Golsdstone-Higgs-Kibble mechanism necessarily to classical fields. Hence in particle physics the existence of classical fields is far from trivial. With chaos the situation is rather inversed. Conventional optical sources are thermal and therefore chaotic. However in particle physics where the wave lengths of particles are of the order of the dimensions of the sources and the lifetime of sources may be small compared with the time necessary for equilibration, one would expect coherence as a rule and thermal equilibrium as exceptional.

2.2 Condensates

One of the most important effects of quantum optics which is based on coherence is the phenomenon of *lasing*. Lasers are Bose condensates and it has been speculated that such condensates, in particular pion condensates, may exist also in nuclei (cf. e.g. [5]) or be created in heavy ion reactions (cf. e.g. [6], [7]).

However there exist important differences between photon condensates i.e. lasers and pion condensates. Furthermore there are different theoretical approaches to the problem of pion condensates and some confusing statements as to how pion condensates are produced. In the following we shall discuss briefly these issues.

2.2.1 Lasers versus pion condensates; pasers?

BEC for inclusive processes, which constitute by far the most interesting and most studied reactions both with hadrons and photons have to be treated by quantum field theory, which is the appropiate formalism when the number of particles is not conserved. For certain purposes however, sometimes one is interested in considering events with a fixed number of particles. Thus the number of particles in a given event can help selecting central collisons with small impact parameter. Theoretically this situation can be handled within field theory, using the methods of quantum statistics [8]. On the other hand for the construction of event generators wave functions appear so far to be a convenient tool and therefore, and also for historical reasons, some theorists have continued to use the "traditional" method of wave function (wf), as introduced in the original GGLP paper. This implies the explicit symmetrization of the products of single particle wf, while in field theory the symmetrization (of amplitudes) is authomatically achieved through the commutation relations of the field operators. When the multiplicities are large, the symmetrization of the wf becomes tedious. This lead Zajc [9] to use numerical Monte Carlo techniques for estimating n particle symmetrized probabibilities, which he then applied to calculate two-particle BEC. He was thus able also to study the question of the dependence of BEC parameters on the multiplicity n. Using as input a second order BEC function parametrized in the form

$$C_2 \sim 1 + \lambda \exp(-\mathbf{q}^2\mathbf{R}^2), \tag{2.1}$$

where $\mathbf{q}$ is the momentum transfer and $\mathbf{R}$ the radius, Zajc found, and this was confirmed in [8], that the "incoherence" parameter λ decreased with increasing n [4].

However Zajc did not consider that this effect means that events with higher pion multiplicities are denser and more coherent. On the contrary he warned against such an interpretation and concluded that his results have to be used in order to eliminate the *bias* introduced by this effect into experimental observations. [5]

This warning apparently did not deter the authors of [6] and [10] to do just that. Ref.[6] went even so far to derive the possible existence of pionic lasers (pasers) from considerations of this type.

Ref.[6] starts by proposing an algorithm for symmetrizing the wf which presents the advantages that it reduces very much the computing time when using numerical techniques, which is applicable also for Wigner type source functions and not only plane wave functions, and which for Gaussian sources provides even analytical results.

Subsequently in ref.[11] wave packets were symmetrized and in special cases the matrix density at fixed and arbitrary n was derived in analytical form. This algorithm was then applied to calculate the influence of symmetrization on BEC and multiplicity distributions. As in [9] it was found that the symmetrization produces an effective decrease of the radius of the source, a broadening of the multiplicity distribution $P(n)$ and an increase of the mean multiplicity as compared to the non-symmetrized case. What is new in [6] is (besides the algorithm) mainly the meaning the author attributes to these results.

In a concrete example Pratt considers a non-relativistic source distribution S in the absence of symmetrization effects:

$$S(k,x) = \frac{1}{(2\pi R^2 mT)^{3/2}} \exp\left(-\frac{k_0}{T} - \frac{x^2}{2R^2}\right)\delta(x_0) \tag{2.2}$$

where

$$k_0/T = k^2/2\Delta^2 \tag{2.3}$$

Here T is an effective temperature, R an effective radius, m the pion mass, and Δ a constant with dimensions of momentum.

Let η_0 and η be the number densities before and after symmetrization, respectively. In terms of $S(k,x)$ we have

$$\eta_0 = \int S(k,x)d^4k d^4x \tag{2.4}$$

and a corresponding expression for η with S replaced by the source function after symmetrization.

Then one finds [6] that η increases with η_0 and above a certain crtitical density η_0^{crit}, η diverges. This is interpreted by Pratt as *pasing*.

The reader may be rightly puzzled by the fact that while η has a clear physical significance the number density η_0 and a fortiori its critical value have no physical significance, because in nature there does not exist a system of bosons the wf of which is not symmetrized. Thus contrary to what is alluded to in ref.[6], this paper does not does address really the question how a condensate is reached. Indeed, the physical factors which induce

---

[4] In [9] the clumping in phase space due to Bose symmetry was also illustrated;

[5] The same interpretation of the multiplicity dependence of BEC was given in [8]. In this reference the nature of the "fake" coherence induced by fixing the multiplicity is even clearer, as one studies there explicitely partial coherence in a consistent quantum statistical formalism.

condensation are, for systems in (local) thermal and chemical equilibrium, [6], pressure and temperature and the symmetrization is contained automatically in the form of the distribution function

$$f = \frac{1}{\exp[(E-\mu]/T]-1} \tag{2.5}$$

where E is the energy and μ the chemical potential.

To realize what is going on it is useful to observe that the increase of η_0 can be achieved by decreasing R and/or T. Thus η_0 can be substituted by one or both of these two physical quantities. Then the blow-up of the number density η can be thought of as occuring due to a decrease of T and/or R. However this is nothing but the well known Bose-Einstein condensation phenomenon.

While from a purely mathematical point of view the condensation effect can be achieved also by starting with a non-symmetrized wf and symmetrizing it afterwards "by hand" , the causal i.e. physical relationship is different: one starts with a bosonic i.e symmetrized system and obtains condensation by decreasing the temperature or by increasing the density of this *bosonic* system. To obtain a pion condensate e.g., the chemical potential has to equate the pion mass.

A scenario for such an effect in heavy ion reactions has been proposed in [7]. It is based on the decay of short lived resonances which leads to an accumulation of pions and takes place if the hadronic (dense) matter decouples from chemical equilibrium earlier than from thermal equilibrium. In [7] it was found that if a pionic Bose condensate is formed at any stage of the collision, it can be expected to survive until pions decouple from the dense matter, and thus it can affect the spectra and correlations of final state pions.

This effect was then studied quantitatively by solving the equations of relativistic hydrodynamics for a fluid which contains also a superfluid component, corresponding to the pion condensate. From the results obtained in this way we quote: in the single inclusive transverse momentum distribution the signature of a maximum velocity appears, which is specific for a superfluid system. The second order correlation function C_2 presents the typical features of a partially coherent system i.e. a lowering of the intercept and a double structure, which in principle could be quite dramatic (up to a given value of q, C_2 vanishes). These features are rather specific for a pion condensate and distinguish such a system from optical condensates. [7].

To conclude the "paser" topic, one must correct another confusing interpretation which relates to the observation made also in [9] that the symmetrization produces a broadening of the multiplicity distribution (MD). In particular starting with a Poisson MD for the non-symmetrized wf one ends up after symmetrization with a negative binomial. While Zajc correctly considers this as a simple consequence of Bose statistics, ref.[6] goes further and associates this with the so called pasing effect. That such an interpretation is incorrect is obvious from the fact that for true lasers the opposite effect takes place. Before "condensing" i.e. below threshold their MD is in general broad and of negative binomial form corresponding to a chaotic (thermal) distribution while above threshold the

[6]For lasers the determining dynamical factor is among other things the inversion of the occupation of atomic levels.

[7]None of the "paser" papers [6] - [11] address the crucial issue of directional coherence which is an essential characteristic of optical lasers. This casts doubts whether the terminology of "paser" is appropiate. For a model of directional coherence, not necessarily related to pion condensates, cf.[12]; experimental hints of this effect have possibly been seen in [13].

laser condensate is produced and as such corresponds to a coherent state and therefore is characterized by a Poisson MD.

2.3 Photon interferometry. Photon spin and bounds of BEC.

In this section we discuss the difference between BEC for photons and for pions. Certain erroneous results and statements in the recent literature will be corrected.

Following [14] and [15] we consider a heavy ion reaction where photons are produced through bremsstrahlung from protons in independent proton-neutron collisions[8]. The corresponding elementary dipole currents are

$$j^\lambda(k) = \frac{ie}{mk^0}\mathbf{p}.\epsilon_\lambda(k) \tag{2.6}$$

where $\mathbf{p} = \mathbf{p}_i - \mathbf{p}_f$ is the difference between the initial and the final momentum of the proton, ϵ_λ is the vector of linear polarization and k the photon 4-momentum; e and m are the charge and mass of the proton respectively. The total current is written

$$J^\lambda(k) = \sum_{n=1}^{N} e^{ikx_n} j_n^\lambda(k). \tag{2.7}$$

For simplicity we will discuss in the following only the case of pure chaotic currents $< J^\lambda(k) >= 0$. The index n labels the independent nucleon collisions which take place at different space-time points x_n. These points are assumed to be randomly distributed in the space-time volume of the source with a distribution function $f(x)$ for each elementary collision. The current correlator then reads

$$\begin{aligned} < J^{\lambda_1}(k_1)J^{*\lambda_2}(k_2) > &= < J^{\lambda_1}(k_1)J^{\lambda_2}(-k_2) >\equiv C^{\lambda_1\lambda_2}(k_1,k_2) \\ &= \sum_{n,m=1}^{N} \int \prod_{l=1}^{N} d^4x_1 f(x_1) \exp(ik_1x_n - ik_2x_m) < j_n^{\lambda_1}(k_1) j_m^{\lambda_2}(-k_2) > \\ &= \sum_{n=1}^{N} [\tilde{f}(k_1-k_2) < j_n^{\lambda_1}(k_1) j_n^{\lambda_2}(-k_2) > -\tilde{f}(k_1)\tilde{f}(-k_2) < j_n^{\lambda_1}(k_1) > \\ &\quad < j_n^{\lambda_2}(-k_2) >] + < J^{\lambda_1}(k_1) >< J^{\lambda_2}(-k_2) > . \end{aligned} \tag{2.8}$$

Here $\tilde{f}(k)$ is the Fourier transform of $f(x)$ with the normalization $\tilde{f}(k=0) = 1$. The function $\tilde{f}$ has a maximum at $k = 0$ and becomes usually negligible for $kR \gg 1$ where R is the effective radius of the source.

We will limit further the discussion to the important case from the experimental point of view of unpolarized photons. The corresponding cross sections are obtained by summing over the the polarization indexes and the elementary currents j_n. Thus the correlator defined above will be proportional to products of the form

$$< J^{\lambda_1}(k_1)J^{\lambda_2}(-k_2) >= \epsilon_{\lambda_1}^i(k_1)\left(\sum_{n=1}^{N} < \mathbf{p}_n^i\mathbf{p}_n^j >\right)\epsilon_{\lambda_2}^j(k_2) \tag{2.9}$$

Due to the axial symmetry around the beam direction one has for the momenta the tensor decomposition

$$< \mathbf{p}_n^i\mathbf{p}_n^j >= \frac{1}{3}\sigma_n\delta^{ij} + \delta_n l^i l^j, \tag{2.10}$$

[8]Photon emission from proton-proton collisions is suppressed because it is of quadrupole form.

where l is the unit vector in the beam direction and σ_n, δ_n are real positive constants. In [14] an isotropic distribution of the momenta was assumed. This corresponds to the particular case $\delta_n = 0$. The generalization to the form (2.10) is due to [15]. The summation over polarization indexes is performed by using the relations

$$< (\epsilon^i.\mathbf{p}_l)(\epsilon^j.\mathbf{p}_{l'}) >= \frac{1}{3}(\epsilon^i.\epsilon^j)\delta_{ll'} \tag{2.11}$$

and

$$\sum_{\lambda=1}^{2} \epsilon^i_\lambda(k).\epsilon^j_\lambda(k) = \delta^{ij} - \mathbf{n}^i\mathbf{n}^j, \tag{2.12}$$

where $\mathbf{n} = \mathbf{k}/|\mathbf{k}|$.

We write below the results for the second order correlation function

$$C_2(k_1, k_2) = \frac{\rho_2(k_1, k_2)}{\rho_1(k_1)\rho_1(k_2)} \tag{2.13}$$

for two extreme cases: (1) Uncorrelated elementary currents (isotropy) ($\sigma \gg \delta$)

$$C_2(k_1, k_2;\ \sigma \neq 0,\ \delta = 0) = 1 + \frac{1}{4}[1 + (\mathbf{n}_1.\mathbf{n}_2)^2]\left[|\tilde{f}(k_1 - k_2)|^2 + |\tilde{f}(k_1 + k_2)|^2\right], \tag{2.14}$$

leading to an intercept

$$C_2(k, k) = \frac{3}{2} + \frac{1}{2}|\tilde{f}(2k)|^2 \tag{2.15}$$

limited by the values (3/2,2). (2) Strong anisotropy ($\sigma \ll \delta$):

$$C_2(k_1, k_2;\ \sigma = 0,\ \delta \neq 0) = 1 + |\tilde{f}(k_1 - k_2)|^2 + |\tilde{f}(k_1 + k_2)|^2 \tag{2.16}$$

with an intercept

$$C_2(k, k) = 2 + |\tilde{f}(2k)|^2 \tag{2.17}$$

limited this time by the values (2,3).

These results are remarkable among other things because they illustrate the specific effects of photon spin on BEC. Thus while for (pseudo-)scalar pions the intercept is a constant (2 for charged pions and 3 for neutral ones) even for unpolarized photons the intercept is a function of k. One thus finds that, while for a system of charged pions (i.e. a mixture of 50% positive and 50% negative) the maximum value of the intercept Max$C_2(k,k)$ is 1.5, for photons Max$C_2(k,k)$ exceeds this value and this excess reflects the space-time properties of the source represented by $\tilde{f}(k)$, the degree of (an)isotropy of the source represented by the quantities σ and δ, and the supplimentary degree of freedom represented by the photon spin. The fact that the differences between charged pions and photons are enhanced for soft photons reminds us of a similar effect found with neutral pions (cf. ref.[16]). Neutral pions are in general more bunched than identically charged ones and this difference is more pronounced for soft pions. This similarity is not accidental, because photons as well as π^0 particles are neutral and this circumstance has quantum field theoretical implications which will be mentioned also below.

We see thus that in principle photon BEC can provide information both about the space-time form of the source and the dynamics.

These results on photon correlations refer to the case that the sources are "static" i.e. not expanding. Expanding sources were considered in [17] within a covariant formalism.

The results quoted above, in particular eqs. (2.14,2.15), which had been initially derived by Neuhauser, were challenged by Slotta and Heinz [18]. Among other things, these authors claim that for photon correlations due to a chaotic source "the only change relative to 2-pion interferometry is a statistical factor $\frac{1}{2}$ for the overall strength of the correlation which results from the experimental averaging over the photon spin". In [18] an intercept $\frac{3}{2}$ is derived which is in contradiction with the results presented above and in particular with eq.(2.15) where besides the factor $\frac{3}{2}$ there appears also the k dependent function $\frac{1}{2}|\tilde{f}(2k)|^2$.

We would like to point out here that the reason for the difference between the results of [14],[15] on the one hand and those of ref.[18] on the other is mainly due to the fact that in [18] a formalism was used which is less general than that used in [14] and [15] and which is inadequate for the present problem. This implies among other things that unpolarized photons cannot be treated in the naive way proposed in [18] and that the results of [14] and [15] are correct, while the results of [18] are not.

In [18] the following formula for the second order correlation function is used:

$$C(\mathbf{k}_1, \mathbf{k}_2) = 1 + \frac{\tilde{g}_{\mu\nu}(\mathbf{q}, \mathbf{K})\tilde{g}^{\nu\mu}(-\mathbf{q}, \mathbf{K})}{\tilde{g}^{\mu}_{\mu}(\mathbf{0}, \mathbf{k}_1)\tilde{g}^{\mu}_{\mu}(\mathbf{0}, \mathbf{k}_2)} \tag{2.18}$$

Here $\tilde{g}$ is the Fourier transform of a source function $(g(x, K)$ and $\mathbf{q} = \mathbf{k}_1 - \mathbf{k}_2$, $\mathbf{K} = \mathbf{k}_1 + \mathbf{k}_2$.

This formula is a particular case of a more general formula for the second order correlation function derived by Shuryak [19] using a model of uncorrelated sources, when emission of particles from the same space-time point is negligible.

As is clear from this derivation there exists also a third term, neglected in eq.(2.18) and which corresponds to the simultaneous emission of two particles from the same point (cf. [16]). While for massive particles this term is in general suppressed, this is not true for massless particles and in particular for soft photons. Indeed in [14] and [15] this additional term had not been neglected as it was done subsequently in [18] and therefore it is not surprising that ref.[18] could not recover the results of refs.[14] and [15]. The neglect of the term corresponding to emission of two particles from the same space-time point is not permitted in the present case. Emission of particles from the same space-time point corresponds in a first approximation to particle-antiparticle correlations and this type of effect leads also to the difference between BEC for identical charged pions and the BEC for neutral pions. This is so because neutral particles coincide with the corresponding antiparticles. (As a consequence of this circumstance e.g. while for charged pions the maximum of the intercept is 2, for neutral pions it is 3 (cf. [16] and Table 1). Photons being neutral particles, similar effects like those observed for π^0-s are expected and indeed found.

This misapplication of the current formalism invalidates completely the conclusions of ref.[18].

Intuitively the fact that for unpolarized photons Max$C_2(k, k)$ is 2 and not 1.5 as stated in [18], can be explained as follows: a system of unpolarized photons consists on the average of 50% photons with the same helicities and 50% photons with opposite helicities. The first ones contribute to the maximum intercept with a factor of 3 and the last ones with a factor of 1 (coresponding to unidentical particles).

For the sake of clarification it must be mentioned that ref.[18] contains also other incorrect statements. Thus the claim in [18] that the approach by Neuhauser "does not correctly take into account the constraints from current conservation" is completely unfounded as can be seen from eq.(2.11) which is a an obvious consequence of current conservation. Last but not least the statement that because the tensor structure in eq.(20) of ref.[17]

is parametrized in terms of k_1 and k_2 separately "instead of only in terms of K, leading to spurious terms in the tensor structure which eventually result in their spurious momentum-dependent prefactor", has also to be qualified. Indeed the additional term, unduely neglected in [18], depends not only on K but also on k_1 and k_2 separately and this contradicts the entire argumentaton of [18] regarding the "spurious terms".

The considerations presented above refer to the effects of photon spin on the upper bounds of the correlation function. Similar specific effects exist also for the lower bounds [16] (cf. Table 1): $C_2^{--}(k_1, k_2) \geq 2/3$ and $C_2^{00}(k_1, k_2) \geq 1/3$. Here the indeces $--$ and 00 refer to charged and neutral pions (photons) respectively. These lower bounds have also lead to confusion in the literature and this issue was clarified and corrected in [20]. For further details of the topics discussed here cf. [21].

References

[1] R. Hanbury-Brown and R. Q. Twiss, Nature, 178 (1956) 1046.

[2] G. Goldhaber, S. Goldhaber, W.Lee and A.Pais, Phys. Rev. 120 (1960) 300.

[3] V.G.Grishin, G. I. Kopylov, M. I. Podgoretskiĭ, Sov. J. Nucl. Phys. 13 (1971) 638.

[4] International Workshop on Correlations and Multiparticle Production (CAMP)(LESIP IV), Marburg 1990; World Scientific 1991, editors M. Plümer, S. Raha, and R. M. Weiner.

[5] A. B. Migdal, Rev. Mod. Phys. 50 (1978) 107.

[6] S.Pratt, Phys.Lett. B301 (1993) 159.

[7] U. Ornik, M.Plümer and D. Strottman; Phys.Lett. B314(1993)401; U.Ornik et al., Phys.Rev.C56 (1997)412.

[8] G.N.Fowler et al., Phys.Lett.B253 (1991) 421.

[9] W.A.Zajc, Phys.Rev. D35 (1987) 3396.

[10] W.Q.Chao, C.S.Gao, and Q.H.Zhang, J.Phys. G21 (1994) 847.

[11] T. Csörgö and J. Zimanyi, Phys. Rev. Lett. 80 (1998) 916.

[12] G. N. Fowler and R. M. Weiner, Phys. Rev. Lett. 55 (1985) 1373.

[13] W. A. Zajc et al., Phys. Rev. C 29 (1984) 2173; H. Bossy et al., Phys. Rev. C 47 (1993) 1659.

[14] D. Neuhauser, Phys. Lett. B 182 (1986) 289.

[15] L. V. Razumov and R. M. Weiner, Phys. Lett. B 319 (1993) 431.

[16] I. Andreev, M. Plümer, R. Weiner, Phys. Rev. Lett. 67 (1991) 3475; Int.J.Mod.Phys. 8A (1993) 4577.

[17] L.Razumov and H. Feldmeier, Phys. Lett. B 377 (1996) 129.

[18] C. Slotta , U. Heinz, Phys. Lett. B 391 (1997) 469.

[19] E. V. Shuryak, Sov. J. Nucl. Phys. 18 (1974) 667.

[20] A. Timmermann et al., Phys. Rev. C50 (1994) 3060.

[21] R. M. Weiner, Physics Reports, to be published.

HOW TO TEST THE EXISTENCE OF THE EARLY PARTON CASCADE USING PHOTON HBT CORRELATIONS?

DANIEL FERENC
University of Regensburg,
Physics Dep.,
93040-Regensburg,
Germany

Div. PPE,
CERN,
1211-Geneva,
Switzerland
E-mail: Daniel.Ferenc@CERN.ch

We report on a possible application of the HBT phenomenon in testing the existence of two hypothetical phenomena. First, it is argued that the existence of a rapidly developing parton cascade in the earliest stages of a high energy nuclear collision process can be tested by studying two–photon HBT correlations over a *wide* longitudinal momentum scale – corresponding to the *early* photon emission time from the hypothetical parton system. This method provides the selectivity for the early emitted photons, since the photons emitted at later times correlate over progressively narrower momentum scales. Second, in a similar way we argue that the existence of a hypothetic dark matter candidate, the Weakly Interacting Massive Particle (WIMP), may be tested by studying HBT correlations of cosmic gamma rays at a relatively *long* detection time scale – corresponding to the very *narrow* spectral line of the photons emerging from WIMP annihilations. Background photons leave no signature since they do not correlate.

1 Introduction

The discovery of the photon intensity interference phenomenon by Hanbury Brown and Twiss (HBT) has initiated, in a rather dramatic way, a new field of physics – the field of quantum optics [1]. The phenomenon has ever since been also utilized in different fields of science as a unique tool capable of revealing important properties of various physical systems. While in astronomy the "HBT tool" has been originally used to study angular sizes of stellar objects, in high energy physics it has provided information on the space-time evolution of expanding hadronic systems created in particle collisions.

In this paper we report on another possible application of the HBT phenomenon, namely in testing the existence of certain hypothetical phenomena. Two applications are proposed. First, the study of the existence of a rapidly developing parton cascade in the earliest stages of a nuclear collision process at very high collision energies (at RHIC and LHC). According to parton cascade models [2] a rapid partial parton thermalisation establishes already within a fraction of one fm/c time. Very strong photon radiation – a flash [3] – should be among the the most prominent consequences and signatures of the phenomenon. In practice, the problem is that the photons are radiated also during all the subsequent evolution stages of the system, most of them from the decays of neutral pions and other hadrons, and in principle there is no way to distinguish between the interesting early photons and all other photons, as long as single–photon momentum spectra are studied. In that case, the most one can do is to estimate the *time-integrated* yield of *all* the direct photons, by subtracting the estimated yield of photons emerging from hadron decays from the total measured contribution.

However, it is possible to distinguish the early photon component from the rest by exploring the emission time information which is imprinted into the HBT correlation pattern [4]. The early emitted photons are HBT correlated over a very *wide longitudinal momentum scale* – thanks to their very *early emission time*. The HBT effect appears, roughly speaking, when the product of these two quantities becomes as small as $\hbar$, or in other words when the photons are found to occupy the same phase space cell in the phase space (in fact in all three directions simultaneously). For example, photons radiated at a time of 0.5 fm/c after collision will correlate over longitudinal momentum difference of about 0.4 GeV/c. All the other photons that emerge later are correlated at progressively narrower longitudinal momentum scales, which provides a possibility to distinguish the contribution of the early photons, and to verify the existence of the hypothetical rapidly developing parton cascade.

Another example, discussed here only briefly and qualitatively, is the search for the Weakly Interacting Massive Particle (WIMP, possibly the lightest SUSY particle) as a dark matter candidate. Abundant annihilation of WIMPs into pairs of photons may be expected in the vicinity of very massive cosmic objects that would assure high annihilation rates due to high dark matter densities. Again, the annihilation photons will be detected together with photons originating from all the other sources, and the phenomenon will be hard to detect, unless a very strong annihilation peak existed above the background (at an energy equal to the WIMP's mass, presently expected at around 100 GeV – in the so far unexplored energy region). The HBT correlation tool can in principle again provide selectivity, because photons emerging from the annihilations could have a very narrow frequency spectrum around the WIMP mass, the spectral width being determined by Doppler smearing due to WIMPs' motion, which is probably very slow. Due to the narrow frequency spectrum, photons from WIMP annihilations will be HBT correlated over a very long time, while the background photons will not be HBT correlated at all. A possible way to search for the annihilation process is therefore by searching for the HBT temporal photon correlations over a relatively long time scale. This time scale depends on the actual WIMP's kinetic energies. In case that the effective temperature of WIMPs would be as high as the microwave background temperature ($\sim$ 3 K), the spectral width would be $\sim 10^{-4}$ eV, and the HBT coherence time would be $\sim 10^{-11}$ seconds. One of the lessons we have learned from the original table-top photon HBT experiment [1] is that the actual time-resolution of the apparatus may be by several orders of magnitude larger than the coherence time. Taking care of the other coordinates in phase space, one also has to make sure that the spread in transverse momentum of the detected photons be sufficiently small (small angular acceptance) to fit a single phase space cell.

The paper is structured as follows: the HBT measurement tool is discussed in Sect.2., momentum variables are defined in Sect.3, photon detection techniques are discussed in Sect.4, and the results of our simple considerations are presented and discussed in Set.5.

2 The HBT Measurement Tool

Statistical fluctuations of a chaotic system consisting of noninteracting identical bosons in the six-dimensional phase space are not Poissonian, like for macroscopic particles, but of a Bose–Einstein type [5]. When a phase–space volume smaller than a single phase–space cell is considered, boson multiplicities fluctuate according to the so called geometric count probability function

$$P(n) = \frac{<n>^n}{(<n>+1)^{n+1}} \tag{1}$$

where $<n>$ is the average count number. This function is very wide – a small number of counts is most probable, but there is also a rather high probability to count a large number of bosons. The count probability distribution for a large phase–space volume which contains many phase–space cells, approaches the Poissonian law, but *never* reaches the Poissonian limit – fluctuations within single phase space cells always remain geometric and only get diluted in bulk as the number of independent cells increases.

The general idea of all HBT measurements is to scan the statistical behavior of identical bosons as a function of the size of the considered phase–space volume. Typically one or more dimensions of the phase–space volume are unknown, while the others are accessible to the measurement; in particle and nuclear physics the volume in the configuration space is unknown (i.e. the physical size of the system at the freezeout), while the momentum space for pions is accessible in the measurement. The goal of a correlation measurement is to find the scale in the accessible coordinate of the phase–space for which the fluctuation pattern turns from the Poissonian to the geometric law – this indicates that the overall observed phase–space volume (including the unobservable dimensions) corresponds to a single phase–space cell. Knowing the volume of a single phase–space cell, $(\Delta p_x x \sim \hbar, \Delta p_y y \sim \hbar, \Delta p_z z \sim \hbar)$, it is straightforward to deduce the scale of the system along the unaccessible coordinates.

In contrast to the particle and nuclear physics, in the famous Hanbury–Brown Twiss stellar interferometry [1] the volume in the configuration space is well known and in fact defined by the detector acceptance, while the transverse momentum spread of light coming from a star is the unknown required quantity; this momentum spread Δp_T is related to the opening angle of the star θ as viwed from the Earth $\Delta p_T \simeq p\theta$, which enables one to estimate the so called angular size of a star (not to be confused with the real size!). In the particle and nuclear physics the phase–space cell is placed at the particle source, while in the stellar interferometry it is placed between the photon detectors on the Earth.

The following methodological problem is how to check the fluctuation pattern in real experiments. There are basically two different methods. First, one can perform a direct study of the count probability statistics as a function of the observational phase–space volume; a still better way is to study moments, e.g. factorial moments. Another method is to study the two–particle correlation functions. Essentially, this means to study a distribution of distances between particles in the accessible part of the phase–space (in particle physics in the momentum–space and in astronomy in configuration space) and to compare this distribution to (or in fact divide by) a somehow synthesized distribution corresponding (locally) to a Poissonian statistics. At a certain small scale a correlation peak emerges, indicating where the actual statistics deviates from the Poissonian, and thus indicating the scale corresponding to the single phase–space cell. As long as two–particle correlations are studied, the correlation technique is very suitable, but when higher–order correlations are to be studied, a serious conceptual problem arises with the definition of the fundamental quantity – the distance between several particles – and it is more natural to study count probability moments.

For expanding particle sources, a correlation arises between momenta and emission coordinates of particles. Due to the Doppler shift, particles emitted from distant parts of the source cannot occupy the same phase-space cell $(\Delta x \Delta p_x >> \hbar)$ and they are not mutually correlated. Only particles emitted from a smaller relative distance Δx, and consequently with a smaller Δp_x, demonstrate correlation. This in turn enables us to exploit the Doppler effect in order to study the way how the system expands.

3 Momentum Variables and the Correlation Function

The two single–particle momentum vectors $\mathbf{p}_1$ and $\mathbf{p}_2$ may be decomposed into the average $\mathbf{k} = 0.5\,(\mathbf{p}_1+\mathbf{p}_2)$ and the relative $\mathbf{Q} = (\mathbf{p}_1-\mathbf{p}_2)$ momentum vectors of the pair, each carrying three degrees of freedom. The relative momentum vector may be decomposed into transverse sideward Q_S, transverse outward Q_O, and longitudinal Q_L components ; the longitudinal component is parallel to the collision axis, while the transverse sideward and outward components are perpendicular and parallel to the average transverse momentum vector $\mathbf{k}_T$, respectively:

$$
\begin{aligned}
\mathbf{Q}_L &= (\mathbf{p}_{L1} - \mathbf{p}_{L2}) \\
\mathbf{k}_T &= \frac{\mathbf{p}_{T1} + \mathbf{p}_{T2}}{2} \\
\mathbf{Q}_T &= (\mathbf{p}_{T1} - \mathbf{p}_{T2}) \\
Q_S &= \frac{|\mathbf{Q}_T \times \mathbf{k}_T|}{|\mathbf{k}_T|} \\
Q_O &= \frac{\mathbf{Q}_T \cdot \mathbf{k}_T}{|\mathbf{k}_T|}
\end{aligned}
\qquad (2)
$$

The correlation function used in the simulations presented in this paper is a function of the three relative momentum components and was assumed to be a Gaussian in all three dimensions:

$$C(Q_S, Q_O, Q_L) = 1 + \lambda e^{(-Q_S^2 R_S^2)} e^{(-Q_O^2 R_O^2)} e^{(-Q_L^2 R_L^2)}, \qquad (3)$$

where λ is the correlation intensity (0.5 for photons), and R_S, R_O and R_L are the parameters which characterize the photon source, or the effective interfero metric source sizes.

4 Photon Detection Techniques

The first problem to be discussed is whether and how the high energy photons may be detected in a realistic experiment. In our discussion we shall consider the ALICE experiment[7] at LHC/CERN.

The experimental parameters of the highest relevance for the proposed correlation study are the acceptance of the photon detector, and the photon detection probability.

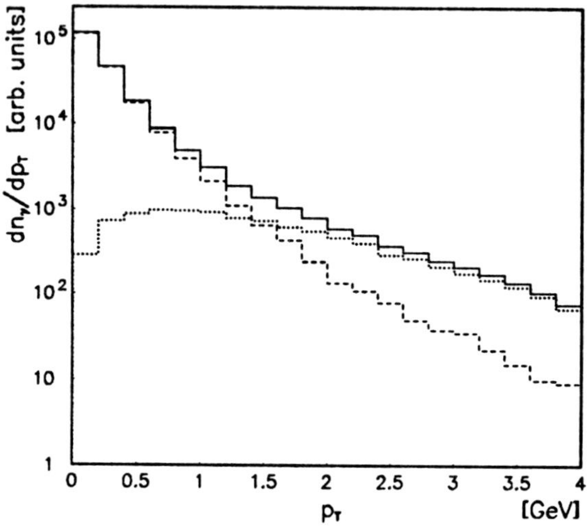

Figure 1: Transverse momentum spectra dn/dp_T for the "early photon component" (dotted line), "late photon component" (dashed line) and the sum of the two (full line).

The acceptance is of twofold importance: first, the statistics increases with the acceptance, and second–more important–acceptance determines the maximal range of relative momenta of particle pairs in correlation analysis. If the acceptance were too narrow, a measured correlation function would not reach its "plateau" (the flat region outside the correlation peak), and the correlation measurement would not be possible. This is of particular importance in the proposed photon correlation measurement, because we are interested in the correlation intensity at a very large longitudinal relative momentum, up to around 2 GeV. For photons of 1 GeV momentum the acceptance in the polar angle should then be around $\pm$ 45 degrees. The dedicated photon detector in ALICE, the Photon Spectrometer (PHOS) (a matrix consisting of lead-tungstenite crystals) is too short since it covers less than a half of the needed acceptance, but fortunately, one can apply another photon measurement technique, namely the photon conversion technique, which has been used in modern satellite borne cosmic gamma ray experiments. In ALICE, photons will convert into electron–positron pairs in the material of the inner tracker system, inner field cage of the Time Projection Chamber (TPC), and in the gas

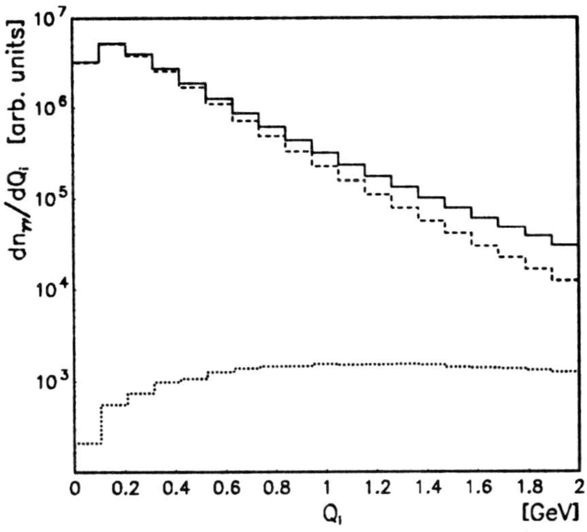

Figure 2: Distribution of photon pairs as a function of Q_i. Pairs of early photons (dotted line) constitute a very small fraction of the total photon contribution (full line); late photons (dashed line). No p_T cuts are applied.

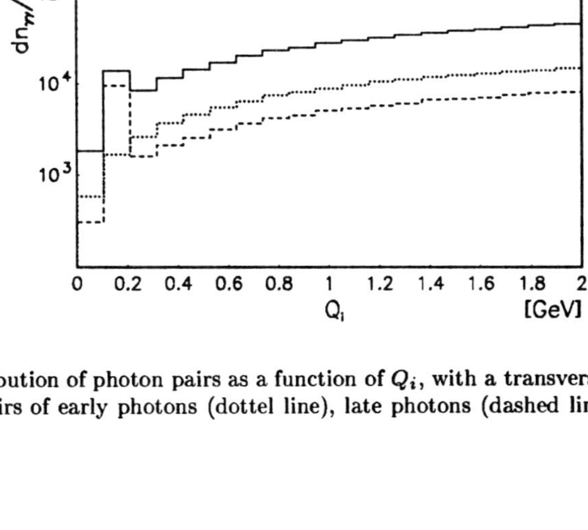

Figure 3: Distribution of photon pairs as a function of Q_i, with a transverse momentum cut p_T >1 GeV. Pairs of early photons (dottel line), late photons (dashed line) and total (full line).

inside the TPC. The created electrons and positrons may be tracked through the TPC, and the information on the original converted photon will be obtained from the electron-positron pair reconstruction. The TPC acceptance for electron-positron pairs is sufficiently wide, and a good resolving power for the two tracks of a conversion pair is confirmed in the simulation.

The integral amount of material contributing to photon conversion will lead to about 5-10% conversion efficiency, and we shall assume 5% overall detection probability.

5 Results

It remains to be seen under which conditions the HBT correlations of the early direct photons will be observable, specifically in the ALICE experiment at LHC. The main problem is the presence of photons emitted later than the early photons, from all other photon sources, particularly from the decays of neutral pions. The HBT correlation at a large relative momentum – the signature of the presence of the early direct photon component – will be therefore strongly suppressed.

For simplicity, in the simulations presented we assume only two photon components: the "early photons", emerging at time 0.3 fm/c, from the rapidly developing parton cascade at 750 MeV effective temperature, and the "late photons", comprising the later direct photons and the photons emerging from the decays of neutral pions (in this simplified approach the late direct photons are not separately simulated, but are effectively assigned as a part of the pion-decay component). We assume further that the fraction of the early direct photons is 5% of the total photon yield.

Assuming a total number of 5000 neutral pions per Pb+Pb central collision in the acceptance of two rapidity units, one arrives at 10,000 photons from pion decays, and 500 early photons per event. Assuming further a detection probability of 5%, one would detect 500 late photons and 25 early photons per event.

Fig. 1 shows the transverse momentum spectrum of the early direct photons, the late photons, and their sum. The high-temperature early component is strongly suppressed at low transverse momenta, below 1 GeV. Fig. 2 shows the distribution of photon pairs as a function of the invariant momentum difference $Q_i^2 = -(p_1^\mu - p_2^\mu)(p_{1\mu} - p_{2\mu})$, where p_1 and p_2 are the four-momenta of the two photons. The contributions shown in Fig. 2 are by early and late photon pairs, while the presented total contribution contains also the mixed early-late photon pairs. The interesting early direct photon component is negligibly small compared to the rest, and in this situation there is no chance to

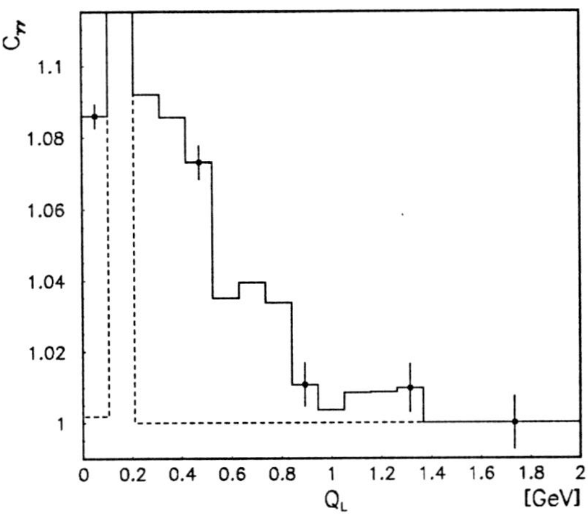

Figure 4: Two–photon correlation function as a function of Q_L, with transverse momentum cut p_T >2 GeV. Full line: R_L=0.3 fm, R_s=3 fm and R_o=3 fm; the correlation enhancement is due to the presence of the early photon component. Dashed line: the early photon component is shifted to later times by setting R_L=6 fm; no correlation enhancement is seen. The π^0 decay peak is also visible. The presented error bars are rescaled and correspond to the statistics expected in the experiment, and not to the actual low statistics used in the simulation.

observe the early photons' correlations. The natural way to proceed is therefore to "enhance" the early direct component by applying a high transverse momentum cut, as suggested by Fig. 1. Fig. 3 shows that already a cut at 1 GeV indeed leads to a significant improvement. In the example presented below we applied a yet "stronger" cut, at 2 GeV.

Two-photon HBT correlations have been simulated with the longitudinal effective length of 0.3 fm, which corresponds to the early appearance of the hypothetical high effective temperature parton gas. The transverse r.m.s. effective radius was set to 3 fm, which corresponds to the r.m.s. transverse size of the colliding Pb nuclei.

Gaussian shape has been assumed for the correlation function in all three dimensions, with an intercept of 1.5 (taking into account [6] that photons are bosons of spin 1). In the results presented we have used an ideal uncorrelated background.

Fig. 4 shows the 3-dimensional correlation function projected onto the Q_L axis, with a cut $Q_S < 0.1$GeV and $Q_O < 0.1$GeV (to reduce "dilution" of the correlation effect). A transverse momentum cut at 2 GeV has been applied. Only 300,000 events have been generated for this analysis, but the error bars presented in Fig. 4 are rescaled to correspond approximately to the expected full statistics of 10^7 events! The expected wide HBT correlation in Q_L due to early photons is indeed seen. In this measurement scheme the correlation intensity at large Q_L is the measure of the yield of the early photon component. The correlation at large Q_L is missing when the early emission time is replaced by a late one (6 fm/c) in the simulation, as shown in Fig. 4, dashed line.

Summary

A method is proposed to test the existence of rapidly developing parton cascades in ultrarelativistic nuclear collisions at RHIC and LHC. An early onset of parton cascades in the collision process should be reflected in abundant emission of photons which are mutually HBT correlated over a wide longitudinal momentum scale. This method provides the necessary selectivity for the early emitted photons, since the late photons correlate only over narrower longitudinal momentum scales. Based on a rather simple approach, our study indicates that such a measurement could be feasible in the ALICE experiment at LHC. The next step is to replace our simplistic assumptions by input from different detailed models.

Acknowledgments

I would like to thank Gordon Baym, Vesa Ruuskanen, Jürgen Schukraft, Josef Sollfrank, Boris Tomašik and Urs Wiedemann for interesting discussions.

References

1. R. Hanbury Brown and R. Q. Twiss, Nature 177 (1956) 27, Nature 178 (1956) 1046, Nature 178 (1956) 1447.
2. K. Geiger, Phys. Rept. 258 (1995) 237, Klaus Geiger, Ron Longacre, Dinesh K. Srivastava, Comput.Phys.Commun. 104 (1997) 70. K. J. Eskola, K. Kajantie, P. V. Ruuskanen, Eur. Phys. J. C1, (1998) 627.
3. Dinesh K. Srivastava and Klaus Geiger, nucl-th/9802034. J. Sollfrank et al., Phys. Rev. C55 (1997) 392.
4. D. Neuhauser, Phys.Lett. B182 (1986) 289. D. K. Srivastava and J. I. Kapusta, Phys. Rev. C48 (1993) 1335. A. Timmermann, M. Plümer, L. V. Razumov and R. Weiner, Phys. Pev. C50 (1994) 3060. L. V. Razumov, H. Feldmeier, Phys.Lett. B377 (1996) 129.
5. R. Loudon, The quantum theory of light, Oxford Univ. Press 1973.
6. C. Slotta and U. Heinz, Phys.Lett. B391 (1997) 469, nucl-th/9610016.
7. Technical proposal for A Large Ion Collider Experiment (ALICE) at the CERN LHC, 1995, CERN/LHCC/95-71, LHCC/P3.

PHOTON-PHOTON INTENSITY INTERFEROMETRY IN HEAVY-ION COLLISIONS AT INTERMEDIATE ENERGIES

A. BADALÀ, R. BARBERA, M. GULINO, A. PALMERI, G. S. PAPPALARDO, F. RIGGI, A. C. RUSSO, G. RUSSO, R. TURRISI
Istituto Nazionale di Fisica Nucleare, Sez. di Catania, and
Dipartimento di Fisica dell'Università di Catania
Corso Italia, 57 – I 95129 Catania, Italy

The correlation between the couples of high-energy photons ($E_\gamma > 30$ MeV) detected in the reactions induced by a ^{36}Ar beam on ^{27}Al, ^{112}Sn, and ^{197}Au targets at 95 MeV/nucleon have been analyzed with the intensity interferometry technique. Both size and lifetime of the emitting source have been quantitatively evaluated. Results support the nucleon-nucleon picture as the dominant hard-photon production mechanism.

1 Introduction

The present availability of large-area and high-granularity detectors for high-energy photons emitted in heavy-ion reactions at intermediate energies has permitted, in these last years, the use of the intensity interferometry technique[1] in the study of the space-time characteristics of the emitting source[2,3,4]. This is not of trivial importance since such a method of analyzing photon-photon correlations can be used as a model independent tool to shed light, in a definitive manner, on the hard-photon production mechanism and time scale. Unfortunately, the very small cross sections involved, associated with the presence of many sources of rather large background, such as neutral pion decays and cosmic ray interactions with the detector, do not allow a direct and comparative study of multi-dimensional correlation functions and the evaluation of the space-time characteristics of the source is not straightforward at all. Furthermore, a very detailed knowledge of the response function of the used detector to the couples of correlated photons and to the sources of noise is of crucial importance to extract reliable values of the emitting source parameters to be compared with the existing theoretical models.

In this contribution we report on an extensive application of the intensity interferometry technique to the pairs of high-energy photons ($E_\gamma > 30$ MeV) emitted in the reactions induced by a ^{36}Ar projectile on ^{27}Al, ^{112}Sn, and ^{197}Au targets at 95 MeV/nucleon. The paper is organized as follows. Section II is devoted to a brief review of the photon-photon correlation function together with a discussion of the problems one has to cope with in the difficult passage from the observed variables to the desired information. Results are presented in Sect. III while summary and conclusions are given in Sect. IV. Due to space limitations, the description of both the experimental setup and the methods of particle detection and identification are missing. The interested reader is addressed to Refs. [2,5] for all details.

2 Photon-photon correlation function

The correlation function between two photons, having wave vectors $\vec{k}_1$ and $\vec{k}_2$, is defined as [6,7]:

$$F^{id}(\vec{k}_1, \vec{k}_2) \equiv \frac{P(\vec{k}_1, \vec{k}_2)}{P(\vec{k}_1)P(\vec{k}_2)} = N \cdot \frac{d^6 n}{d^3\vec{k}_1 d^3\vec{k}_2} / (\frac{d^3 n}{d^3\vec{k}_1}\frac{d^3 n}{d^3\vec{k}_2}) \tag{1}$$

where $P(\vec{k}_{1,2})$ and $P(\vec{k}_1, \vec{k}_2)$ are the one photon and two-photon inclusive distributions, respectively. The factor N in Eq. (1) follows from the relative normalization of these distributions. Taking into account the polarization of the photons and assuming that: (i) the coherent electromagnetic current can be neglected; (ii) all neutron-proton collisions produce the same magnitude of electromagnetic current, and (iii) currents from different collisions have uncorrelated directions, Neuhauser has shown [7] that the two-photon correlation function assumes the form:

$$F(\vec{k}_1, \vec{k}_2) = 1 + \frac{1}{2}\,[1 + (\hat{\mathbf{k}}_1 \cdot \hat{\mathbf{k}}_2)^2] \times \frac{1}{2}\,[|\rho(k_1 + k_2)|^2 + |\rho(k_1 - k_2)|^2] \equiv 1 + C_2(\vec{k}_1, \vec{k}_2) \tag{2}$$

where $\hat{\mathbf{k}}_1$ and $\hat{\mathbf{k}}_2$ are the unit vectors along the directions of the two photons and $\rho(k)$, as a function of the four-vector k, is the four-dimensional Fourier transform of the normalized space-time density $\rho(\vec{r}, t)$ of the photon source. The term $[1 + (\hat{\mathbf{k}}_1 \cdot \hat{\mathbf{k}}_2)^2]$, which reduces the maximum value of the correlation function from 2 to 1.5, is due to the photon polarization [7] and takes into account that orthogonally polarized photons cannot interfere. If the energies of both correlated photons are sufficiently high (E_{γ_1}, $\mathrm{E}_{\gamma_2} \geq 30 - 40$ MeV), it is possible to neglect, in Eq. (2), $|\rho(k_1 + k_2)|$ with respect to $|\rho(k_1 - k_2)|$ and, hence, the correlation function can be re-written as:

$$C_2(\vec{k}_1, \vec{k}_2) = \frac{1}{4}\,[1 + (\hat{\mathbf{k}}_1 \cdot \hat{\mathbf{k}}_2)^2]\,|\rho(k_1 - k_2)|^2 \tag{3}$$

In order to obtain numerical estimates of the photon correlation function, it is usual [7] to assume the space-time density as a spherically symmetric Gaussian function with uncorrelated space and time coordinates:

$$\rho(r,t) = \frac{1}{4\pi^2 R^3 \tau} \exp(-r^2/2R^2 - t^2/2\tau^2) \tag{4}$$

where R and τ characterize the source radius and lifetime, respectively. Taking the four-dimensional Fourier transform of Eq. (4), it is possible to express Eq. (3) in terms of the observables $q_{rel} = |\vec{p}_{\gamma_1} - \vec{p}_{\gamma_2}|$, and $q_0 = |E_{\gamma_1} - E_{\gamma_2}|$ $(c = 1)$:

$$C_2(q_{rel}, q_0) = \frac{1}{4} (1 + \cos^2\theta_{rel}) \exp(-\frac{q_{rel}^2 R^2}{\hbar^2}) \exp(-\frac{q_0^2 \tau^2}{\hbar^2}) \tag{5}$$

(θ_{rel} being the relative angle between the two photons).

Generally, both geometrical and physical limitations of real detectors prevent the complete knowledge of the full six-dimensional photon-photon distribution reported in Eq. (1). Thus, the photon correlation function relative to filtered events can not be constructed from the ratio given in that equation. In practice, the six-dimensional two-particle correlation function $P(\vec{k}_1, \vec{k}_2)$ is projected on an one- or, at most, two-dimensional one, $P(\vec{k}_1, \vec{k}_2) \rightarrow A(q_1), A'(q_1, q_2)$, where q_i are the so-called correlation observables. In our analysis we have used both the relative momentum, q_{rel}, and the relative energy q_0. They are the most natural choices since the emitting source has been parameterized through Eq. (4) allowing a separated study of space and time coordinates. To evaluate the quantity $C(q)$, the correlation term $A(q)$ is then compared to a similar projection of some background sample $B(q)$, so that the experimental correlation function is given by:

$$C(q) = \frac{A(q)}{B(q)} \cdot \tag{6}$$

The uncorrelated term $B(q)$ is usually evaluated either starting from inclusive events, advisable when they are available, or using the event mixing technique[8], i.e. creating artificial pairs by combining single particles coming from randomly chosen two-particle events. In this work we have adopted the first method. This approach may be justified considering that, in reactions between heavy ions producing photons in the intermediate energy regime, the mean value of the inclusive high energy photon multiplicity (leaving out the case of no detected photon) is very close to one and, thus, the residual two-particle correlations are negligible.

In the most general case of whatever-oriented couples of photons (i.e., when using large solid angle multi-detectors) the quantity q_{rel} is different from q_0 and the polarization term $(1 + \cos^2\theta_{rel})$ in Eq. (5) changes on an event-by-event basis in a very complicated manner. To help the understanding,

Fig. 1 shows the simulated two-dimensional correlation function $C_2(q_{rel}, q_0)$ for photons coming from a source with $R = 5$ fm and $\tau = 10$ fm/c and detected by an ideal detector (namely 4π sr solid angle, infinite granularity, infinite angular resolution, no thresholds, perfect photon-charged particle and photon-neutron separation, no neutral pion background, complete cosmic ray rejection, etc.).
The condition $q_{rel}^2 \geq q_0^2$ $(c = 1)$, which must hold for real particles, makes the distribution not symmetric with respect to the $q_{rel} = q_0$ plane. The value of 0.5, expected for the maximum of the correlation function, is well reproduced and the widths of the distribution, from q_0 and q_{rel} directions (see Fig. 1, lower panels), are in agreement with the input values of R and τ. On the contrary, if one reports the quantities $C(q_{rel})$ and $C(q_0)$ (see for example Figs. 7 and 8 of Ref. [2]), through an implicit integration process over the not used variable, both maxima and widths are not reproduced, notwithstanding the shape remain roughly (but not exactly) Gaussian. Both R' and τ' are larger than R and τ, erroneously indicating a source with space-time size larger than the real one. This is a crucial point in the intensity interferometry technique. In order to extract from data reliable values of source sizes and lifetimes, one can either directly fit the bi-dimensional correlation function $C_2(q_{rel}, q_0)$ or the two mono-dimensional distributions $C(q_{rel})$ and $C(q_0)$. The application of the first method needs a very good statistics, often unrealistic in actual experiments. The second method is more suitable, as it is shown in Ref. [2,9], even if it involves nested integrations of the asymmetric $C_2(q_{rel}, q_0)$. In any case, however, the results must be considered with great care. Both the relative angle term in the correlation function and the detector efficiency make R' and τ' different from the real source parameters, although they are correlated to them. Only through a complete simulation it is possible to find the relations between these widths and the true R and τ values. This point will be discussed in a greater detail in the next Section.

3 Results

As an example, the relative momentum distribution of all two-photon events detected in the reaction ^{36}Ar+^{27}Al at 95 MeV/nucleon is reported in Fig. 2. Filled dots refer to those couples of photons coming from π^0 decay while open dots refer to the rest of photon couples (see Ref. [2] for details). These latter pairs of photons, producing the large bump centered around 80-100 MeV/c, are relative both to "badly" detected pions and to $\gamma - \gamma^{\pi^0}$ events where only one photon comes from the source under investigation and the other one from a π^0 decay not completely detected due to the geometrical efficiency of the

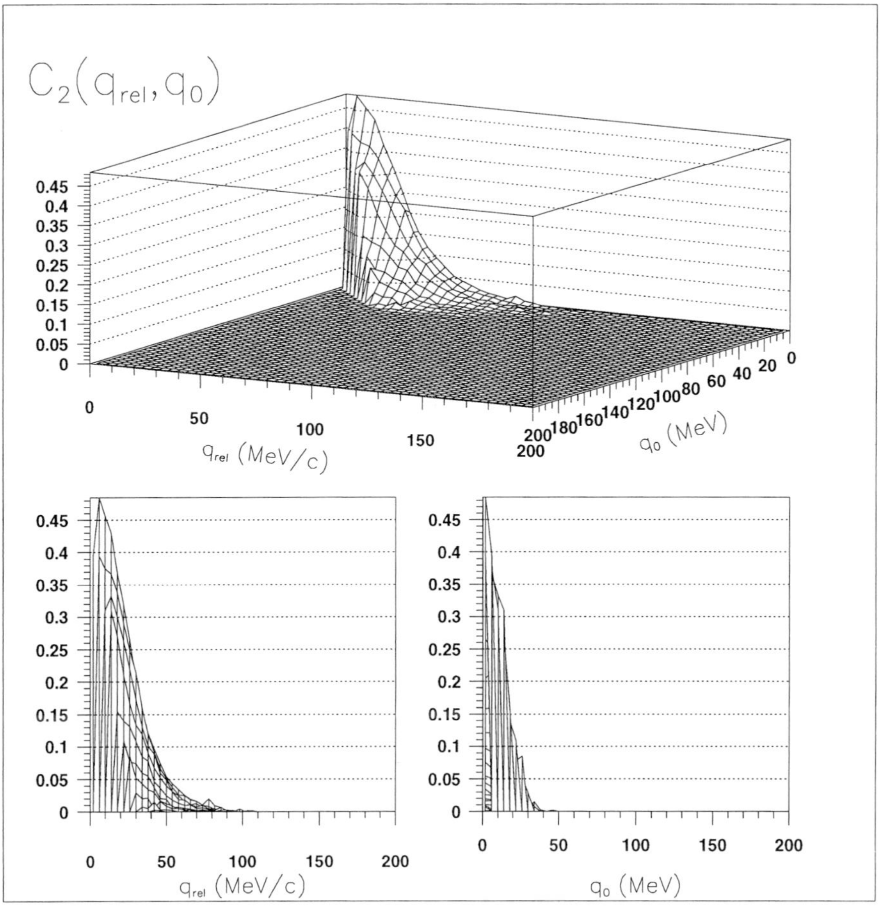

Figure 1: Bi-dimensional correlation function $C_2(q_{rel}, q_0)$ for photons emitted from a source with $R = 5$ fm and $\tau = 10$ fm/c (upper panel). In the lower panels the views from the q_{rel} and q_0 axes are reported. The events are not filtered.

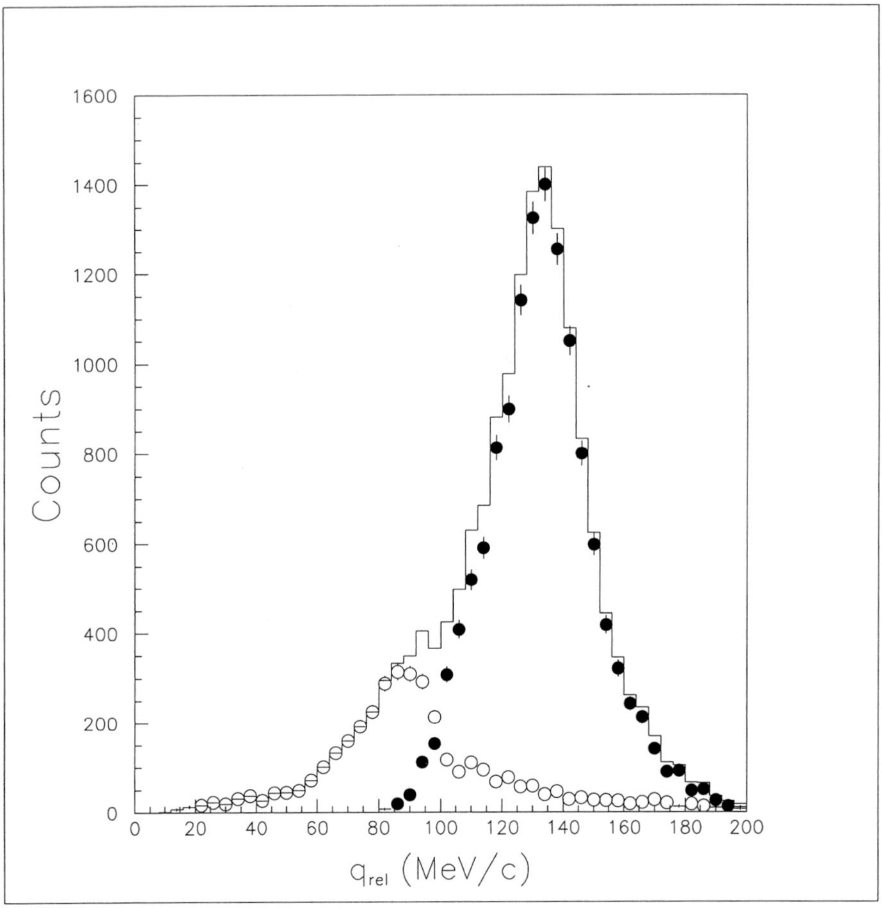

Figure 2: Relative momentum distribution of two-photon events detected in the ^{36}Ar+^{27}Al reaction at 95 MeV/nucleon (continuous histogram). Filled dots refer to those couples of photons coming from π^0 decay while open dots refer to the rest of photon couples.

real detector. True $\gamma - \gamma$ events are expected to place themselves in the lowest part of the relative momentum spectrum. The absence of values of q_{rel} lower than about 10 MeV/c is due to the limitation imposed both in the minimum energy (30 MeV) and relative angle (15°) of the two detected photons.

In order to evaluate the correlation function, the relative momentum distribution plotted in Fig. 2 has been divided by that obtained folding the inclusive photon events through the event mixing technique [8]. The normalization factor has been calculated normalizing to 1 the mean of the correlation function between about 45 and 60 MeV/c and it has been independently verified that it takes into account the different run times of the two sets of events . The correlation distributions $1 + C(q_{rel})$, as a function of the relative momentum, are reported in Fig. 3 for all the used targets.

The three above-cited contributions are visible. A peak, due to neutral pions (centered around 140 MeV/c), a large bump, due to 'badly" measured pions and to $\gamma - \gamma^{\pi^0}$ events (between 50 and 150 MeV/c), and the photon-photon correlation signal (for $q_{rel} < 45$ MeV/c). The continuous line passing through the points refers to the result of a best-fit procedure using the sum of three Gaussian functions ($c = 1$):

$$C(q_{rel}) = \lambda_{R'}\, e^{-\frac{q_{rel}^2 R'^2}{\hbar^2}} + A_1\, e^{-\frac{(q_{rel}-q_1)^2}{2\sigma_1^2}} + A_2\, e^{-\frac{(q_{rel}-q_2)^2}{2\sigma_2^2}}, \qquad (7)$$

one for each component. R' is the width of the spatial correlation signal. For the sake of correctness, the first two (black) points in the correlation distribution, having $q_{rel} < 15$ MeV/c, have been not included in the fitting because the evaluated detector efficiency for this correlation function, reported in the upper panel of Fig. 4, shows a steep fall-off to zero below that value of the relative momentum.

An efficiency greater than one has not to be surprising as it is defined as the ratio between the detector efficiencies for the numerator and the denominator of the correlation function. The extracted values of the width R' are summarized, for all targets, in the second column of Table 1.

The correlation distributions $C(q_0)$, as functions of the relative energy, are plotted, for the various targets, in Figs. 5.

Only those couples of photons having $q_{rel} < 45$ MeV/c have been taken into account to build those plots. The continuous curves are the results of best-fit procedures using the function ($c = 1$):

$$C(q_0) = \lambda_{\tau'}\, e^{-\frac{q_0^2 \tau'^2}{\hbar^2}} \qquad (8)$$

where τ' is the width of the temporal correlation signal. In this case all the points have been included in the fitting due to the shape of the evaluated

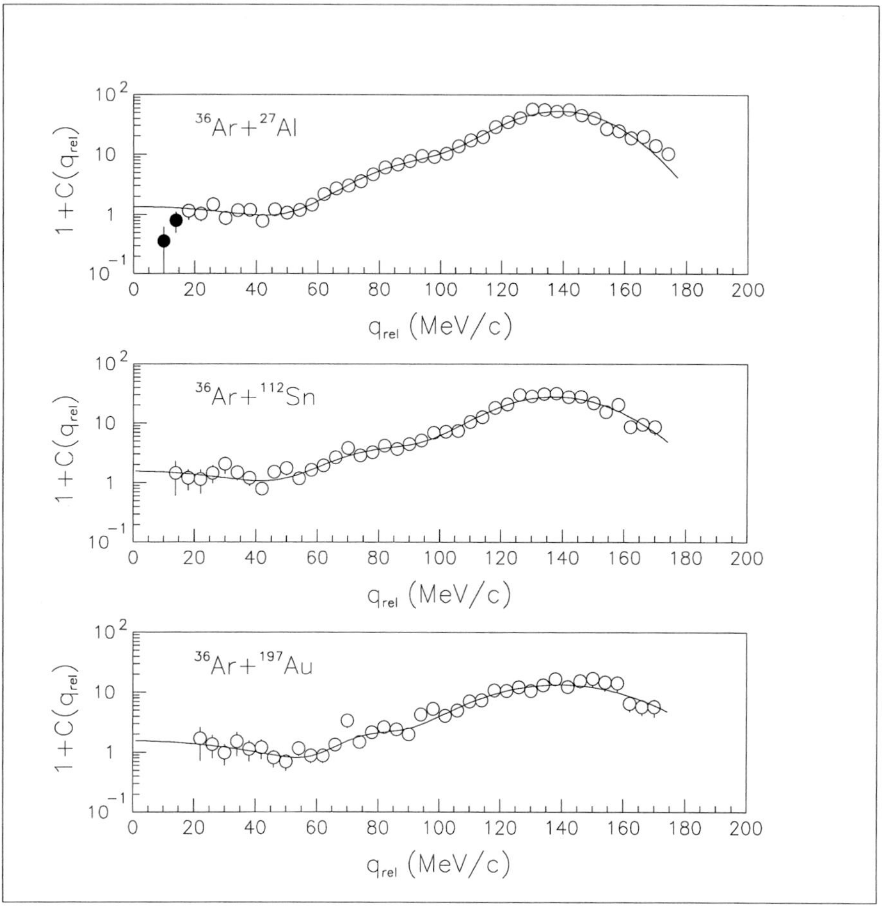

Figure 3: Correlation distributions $1 + C(q_{rel})$, as functions of the relative momentum, for the pairs of photons detected in all studied reactions. Continuous curves are relative to the result of the best-fit procedure discussed in the text.

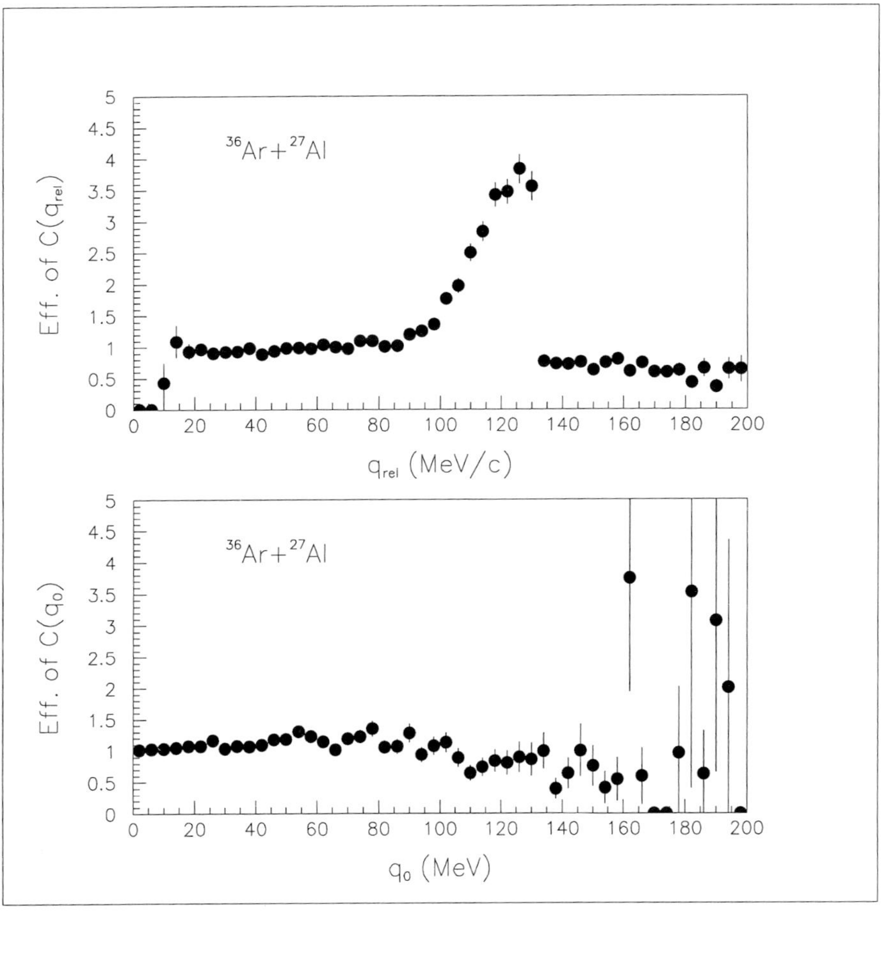

Figure 4: Upper panel: detector efficiency for the relative-momentum photon-photon correlation function. Lower panel: detector efficiency for the relative-energy photon-photon correlation function.

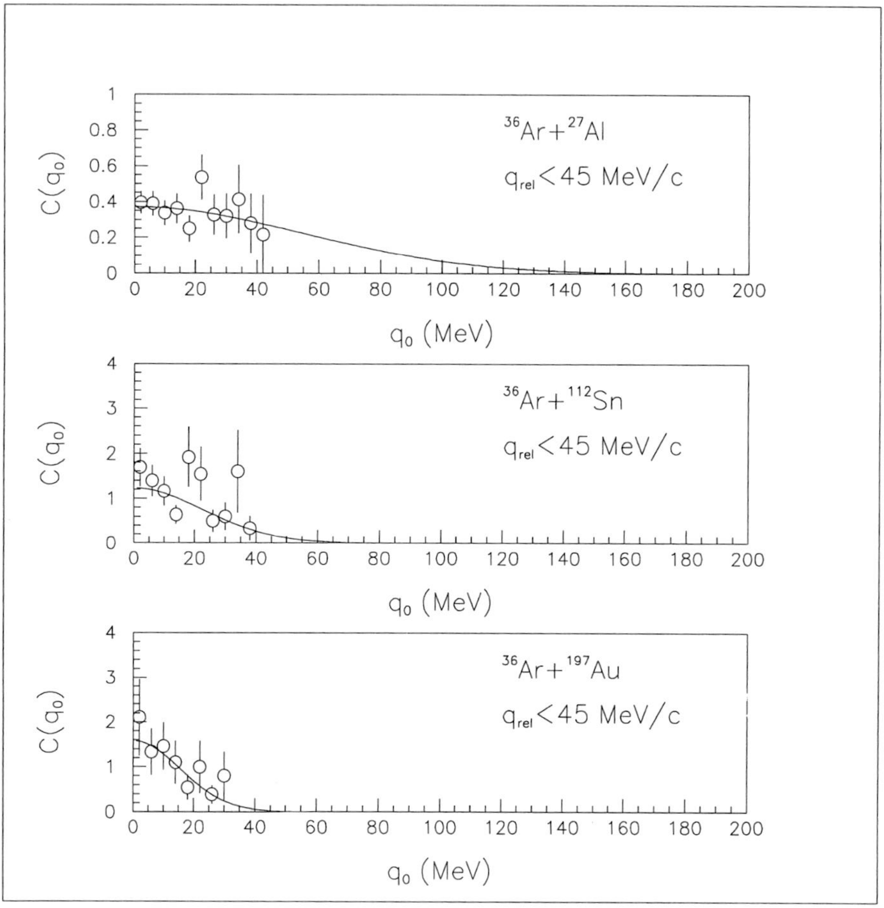

Figure 5: Correlation distributions $1 + C(q_0)$, as functions of the relative energy, for the pairs of photons detected in all studied reactions and having $q_{rel} < 45$ MeV/c. Continuous curves are relative to the result of the best-fit procedure discussed in the text.

Table 1: Values of the measured source parameters R' and τ' and those of the extracted ones R and τ, for the various studied systems.

System	R' (fm)	R (fm)	τ' (fm/c)	τ (fm/c)
^{36}Ar+^{27}Al	3.1 ± 1.2	0.8 ± 0.4	2.6 ± 1.7	2.0 ± 1.0
^{36}Ar+^{112}Sn	4.0 ± 1.4	1.7 ± 0.7	6.1 ± 1.4	6.3 ± 1.8
^{36}Ar+^{197}Au	3.2 ± 1.0	2.2 ± 1.1	9.0 ± 1.9	9.9 ± 1.9

detector efficiency for this correlation function which is reported in the lower panel of Fig. 4. The extracted values of the width τ' are summarized, for all targets, in the fourth column of Table 1. It is worth stressing that changing the value of the upper threshold of q_{rel} in the realistic relative momentum region between 40 and 50 MeV/c the extracted values of τ' resulted all consistent among them within their error bars.

As it has been discussed in detail in Ref.[2] and reported here in the previous Section, the effects deriving from the interference between the polarizations of the two emitted photons, always present in the correlation function, and that due to the detector efficiency, make R' and τ' different from the real source parameters. These effects are also not simply additive and operate at the same time in a very complicated way depending both on the values of the source parameters and on the detector geometry. So, in order to unfold the searched values of R and τ from the measured ones of R' and τ' we performed a lot of full GEANT [10] simulations of the detection of pairs of correlated photons by the used experimental setup and we submitted the simulated events to the same analysis technique adopted for experimental data.

Owing to the fact that the interesting region of the plane (R, τ) is quite small and the chosen grids in this plane are rather close, we expanded the unknown functions $R(R', \tau')$ and $\tau(R', \tau')$ as second-order polynomials whose free parameters have to be fitted to the values of R' and τ' extracted from simulated data. So, the problem of finding R and τ for the three studied systems reduces to an interpolation of $R(R', \tau')$ and $\tau(R', \tau')$. The extracted values of R and τ are reported in the third and fifth column of Table 1 for the various targets.

As it has been pointed out in Ref. [6], however, the numeric value of R represents only one (the other is τ) of the σ-values of the assumed space-time distribution of the source. A more reliable quantity is the root-mean-squared radius r_{rms} that, in the hypothesis of a spherical symmetric source, writes:

$$r_{rms} = \sqrt{\langle \vec{r} \cdot \vec{r} \rangle} = \sqrt{3}R \tag{9}$$

Table 2: The values of the root-mean-squared radius r_{rms} are compared, for the three studied systems, with those of the overlap radius R_{ov} and those of the total radius R_{tot}.

System	r_{rms} (fm)	R_{ov} (fm)	R_{tot} (fm)
^{36}Ar+^{27}Al	1.4 ± 0.7	1.38	7.56
^{36}Ar+^{112}Sn	3.0 ± 1.2	2.12	9.75
^{36}Ar+^{197}Au	3.8 ± 1.9	2.37	10.94

The values of r_{rms} for the three studied system are reported in the second column of Table 2.
In heavy-ion collisions at around 100 MeV/nucleon bombarding energy, hard photons are thought to come from incoherent nucleon-nucleon collisions which take place, in the very first phase of the reaction, within the overlap region between the two nuclei [11]. Using the geometrical estimation [12] of the number (averaged over the impact parameter) of the participant nucleons producing two hard photons

$$\langle N_{pn}\rangle_{\gamma\gamma} \equiv \langle N_{pn}(N_{pn} - 1)\rangle_b \tag{10}$$

(where N_{pn} is the number of first proton-neutron collisions), it is possible to calculate the Gaussian radius [13] of the two-nuclei overlap zone through the formula:

$$R_{ov} = 1.2\langle N_{pn}\rangle_{\gamma\gamma}^{1/3}/\sqrt{5} \tag{11}$$

The values of R_{ov} for the three systems are reported in the third column of Table 2. Within the error bars, the agreement with the r_{rms}'s is good. Furthermore, both series of values are very much different from the sums of projectile and target radii $R_{tot} = 1.2(A_p^{1/3} + A_t^{1/3})$ which are reported for comparison in the last column of Table 2. Our data are also in agreement with much more refined calculations based on the resolution of microscopic transport equations. In a recent paper, in fact, Barz *et al.* [14] have successfully reproduced the experimental correlation functions relative to the ^{27}Al target with a BUU calculation based on first-chance nucleon-nucleon collisions.

4 Summary and conclusions

In this contribution we have exploited the Hanbury-Brown–Twiss effect for an extensive study of two-photon correlation data relative to the reactions ^{36}Ar+^{27}Al, ^{36}Ar+^{112}Sn, and ^{36}Ar+^{197}Au at 95 MeV/nucleon.

A new method, based on computer simulations, has been used to get information both on the source size and lifetime. Experimental results are consistent with the picture of the hard-photon production mechanism due to first, single, and incoherent nucleon-nucleon collisions. The values of the spatial size of the source, for the various systems, are in fact very small and compatible with the radii of the geometric overlap volumes between the two colliding nuclei. The source lifetimes are in agreement, at least for the lightest used target, with BUU calculations based on first-chance proton-neutron collisions. They steadily increase as a function of the target mass possibly due to the presence of projectile stopping effects which have already been observed for the same systems at the same bombarding energy [15].

References

1. For a review, see e.g., D. H. Boal, C. K. Gelbke, and B. J. Jennings, Rev. Mod. Phys. **62**, 553 (1990), and references therein; W. Bauer, C. K. Gelbke, and S. Pratt, Annu. Rev. Nucl. Part. Sci. **42**, 77 (1992), and references therein.
2. A. Badalà, R. Barbera, A. Palmeri, G. S. Pappalardo, F. Riggi, G. Rizza, A. C. Russo, G. Russo, and R. Turrisi, Phys. Rev. C **55**, 2521 (1997) and reference therein.
3. A. Badalà, R. Barbera, A. Palmeri, G. S. Pappalardo, F. Riggi, G. Rizza, A. C. Russo, G. Russo, and R. Turrisi, Phys. Rev. C **57**, 2765 (1998).
4. F. M. Marqués, G. Martinez, T. Matulewicz, R. W. Ostendorf, Y. Schutz, Phys. Rep. **284**, 91 (1997).
5. E. Migneco *et al.*, Nucl. Instrum. Methods Phys. Res., Sect. A **314**, 31 (1992).
6. W. A. Zajc, in *Proceedings of a NATO Advanced Study Institute on Particle Production in Highly Excited Matter*, IL Ciocco, Italy, 1992, edited by H. H. Gutbrod and J. Rafelski (Plenum Press, New York, 1993).
7. D. Neuhauser, Phys. Lett. B **182**, 289 (1986).
8. D. Drijard, H. G. Fischer, and T. Nakada, Nucl. Instrum. Methods Phys. Res., Sect. A **225**, 367 (1984).
9. A. Badalà, R. Barbera, A. Palmeri, G. S. Pappalardo, F. Riggi, A. C. Russo, G. Russo, and R. Turrisi, Nucl. Instrum. Methods Phys. Res., Sect. A **351**, 387 (1994).
10. CERN Application Software Group, GEANT: *Detector Description and Simulation Tool* (CERN, Geneva, 1993); CERN Program Library Long Writeups W5013.
11. W. Cassing, V. Metag, U. Mosel, and K. Niita, Phys. Rep. **188**, 363

(1990).
12. H. Nifenecker, and J. P. Bondorf, Nucl. Phys. A **442**, 478 (1985).
13. W. A. Zajc *et al.*, Phys. Rev. C **29**, 2173 (1984).
14. H. W. Barz, B. Kämpfer, Gy. Wolf, and W. Bauer, Phys. Rev. C **53**, R553 (1996).
15. A. Badalà, R. Barbera, A. Palmeri, G. S. Pappalardo, F. Riggi, A. C. Russo, G. Russo, and R. Turrisi, Phys. Rev. C **53**, 1782 (1996), and references therein.

CONNECTION BETWEEN THE INTERFERENCE CORRELATIONS OF IDENTICAL PARTICLES IN HEAVY ION PHYSICS AND ASTROPHYSICS: ANALOGY AND DIFFERENCE

V. L. LYUBOSHITZ

Joint Institute for Nuclear Research,
Laboratory of High Energies,
141980, Dubna, Moscow Region, Russia
E-mail: lyubosh@sunhe.jinr.ru

The wide-spread method of measuring the space-time characteristics of the region of particle multiple generation by investigating the pair correlations of identical pions with close momenta, and the Hanbury-Brown and Twiss method of determining the angular sizes of stars by studying the intensity correlations of two optical photons are compared. Both the correlation methods are based on the correlation properties of identical particles. However, in spite of the deep analogy between them, the difference of principle in the arrangement and performance of the corresponding experiments exists. In high energy physics (heavy ion physics) we deal with the momentum-energy correlations, when the momenta and energies are measured and fixed experimentally. In this case the correlations contain the direct information on space-time parameters of the source group. In astrophysics we deal with space-time correlations, and sources and detectors are mutually replaced as compared with the situation in high energy physics. The space-time correlations of intensities in two counters of photons are determined directly by the distances between the observation points and by the difference of the registration time moments (these parameters are measured and fixed experimentally), and they contain the information on the angular radius of the star only through the dependence on the difference of transverse momenta of photons emitted by different regions of the star surface.

1. About 45 years have elapsed after the publication of the classical work of Hanbury-Brown and Twiss [1], in which the method of determining the angular sizes of stars by the way of studying the intensity correlations at the registration of two optical photons was elaborated. 6 years later Goldhaber et al [2] showed that in the framework of the statistical model the effect of symmetrization of wave functions of two identical pions leads to the narrowing of the distribution of angles between their flight directions, depending on the radius R of the generation region. In the seventies, in the cycle of papers by Kopylov and Podgoretsky [3-8] the deep analogy between pair correlations of identical particles, produced in nuclear and hadron collisions, and pair correlations of photons, emitted by optical sources, in particular, by stars, was established for the first time. That became the base for the creation of the general correlation method in high energy physics, overstepping the narrow bounds of the stationary statistical model [2] and allowing to measure not only the space sizes of the genera-

tion region of identical particles but also the shape of this region and the duration of the generation processes [8].

During the last 25 years the method of correlations of identical particles with small relative velocities has got wide propagation in the practical activity of many laboratories of the world, and it has been formed as the new perspective trend in nuclear physics and particle physics, including hundreds of experimental and theoretical works.

Both the methods (in high energy physics, as well as in astronomy) are based on the correlation properties of identical particles. However, in the recent years the very important analogy between them has obscured the difference of principle (even opposition!) in the arrangement and performance of the corresponding experiments, indicated clearly in the paper [7].

2. Really, in high energy physics (heavy ion physics) we deal with the momentum-energy correlations. The model of one-particle sources is used. It is assumed that the emission sources have a sufficiently broad momentum spectrum. The sizes of sources themselves and the distances between them are by many orders smaller than the uncertainties in the space localization of the detectors; the times of the source decays are very small in comparison with the time characteristics of detectors. The momenta of two identical pions in the final state are measured and fixed. We will not dwell on details here, restricting ourselves by the description of the structure of principle of these correlations.

The basic formula of the pion interferometry, which should be averaged over the space-time distribution of two sources at the calculation of the correlation function, independently of concrete processes, has the form

$$W = {}^1\!/_2 \,|\, <p_1|x_1><p_2|x_2> + <p_2|x_1><p_1|x_2> \,|^2 =$$

$$= {}^1\!/_2 \,|\, \exp(-ip_1x_1)\exp(-ip_2x_2) + \exp(-ip_1x_2)\exp(-ip_2x_1) \,|^2$$

$$= 1 + \cos\left((p_1-p_2)(x_1-x_2)\right). \tag{1}$$

Formula (1) describes the probability of the event that two noninteracting spinless identical particles, emitted at the definite space points $\mathbf{x}_1$ and $\mathbf{x}_2$ at the definite time moments t_1 and t_2, are registered by detectors in the states with the definite 4-momenta $p_1 = \{E_1, \mathbf{p}_1\}$ and $p_2 = \{E_2, \mathbf{p}_2\}$. This relation corresponds to the Dirac transformation from the space-time representation to the momentum-energy one. In so doing, $x_1 = \{t_1, \mathbf{x}_1\}$ and $x_2 = \{t_2, \mathbf{x}_2\}$ mean the 4-coordinates of sources, or, more exactly, their centres (taking into account the non-zero dimensions of the sources themselves).

By definition, the argument of cosine in Eq.(1) is the following:

$$(p_1-p_2)(x_1-x_2) = (\mathbf{p}_1-\mathbf{p}_2)(\mathbf{x}_1-\mathbf{x}_2)-(E_1-E_2)(t_1-t_2) \ . \tag{2}$$

Due to the equality $(\mathbf{p}_1-\mathbf{p}_2)(\mathbf{p}_1+\mathbf{p}_2)=0$, the difference of energies is:

$$E_1-E_2 = (\mathbf{p}_1-\mathbf{p}_2)\,\mathbf{u} \quad , \tag{3}$$

where $\mathbf{u} = \dfrac{(\mathbf{p}_1+\mathbf{p}_2)}{E_1-E_2}$ is the velocity of the pair of the identical pions. It should be stressed that here we use the unit system in which $\hbar = c = 1$.

3. In the framework of the model of one-particle sources, the double inclusive cross-section of the production of two identical pions with small relative momenta in the collisions of heavy ions has the structure:

$$\frac{d^6\sigma}{d^3\mathbf{p}_1\, d^3\mathbf{p}_2} = R(\mathbf{p}_1,\mathbf{p}_2)\,\frac{d^6\sigma^{(0)}}{d^3\mathbf{p}_1\, d^3\mathbf{p}_2} \quad , \tag{4}$$

$$R(\mathbf{p}_1,\mathbf{p}_2) = 1 + \langle \cos((p_1-p_2)(x_1-x_2)) \rangle =$$

$$= 1 + \int W_P(x_1,x_2)\cos(q(x_1-x_2))\, d^4x_1\, d^4x_2 \quad . \tag{5}$$

Here $\dfrac{d^6\sigma^{(0)}}{d^3\mathbf{p}_1\, d^3\mathbf{p}_2}$ is the "background" cross-section of the pion production without the quantum statistics effect (pion identity) and the final state interaction effect too, $W_P(x_1, x_2)$ is the distribution of 4-coordinates of two sources, normalized by unity, which depends on the 4-momentum sum $P = p_1 + p_2$, in general, and does not depend practically on the small 4-momentum difference $q = p_1 - p_2$ [9,10]. Pair correlations of pions, containing the information on space-time parameters of the multiple generation region, are described by the expression

$$C_P(q) = R(\mathbf{p}_1,\mathbf{p}_2) - 1 = \langle \cos(q(x_1-x_2)) \rangle \quad . \tag{6}$$

It is essential that the correlation function $C_P(q)$ tends to zero at sufficiently large values of the momentum difference $|\mathbf{q}| = |\mathbf{p}_1 - \mathbf{p}_2| >> 1/r$ and the energy difference $|q_0| = |E_1 - E_2| >> 1/\tau$, where r is the characteristic radius of the emission region, τ is the characteristic duration of the particle production process. It should be noted that often one names the ratio $R(\mathbf{p}_1, \mathbf{p}_2)$ itself of the true two-particle spectrum to the "background" one as the correlation function. Concerning the background, in heavy ion collisions we can neglect kinematic constraints and most of the dynamic correlations, and construct the reference two-particle distribution as the product of one-particle spectra by mixing the particles from different events.

The final state interaction is neglected in Eqs. (1), (5) and (6). This approximation is sufficiently good for pions. In general, the final state interaction also influences the character of pair correlations at small relative velocities. It should be emphasized that in the case of narrow pair correlations of non-identical particles only the final state interaction effect plays role [9-11]. We will not discuss this question in detail here.

4. In the case of the emission of unpolarized identical particles with the non-zero spin j the correlations decrease. Neglecting the final state interaction, one should write for the ratio $R(\mathbf{p}_1, \mathbf{p}_2)$ of the true two-particle spectrum to the "background" one the following expression [9,10,12]:

$$R(\mathbf{p}_1, \mathbf{p}_2) = 1 + \frac{(-1)^{2j}}{2j+1} < \cos(q(x_1 - x_2)) > \quad , \tag{7}$$

which coincides with Eq. (5) at $j = 0$. Let us note that the factor $N(j) = 2j + 1$ in the denominator of Eq. (7) means the number of spin states of a particle with the spin j; the sign factor $(-1)^{2j}$ is equal to (+1) for bosons and to (–1) for fermions.

Formula (7) is related to the identical particles with the non-zero mass. When the unpolarized photons are emitted, one should introduce in Eq.(7), instead of the factor $\frac{(-1)^{2j}}{2j+1}$, the factor

$$K = {}^1/_4\,(1 + \cos^2\theta) \quad , \tag{8}$$

in which θ is the angle between the directions of photon flights, taking into account the transversality of the electromagnetic field of radiation [12]. Really, the angle θ is very small at small values of the momentum difference, so that $K = {}^1/_2$. In so doing, the correlation function of two photons with small relative momenta is described by the formula

$$C_P(q) = {}^1/_2 < \cos(q(x_1 - x_2)) > \quad (9),$$

differing from Eq.(6) by the factor ${}^1/_2$. This factor corresponds to the presence of two possible polarization states of a photon (although the photon has the spin 1, the photon helicity can take only two values (+1) and (−1) due to the zero mass of photon).

5. We discussed above the momentum-energy correlations of pions and photons in nuclear and particle physics (heavy ion physics). In optics and astronomy we deal with the space-time correlations of photons [7]. In this situation the sizes of the emission region, the distances between sources and the duration of the emission process are by many orders larger than the corresponding characteristics of detectors. Let us denote the distance between two sources as $r_{12} = |\mathbf{x}_1 - \mathbf{x}_2|$, the distance between two detectors as $r_{34} = |\mathbf{x}_3 - \mathbf{x}_4|$ and the distance between the observation region and the emission region as L. The space-time correlations are observed when the conditions

$$r_{34} << r_{12} << L \quad (10)$$

are satisfied (i.e. the emission region is very large, and it is situated very far from the observation region). Then, the amplitude of the registration of two identical particles (concretely, two photons with the same polarizations) by two detectors is proportional to

$$A = \frac{1}{\sqrt{2}} (< x_3 | p_1 > < x_2 | p_2 > + < x_3 | p_2 > < x_4 | p_1 >) =$$

$$= \frac{1}{\sqrt{2}} (\exp(-i p_1 x_3) \exp(-i p_2 x_4) + \exp(-i p_1 x_4) \exp(-i p_2 x_3)) \quad , \quad (11)$$

and it corresponds to the Dirac transformation from the momentum-energy representation to the space-time one. In accordance with this, the equation

$$W = |A|^2 = 1 + \cos((p_1 - p_2)(x_3 - x_4)) =$$

$$= 1 + \cos[(\mathbf{p}_1 - \mathbf{p}_2)(\mathbf{x}_3 - \mathbf{x}_4) - (E_1 - E_2)(t_3 - t_4)] \tag{12}$$

describes the probability of the event that two photons, emitted in the states with the definite 4-momenta $p_1 = \{E_1, \mathbf{p}_1\}$ and $p_2 = \{E_2, \mathbf{p}_2\}$, will be registered by two detectors at the definite space points $\mathbf{x}_3$ and $\mathbf{x}_4$ at the definite time moments t_3 and t_4. In so doing, $x_3 = \{t_3, \mathbf{x}_3\}$ and $x_4 = \{t_4, \mathbf{x}_4\}$ in Eqs. (11) and (12) are already the 4-coordinates of detectors, but not sources; the detectors and the sources are mutually replaced as compared with the situation in high energy physics.

6. When the unpolarized photons with close momenta are emitted, it should be written instead of Eq. (12), passing to the usual unit system:

$$W = 1 + {}^1\!/_2 \cos[(\mathbf{k}_1 - \mathbf{k}_2)(\mathbf{x}_3 - \mathbf{x}_4) - (\omega_1 - \omega_2)(t_3 - t_4)] \quad , \tag{13}$$

where $\omega_1 = E_1 / \hbar$ and $\omega_2 = E_2 / \hbar$ are the emission frequencies, $\mathbf{k}_1 = \mathbf{p}_1 / \hbar$ and $\mathbf{k}_2 = \mathbf{p}_2 / \hbar$ are the wave vectors. Let $\mathbf{n}_1$ and $\mathbf{n}_2$ be the unit vectors along the directions from two sources to the observation region, which can be considered here formally as a point, in accordance with the condition (10). Then

$$\mathbf{k}_1 = \frac{\omega_1}{c}\mathbf{n}_1 \quad , \qquad \mathbf{k}_2 = \frac{\omega_2}{c}\mathbf{n}_2 \quad . \tag{14}$$

Taking into account the small values of the angular distance between sources and the relatively small values of the frequency difference, the approximate relation

$$\mathbf{k}_1 - \mathbf{k}_2 = (\omega_1 - \omega_2)\mathbf{l} + \frac{\omega}{c}(\boldsymbol{\theta}_1 - \boldsymbol{\theta}_2) \tag{15}$$

is valid. Here $\omega = \dfrac{\omega_1 + \omega_2}{2}$, $\mathbf{l}$ is the unit vector along the direction from the centre of the source group (concretely, the star centre) to the observation region, $\boldsymbol{\theta}_1$ and $\boldsymbol{\theta}_2$ are the transverse vectors, which are perpendicular to the vector $\mathbf{l}$, their moduli $|\boldsymbol{\theta}_1|$ and $|\boldsymbol{\theta}_2|$ are the angles between the vectors $\mathbf{l}$ and $\mathbf{n}_1$,

$\mathbf{l}$ and $\mathbf{n}_2$, respectively (i.e., the angular distance between the centre of the star and the sources on its disk). Using Eq. (15), we can rewrite the formula (13) for the probability of coincidences (double counts) in the form:

$$W = 1 + {}^1/_2 \cos\left[(\omega_1 - \omega_2)\left(t_{34} - \frac{r_{34}^{(||)}}{c}\right) - \frac{\omega}{c}(\boldsymbol{\theta}_1 - \boldsymbol{\theta}_2)\,\mathbf{r}_{34}^{(\perp)}\right], \qquad (16)$$

where $t_{34} = t_3 - t_4$ is the difference of the time moments of registration, $r_{34}^{(||)} = \mathbf{r}_{34}\mathbf{l}$ is the longitudinal component of the vector $\mathbf{r}_{34} = \mathbf{x}_3 - \mathbf{x}_4$ and $\mathbf{r}_{34}^{(\perp)}$ is the transverse component of the same vector with respect to the direction of $\mathbf{l}$. Performing the integration of the expression (16) over the star disk with the radius $r = L\theta_0$ and the frequency interval $(\omega_0 - \Delta\omega/2, \omega_0 + \Delta\omega/2)$ at the fixed parameters of the observation region, we find that the number of double counts is proportional to

$$N_{34} = 1 + {}^1/_2\left[\frac{2J_1(\rho)}{\rho}\,\frac{\sin\left({}^1/_2\,\Delta\omega\,(r_{34}^{(||)}/c - t_{34})\right)}{{}^1/_2\,\Delta\omega\,(r_{34}^{(||)}/c - t_{34})}\right]^2, \qquad (17)$$

where J_1 is the first-order Bessel function,

$$\rho = \frac{\omega_0}{c}\,|\mathbf{r}_{34}^{(\perp)}|\,\theta_0 . \qquad (18)$$

We see that the space-time correlations of intensities in two counters of photons are determined directly by the distance between the observation points and by the difference of the registration time moments. Just these parameters are measured and fixed experimentally. The information on angular sizes of the source group (concretely, star) appears only due to the dependence of the intensity correlations on the difference of transverse momenta of photons emitted by different regions of the star surface. The frequency ω_0 can be measured independently by the methods of spectral analysis.

Thus, the comparison of correlation methods in high energy physics and astronomy shows that the use of the term "HBT-interferometry" in nuclear and particle physics (heavy ion physics) seems to be not quite adequate.

Acknowledgments

I am grateful to R.Lednicky for fruitful discussions and to V.V.Lyuboshitz for the assistance in work.

The present work has been performed under the support of Russian Foundation of Fundamental Investigations (grant No. 97-02-16699).

References

1. R. Hanbury–Brown, R.Q.Twiss. *Phil. Mag.* **45**, 663 (1954).
2. G.Goldhaber, S.Goldhaber, W.Lee, A.Pais. *Phys. Rev.* **120**, 300 (1960).
3. V.G.Grishin, G.I.Kopylov, M.I.Podgoretsky. *Yad. Fiz.* **13**, 1116 (1971) (*Sov. J. Nucl. Phys* **13**, 638 (1971)).
4. G.I.Kopylov, M.I.Podgoretsky. *Yad. Fiz.* **15**, 392 (1972) (*Sov. J. Nucl. Phys* **15**, 219 (1972)).
5. G.I.Kopylov, M.I.Podgoretsky. *Yad. Fiz.* **18**, 656 (1974) (*Sov. J. Nucl. Phys* **18**, 336 (1974)).
6. G.I.Kopylov. *Phys. Lett.* B **50**, 472 (1974).
7. G.I.Kopylov, M.I.Podgoretsky. *Zh. Exp. Teor. Fiz.* **69**, 414 (1975) (*Sov. Phys. JETP* **42**, 211 (1975)).
8. M.I.Podgoretsky. *Fiz. Elem. Chast. At. Yadra* **20,** 628 (1989) (*Sov. J. Part. Nucl.* **20**, 266 (1989)).
9. R. Lednicky, V.L.Lyuboshitz. *Yad. Fiz.* **35**, 1316 (1982) (*Sov. J. Nucl. Phys* **35**, 770 (1982)).
10. V.L.Lyuboshitz. *Yad. Fiz.* **48**, 1501 (1988) (*Sov. J. Nucl. Phys* **48**, 956 (1988)).
11. V.L.Lyuboshitz. *Yad. Fiz.* **53**, 823 (1991) (*Sov. J. Nucl. Phys* **53**, 514 (1991)
12. V.L.Lyuboshitz, M.I.Podgoretsky. *Yad. Fiz.* **58**, 33 (1995) (*Phys. At. Nucl.* **58**, 30 (1995)).

Multiple Sources for Meson– and Baryon–Radiation in High Energy Nucleus–Nucleus Collisions

Sven Soff[a,b], S. A. Bass[c], S. Schramm[b], M. Belkacem[a,b], M. Bleicher[a], M. Brandstetter[a], L. Bravina[a], C. Ernst[a], L. Gerland[a], C. Spieles[d], H. Weber[a], H. Stöcker[a], W. Greiner[a].

[a] *Institut für Theoretische Physik, Johann Wolfgang Goethe–Universität Frankfurt, Postfach 111932, 60054 Frankfurt am Main, Germany.*

[b] *Gesellschaft für Schwerionenforschung Darmstadt, Postfach 110552, 64220 Darmstadt, Germany.*

[c] *Department of Physics, Duke University, Durham, NC 27708-0305, USA.*

[d] *Nuclear Science Division, Lawrence Berkeley National Laboratory, Berkeley, CA 94720, USA.*

The source characteristics of different particle species (pions, kaons, lambdas, ...) are investigated in central Pb (158 A GeV) Pb collisions within the microscopic Ultrarelativistic–Quantum–Molecular–Dynamics (UrQMD) model. The different feeding contributions (resonances, strings) to the final particle yields are discussed. The transverse radius and time distributions of the last strong interaction points (freeze-out) depend strongly on the feeding mechanism for a given particle species. This multisource behaviour can be studied in the correlation functions of identical particle pairs.

1 Introduction

The investigation of excited nuclear matter by means of heavy ion collisions addresses questions as the (nuclear) equation of state or phase transitions between different forms of nuclear matter.

At relativistic incident energies experiments have been performed at the SPS accelerator at Cern with the massive system Pb (158 A GeV) Pb. The space-time evolution of the system plays an essential role for the interpretation of the observables and the applicability of thermodynamic concepts in heavy ion collisions.

In contrast, non-equilibrium microscopic transport models offer the unique possibility to study the full space-time evolution of the collision system without making assumptions on thermalization or emission profiles. Additionally, they also allow for an analysis of the various processes and their relative importance in such complex reactions.

In order to analyse the predicted source characteristics several questions are listed in the following. The first question is how many particles does the source emit ? Secondly, which of the various reaction channels do determine the final state ? How important are the various particles incorporated in the microscopic model and, how do the freeze-out dynamics look like ? The last generation sources are separated, i.e. the resonance types and strings in the UrQMD model, and strong differences are observed among them.

Finally, consequences for the resulting two-particle correlation functions of identical particle pairs and the HBT-radius parameters are discussed.

2 UrQMD model

The UrQMD model is based on analogous principles as (Relativistic) Quantum Molecular Dynamics. Hadrons are represented by Gaussians in phase space and a nucleus A is initialized by distributing nucleons in a sphere of radius $R = 1.12A^{1/3}$ fm. Momenta are chosen according to a non-interacting Fermi-gas ansatz. Hadrons are then propagated according to Hamilton's equations of motion.

The microscopic evolution of the hadrochemistry in heavy-ion reactions requires the solution of a set of hundreds of coupled (Boltzmann-type) integro-differential equations. This means that all (known) hadrons need to be included into the model as realistically as possible. The collision term of the UrQMD model treats 55 different isospin (T) degenerate baryon (B) species (including nucleon-, delta- and hyperon- resonances with masses up to 2.25 GeV) and 32 different T-degenerate meson (M) species, including (strange) meson resonances as well as the corresponding anti-particles, i.e. full particle/antiparticle symmetry is incorporated in the model. The number of implemented baryons therefore defines the number of antibaryons in the model and the antibaryon-antibaryon interaction is defined via the baryon-baryon interaction cross sections. Isospin is explicitly treated (although the SU(2) multiplets are assumed to be degenerate in mass). The baryons and baryon-resonances which can be populated in UrQMD are listed in table 1, the respective mesons in table 2. The states listed can either be produced in string decays, s-channel collisions or resonance decays. For excitations with masses > 2 GeV (B) and 1.5 GeV (M) a string model is used. All (anti-)particle states can be produced – in accordance with the conservation laws – both, in the string decays as well as in s-channel collisions or in resonance decays.

Tabulated or parametrized experimental cross sections are used when available. Resonance absorption and scattering is handled via the principle of detailed balance. If no experimental information is available, the cross section is either calculated via an OBE model or via a modified additive quark model, which takes basic phase space properties into account. The baryon-antibaryon annihilation cross section is parameterized as the proton-antiproton annihilation cross section and then rescaled to equivalent relative momenta in the

N_{938}	Δ_{1232}	Λ_{1116}	Σ_{1192}	Ξ_{1315}	Ω_{1672}
N_{1440}	Δ_{1600}	Λ_{1405}	Σ_{1385}	Ξ_{1530}	
N_{1520}	Δ_{1620}	Λ_{1520}	Σ_{1660}	Ξ_{1690}	
N_{1535}	Δ_{1700}	Λ_{1600}	Σ_{1670}	Ξ_{1820}	
N_{1650}	Δ_{1900}	Λ_{1670}	Σ_{1750}	Ξ_{1950}	
N_{1675}	Δ_{1905}	Λ_{1690}	Σ_{1775}	Ξ_{2030}	
N_{1680}	Δ_{1910}	Λ_{1800}	Σ_{1915}		
N_{1700}	Δ_{1920}	Λ_{1810}	Σ_{1940}		
N_{1710}	Δ_{1930}	Λ_{1820}	Σ_{2030}		
N_{1720}	Δ_{1950}	Λ_{1830}			
N_{1900}		Λ_{1890}			
N_{1990}		Λ_{2100}			
N_{2080}		Λ_{2110}			
N_{2190}					
N_{2200}					
N_{2250}					

Table 1: Baryons and baryon-resonances included into the UrQMD model. Through baryon-antibaryon symmetry the respective antibaryon states are included as well.

incoming channel. For a detailed overview of the elementary cross sections and string excitation scheme included in the UrQMD model, see ref. [1].

The UrQMD model allows for systematic studies of heavy-ion collisions over a wide range of energies in a unique way [1,2,3,4]: the basic concepts and the physics input used in the calculation are the same for all energies. A relativistic cascade is applicable over the entire range of energies from 100 MeV/nucleon up to 200 GeV/nucleon. However, UrQMD can also perform infinite matter calculations by evolving the initial state in a box with periodic boundary conditions. Thus, the equilibrium limit of the UrQMD transport model may be investigated in a unique fashion [3].

0^{-+}	1^{--}	0^{++}	1^{++}	1^{+-}	2^{++}	$(1^{--})^*$	$(1^{--})^{**}$
π	ρ	a_0	a_1	b_1	a_2	ρ_{1450}	ρ_{1700}
K	K^*	K_0^*	K_1^*	K_1	K_2^*	K_{1410}^*	K_{1680}^*
η	ω	f_0	f_1	h_1	f_2	ω_{1420}	ω_{1662}
η'	ϕ	f_0^*	f_1'	h_1'	f_2'	ϕ_{1680}	ϕ_{1900}

Table 2: Mesons and meson-resonances, sorted with respect to spin and parity, included into the UrQMD model.

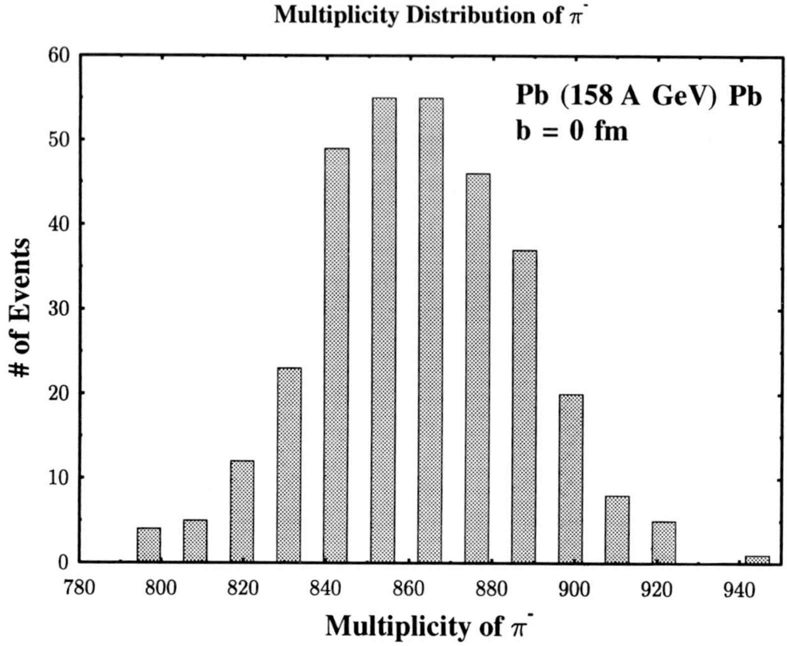

Figure 1: Multiplicity distribution of π^- as obtained from 320 collisions of two lead nuclei at 158 A GeV and zero impact parameter as calculated with UrQMD.

3 Microscopic Investigation of the Particle Emitting Source

The collision system Pb+Pb is studied at an incident energy of 158 A GeV and zero impact parameter $b = 0$ fm.

The first question is: How many particles does the predicted source emit? Fig. 1 shows the multiplicity distribution $1/N_{\mathrm{tot}}\, dN_{\mathrm{events}}/dM$ of negative pions π^- for an event sample of $N_{\mathrm{tot}} = 320$ events. The mean value for the number of emitted π^- per event is $\langle M_{\pi^-}\rangle \approx 860$. Nevertheless, fluctuations from this mean value are obvious. The shape of the distribution can be fitted by a Gaussian (squares) with a width of $\sigma \approx 27$.

Where do the final observed pions come from, i.e. what is the last interaction they have encountered before they were produced as asymptotically free particles? In the microscopic description the space-time evolution of the collision is traced. Newly produced particles either stem from resonance decays, i.e. from baryon or meson resonances, or they are created out of longitudinally

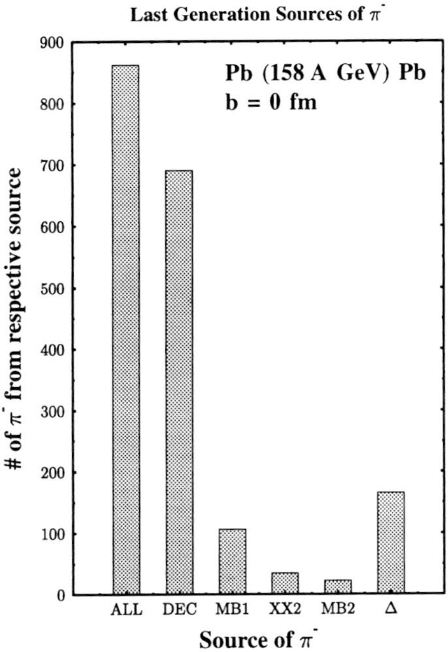

Figure 2: Rough classification of the last generation sources of π^-'s. Baryon and meson resonance decays (DEC) and string decays (MB1,XX2,MB2) whith emission of pions are displayed as the different bars. The string part is divided with respect to the creation channel of the string. Meson-baryon collisions producing pion emitting strings (MB1), baryon baryon or meson meson collisions (XX2) and meson baryon collisions (MB2) which lead to two strings. Clearly, the total pion yield is dominated by resonance decays. The right bar (Δ) are Δ_{1232}-decays and are also contained in the (DEC) bar.

excited parton states, i.e. from strings. Additionally, particles may also scatter elastically.

How many pions are radiated from the various channels? Fig. 2 shows the origin of π^-'s. The contributions of the different last generation sources to the pions after freeze-out are plotted. Elastic collisions are not taken into account when tracing back the individual history of one pion to the last decay (from a resonance or string) and before it became a free particle. However, only 5% of the pions scatter elastically after this last decay.

Which of the many incorporated resonance-species are most relevant to the final decay channels which emit pions in such a complex heavy ion collision

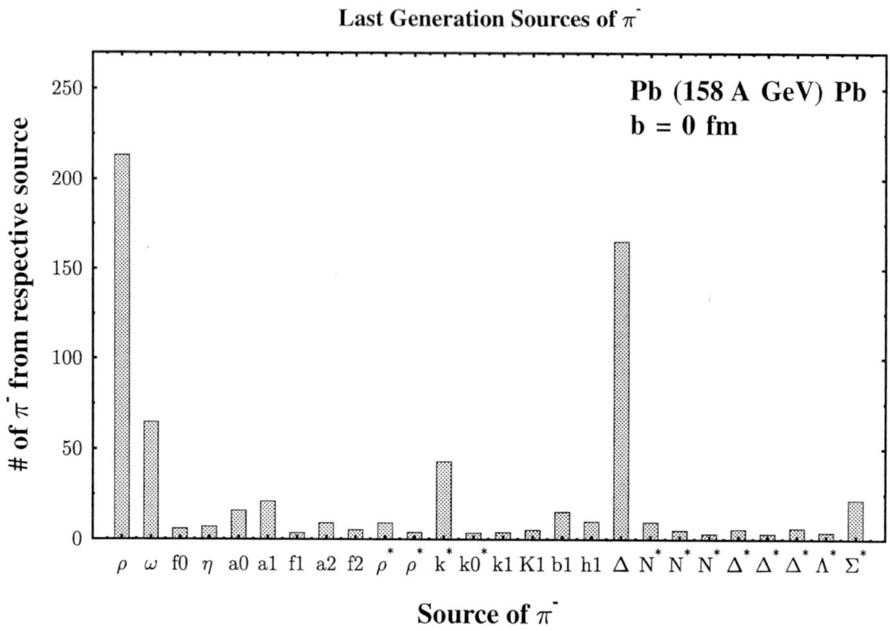

Figure 3: Different resonance contributions to the frozen-out pions. Most important are ρ-, Δ-, ω-, K^*-decays leading to π-emission. Various other sources are displayed as well (e.g. higher mass Δ^*, N^* or ρ^* resonances with increasing mass from the left to the right)

system? To investigate this point we study the 80% contribution from decay processes (Fig. 2) in more detail. The different resonance sources for pions are shown in Fig. 3. The largest contribution is given by the ρ meson ($\approx 25\%$ of all pions come from ρ's), followed by the Δ_{1232} ($\approx 20\%$), the ω ($\approx 7\%$) and the k^*_{892} meson ($\approx 5\%$).

The rather large ρ-contribution to the final pion yield is due to the various processes producing ρ's. The analysis of these ρ–sources as a function of time shows that at early times ($t_{\rm cms} \leq 5\,{\rm fm/c}$) high energetic baryon-baryon– and meson-meson–collisions which produce strings are an important ρ–source. Approximately 50% of the initial ρ's are produced in such collisions. Strings, produced in high energetic meson-baryon collisions, represent the other 50% of the early ρ–source. The process $\pi^+\pi^- \to \rho$ increases approximately linearly as a function of time and reaches a plateau at $t_{\rm cms} \approx 7\,{\rm fm/c}$ where this ρ–source is equally important as ρ's emitted from higher mass resonances (N^*'s,

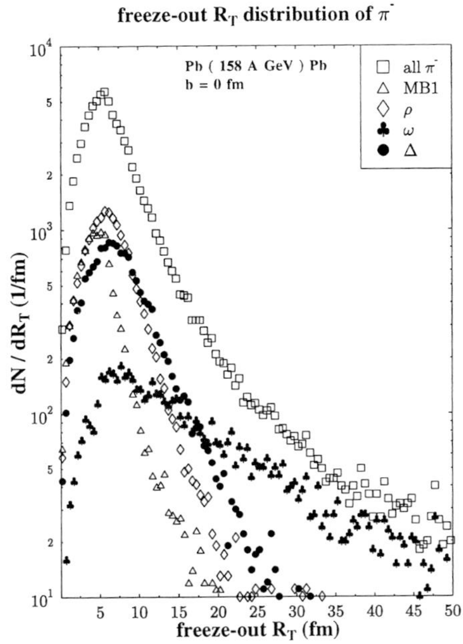

Figure 4: Transverse freeze-out radius distributions dN_{π^-}/dR_T of π^- in the collision of Pb (158 A GeV) Pb at $b = 0$ fm. Shown are the total integrated distribution (squares) and the contributions from ρ- (triangle), Δ_{1232}- (star), ω- (diamond), and (MB1-)string-decays.

ρ^*'s, ...) and ρ's from string decays (MB1). For $t \geq 12$ fm/c the $\pi^+\pi^- \to \rho$ channel dominates the ρ production. So it is obvious that a major fraction of the ρ source is due to pions which were produced in earlier stages of the reaction and rescatter in this meson-dominated environment, where the baryon to meson ratio is of the order of 10%.

Δ_{1232} resonance decays generate nearly as many pions as the decay of the ρ meson. Here, the pion-nucleon-delta cycle plays a major role as is already known from the 1 GeV/nucleon [5] or 10 GeV/nucleon [6] energy regime. Delta-resonance production via pion-nucleon and nucleon-nucleon scattering, reabsorption and elastic scattering allow for delta resonance matter creation [6].

Due to different lifetimes and reaction cross sections of ρ's and Δ's the question arises if the freeze-out emission pattern of the respective pions shows also characteristic differences. This point will be investigated later.

The ω-meson decays dominantly ($\approx 89\%$) into three pions ($\omega \to \pi^+\pi^-\pi^0$)

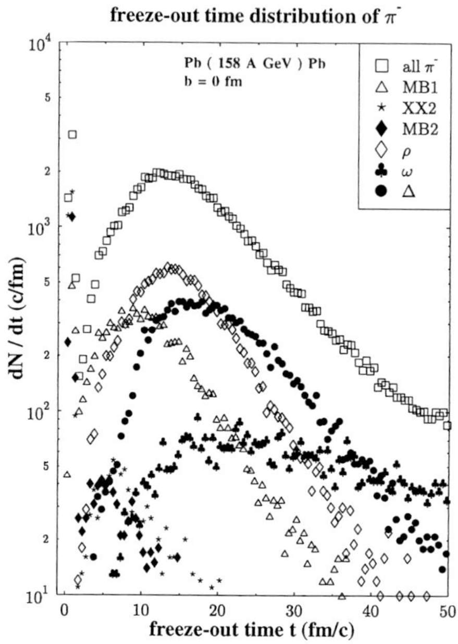

Figure 5: Freeze-out time dN/dt_f-distribution of π^-'s in Pb (158 A GeV) at $b = 0$ fm as calculated with UrQMD1.0. Depicted is the total distribution and single contributions from Δ-, ρ-, ω- and various string-decays (MB1,XX2,MB2).

and contributes about 7% to the total pion yield. Decays of strange resonances contribute to the total pion yield mainly through the $K^*(892)$ resonance ($K^*(892) \to K\pi$) ($\approx 5\%$) and the Σ^* resonance ($\Sigma^* \to \Lambda\pi$ or $\Sigma^* \to \Sigma\pi$) ($\approx 2.5\%$). The model offers ≈ 100 other possible origins to the pion production, e.g. higher mass nucleon and delta resonances ($\Delta^*, N^*, \ldots,$) and other mesonic resonances. Each of them contributes to less than 2% to the total pion yield but cannot be neglected when summed up.

Fig. 4 shows the predicted transverse freeze-out radius distribution dN_{π^-}/dR_T of π^-'s at midrapidity, $-1 \leq y \leq 1$.

Clearly, the total distribution dN_{π^-}/dR_T is rather broad. Although the most probable value is located around 5.5 fm, the distribution drops rather slowly with transverse radius, with a decay constant $\lambda \approx 1/3.8$ fm for 6 fm $\leq R_T \leq 15$ fm and $\lambda \approx 1/10.5$ fm for 25 fm $\leq R_T \leq 35$ fm when the distribution is approximated by $dN_{\pi^-}/dR_T = N_0 \exp(-\lambda R_T)$ in the respective R_T interval.

The profile is not compatible with a Gaussian source. The fall-off resembles an exponential or even weaker fall off.

Do the various sources of pions show different transverse freeze-out distributions? The ρ source of pions, i.e. the freeze-out distribution for pions emitted from ρ's, is narrower than the total distribution which is obtained if one integrates over all sources which was discussed before. The centroid of the Δ source of pions is located at a larger transverse radius than that of the ρ source and the distribution is also broader for the Δ source.

On the other hand, the $\pi(\omega)$-distribution is rather flat. The ω source dominates the total distribution for large radii ($\geq$ 25fm) due to the long lifetime of the ω ($\tau \approx 25\,\mathrm{fm/c}$).

The R_T-distribution for pions originating from strings is much narrower and the centroids are located closer to the beam axis. The R_T-distribution for π^-'s originating from strings which where produced in meson-baryon collisions is in Fig. 6 indicated by (MB1). It peaks around $R_T \approx 5\,\mathrm{fm}$ and the emissivity has already dropped to one-tenth of the maximum emissivity at $R_T \approx 10\,\mathrm{fm}$. For $R_T \geq 20\,\mathrm{fm}$ this string source of pions can be neglected. So it is evident that the various contributions to the total π^- transverse radius distribution show completely different freeze-out distributions. For these sub-sources an average freeze-out order $\bar{R_T}_\pi(string) < \bar{R_T}_\pi(\rho) < \bar{R_T}_\pi(\Delta) < \bar{R_T}_\pi(\omega)$ is found.

Qualitatively this behaviour is also reflected in the freeze-out time (t_f) distributions dN_{π^-}/dt_f shown in Fig.5. Here t_f is the c.m.s freeze-out time. The emission of the midrapidity pions is peaked at $t_f \approx 12 - 15\,\mathrm{fm/c}$ and the width is characterized by a FWHM of $\Delta t_{f,FWHM} \approx 17\,\mathrm{fm/c}$, thus showing a rather long emission duration. The total t_f-distribution can be decomposed into ≈ 100 sources, out of which, the string and the ρ, ω, Δ are most important. A peak is visible in the total dN_{π^-}/dt_f-distribution at the very beginning of the collision ($t_f \leq 2\,\mathrm{fm/c}$) which is mainly due to first nucleon nucleon collisions creating (baryonic) strings which then promptly emit pions that do not rescatter, but directly enter the detector. Such pions correspond to the output of all cascade models which neglect rescattering.

The freeze-out time t_f-distributions $dN_{\pi(\rho)}/dt_f$ show that π's stemming from ρ's freeze out earlier (corresponding to the observed smaller freeze-out radii R_T), whereas π's from Δ's freeze out later and at larger freeze-out radii R_T. Again π's from the ω-meson determine the total distribution at late freeze-out times which is clear from the relatively long lifetime of the ω. So the sub-sources also fullfill a average time-like freeze-out order $\bar{t}_\pi(string) < \bar{t}_\pi(\rho) < \bar{t}_\pi(\Delta) < \bar{t}_\pi(\omega)$.

The dependence of the freeze-out distribution dN/dR_T on the particle species (π^-'s, $K^\pm$'s, p's and Λ's) is compared in Fig.6. The peak of the freeze-

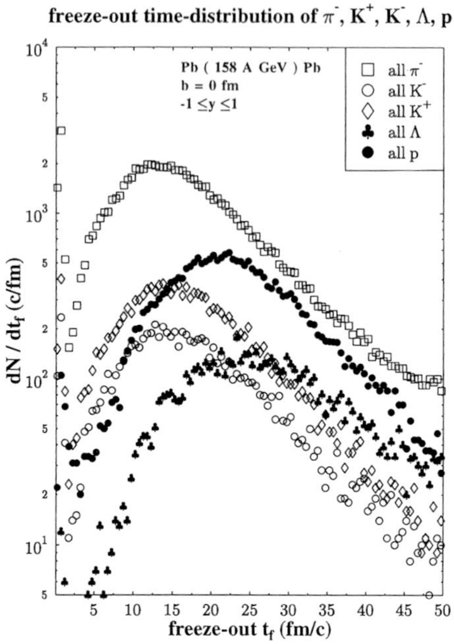

Figure 6: Freeze-out time distributions dN/dt_f of π^-'s, $K^{\pm}$'s, p's and Λ's in the collision of Pb (158 A GeV) Pb at $b = 0$ fm.

out distributions of mesons is approximately located at 5 fm, thus indicating similar freeze-out radii for pions and kaons. The Λ's and p's freeze out at larger transverse radii, peaking both around ≈ 8 fm. Accordingly, also the freeze-out time distributions, depicted in Fig. 9, show later times for the maximum emissivity of Λ's and p's ($\approx 20-25$ fm/c) compared to π's and K's ($\approx 12-15$ fm/c). Thus, a different freeze out scenario for baryons and mesons is observed.

All distributions exhibit non-Gaussian shapes and have rather long tails at larger transverse freeze-out radii and freeze-out times.

The question arises whether one can get access to this complicated space-time-structure of such a multi-source behaviour, e.g. by means of many-particle quantum correlations. Therefore, we have calculated the correlation function C_2 of identical particle pairs as a function of the relative four-momentum

$$\Delta q = Q_{\text{inv}} = \sqrt{(E_1 - E_2)^2 - (\vec{p}_1 - \vec{p}_2)^2}.$$

In fig. 7 the $\pi\pi$ correlation function is plotted (open squares). Addition-

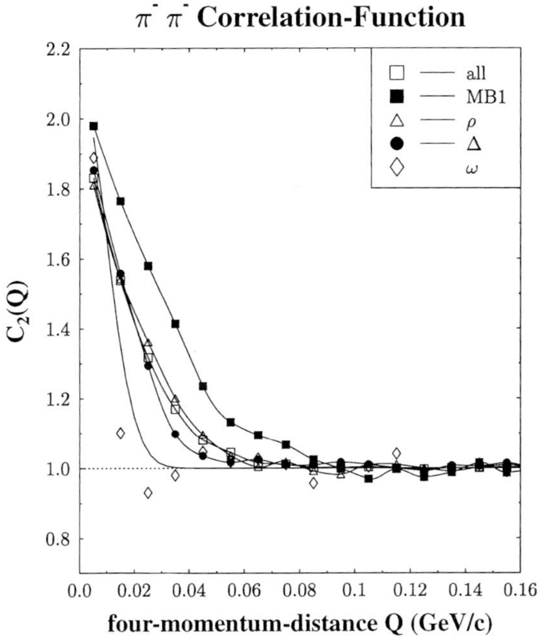

Figure 7: $\pi^-\pi^-$-correlation function for Pb (158 A GeV) collisions at $b = 0$ fm as calculated with UrQMD1.0. There are dipicted the full correlation function, which is calculated from all π^--pairs and the "spectral" components, i.e. the correlation functions, which are constructed from π^-'s which origin from Δ-, ρ-, ω- or string-decays, respectively.

ally the correlation functions for π^-'s separated for their sources are drawn as well. This means for each of these curves only pions which stem from a particular source have been chosen. In other words we add to source function $S_{\pi^-}(x_1, x_2, p_1, p_2)$ an index which depicts the last generation source of the pions, thus having $S_{\pi^-(\rho)}$, $S_{\pi^-(\omega)}$, $S_{\pi^-(\Delta)}$, $S_{\pi^-(k^*)}$, $S_{\pi^-(\text{string})}, \ldots$. Clearly the correlation functions C_2 show different widths depending on which source you look at. The correlation function $C_2(\pi^-(\rho))$ has a smaller width than the overall integrated one $C_2(\text{all}\pi^-)$ whose width is again smaller than the width of $C_2(\pi^-(\Delta))$. The two extreme cases are $C_2(\pi^-(\omega))$ exhibiting a rather narrow width and $C_2(\pi^-(\text{string}))$ showing a broad width. Consequently, the observations from fig.4 are reflected in the two-particle correlation function, where the inverse of the width is proportional to the radius. Experimentally it will be the (difficult) task to disentangle these different components of this - in the first step theoretical - correlation function e.g. by reconstruction of the resonances by means of invariant mass spectra. So one could be able to compare $C_2(\mathrm{all}\pi^-)$ with $C_2(\pi^-(\Delta))$ or $C_2(\pi^-(\omega))$ and get clear indications for the multisource-behaviour.

4 Conclusions

The source characteristics of the collision system Pb (158 A GeV) Pb at $b = 0$ fm were investigated with the microscopic transport model UrQMD1.0.

The multiplicity distributions for π's at fixed impact parameter show a broad symmetric distribution which is compatible with a Gaussian. The final π yield is dominated (80%) by resonance contributions, where the ρ, Δ, ω and K^* are the most important sub-sources among several others. Only 20% of the π's are directly emitted from strings.

The freeze-out distributions of pions differ with respect to their last-generation sources. The average freeze-out times for the various decay channels are ordered like follows: $\bar{t}_\pi(string) < \bar{t}_\pi(\rho) < \bar{t}_\pi(\Delta) < \bar{t}_\pi(\omega)$. Pions emitted from strings freeze out earlier (and at smaller transverse radii, on the average) than pions emitted from ρ's followed by pions from Δ's and finnaly pions from ω's. The $\pi(\omega)$ contribution dominates the late and outer freeze-out region.

The overall integrated freeze-out distributions exhibit rather broad distributions in time as well as in transverse radius with long tails at large radii and times. The shape is not compatible with a Gaussian source, even an exponential fall-off seems to be too strong. The peak values for the produced mesons (π's and K's) are located at the same point in time and transverse radius whereas the baryons Λ's and p's freeze out later and more further outside.

This multisource behaviour, the complex substructure of the emitting particle source, opens up a more detailed view into the dynamics and space-time evolution of heavy-ion collisions. When concentrating on different source contributions, which may be possible by reconstruction via invariant-mass pair distributions, different space-time regions are scanned.

If only pion pairs from one specific contribution of the multisource are used to construct the two-pion correlation function these were shown to differ completley for the different contributions. Thus, the different freeze-out distributions are reflected in the correlation functions, e.g. their widths, as well. This opens the possibility for a new kind of spectroscopy when analysing the particle-emitting source.

Acknowledgments

This work was supported by DFG, BMBF, Graduiertenkolleg "Theoretische und Experimentelle Schwerionenphysik", J. Buchmann Stiftung and A. v. Humboldt Stiftung.

References

1. S. A. Bass, M. Belkacem, M. Bleicher, M. Brandstetter, C. Ernst, L. Gerland, M. Hofmann, S. Hofmann, J. Konopka, G. Mao, L. Neise, S. Soff, C. Spieles, H. Weber, L. A. Winckelmann, H. Stöcker, W. Greiner, C. Hartnack, J. Aichelin, N. Amelin, Prog. Nucl. Part. Phys. **41** (1998).
2. S. A. Bass, M. Belkacem, M. Brandstetter, M. Bleicher, L. Gerland, J. Konopka, L. Neise, C. Spieles, S. Soff, H. Weber, H. Stöcker, W. Greiner. Phys. Rev. Lett. in press (1997).
3. M. Belkacem, M. Brandstetter, S. A. Bass, M. Bleicher, L. Bravina, M. I. Gorenstein, J. Konopka, L. Neise, C. Spieles, S. Soff, H. Weber, H. Stöcker, W. Greiner. Phys. Rev. C **58** (1998) 1727.
4. L. V. Bravina, M. I. Gorenstein, M. Belkacem, S. A. Bass, M. Bleicher, M. Brandstetter, M. Hofmann, S. Soff, C. Spieles, H. Weber, H. Stöcker, W. Greiner, Phys. Lett. **B** in press (1998).
5. S. A. Bass, M. Hofmann, C. Hartnack, H. Stöcker, W. Greiner, Phys. Lett. **B335**, 289 (1994).
6. M. Hofmann, R. Mattiello, H. Sorge, H. Stöcker, W. Greiner, Phys. Rev. **C51**, 2095 (1995).
7. S. Soff, D. Ardouin, C. Spieles, S. A. Bass, H. Stöcker, D. Gourio, S. Schramm, C. Greiner, R. Lednicky, V. L. Lyuboshitz, J.-P. Coffin, C. Kuhn. J. Phys. G: Nucl. Part. Phys. **23** (1997) 2095.

GIANT RESONANCES AND LOW-MASS (e^+e^-) PAIR PRODUCTION IN RELATIVISTIC HEAVY ION COLLISIONS.

R. ALZETTA, T. BUBBA, R. LE PERA, G. LIBERTI [a] and D. TARANTINO
Dipartimento di Fisica, Università della Calabria.
INFN, Cosenza, Italia.
e-mail:Alzetta@fis.unical.it

G. PREPARATA
Dipartimento di Fisica, Università di Milano.
INFN, Sezione di Milano, Italia.
e-mail:Preparata@mi.infn.it

We show that the coherent fluctuations of the e.m. field induced by the collective quantum fluctuations of the nuclear matter field associated to the giant resonances, with frequencies $\omega_A = 78A^{-\frac{1}{3}}$ MeV give rise to a significant e^+e^- pair production in high energy Heavy Ion collisions. The parameterless calculation of such yield is in good agreement with recent experimental observations.

1 Introduction

Motivated by the search of a new state of hadronic matter, the Quark-Gluon-Plasma (QGP) predicted by some lattice Montecarlo studies [1], an impressive research program has been recently launched in the field of Heavy-Ion collisions at energies of a few hundred GeV per nucleon [b]. In particular three experiments has been conducted at the CERN SpS in the last few years [3,4,5] to measure the yield of low mass (M<2 GeV) e^+e^- pairs in such collisions in search for some tipical signal of the QGP. In Fig.1 we report a typical observation of such experiments, showing a large yield of e^+e^- pairs in the mass region (0.2<M<1 GeV), well in excess of that predicted as arising from conventional sources. Furthermore in experiments with a high energy proton beam impinging on a heavy nucleus [2] no such excess has been observed, suggesting that the source of the excess pairs lies in some peculiar aspect of the nucleus-nucleus interaction. A wide consensus has formed so far around the hypothesis that the e^+e^- excess is due to some special manifestation of the QGP, and the important question is thus which peculiar QGP feature allows this hypothetical new state of matter to reveal its footprints in the form of abundant low mass e^+e^- pairs. As shown in Ref.2, the situation is quite subtle and complicated, and the theory that appear now generally accepted [6] requires not only the production

[a] INFM, Unità di Cosenza, Italia.
[b] For a recent review consult Ref.2

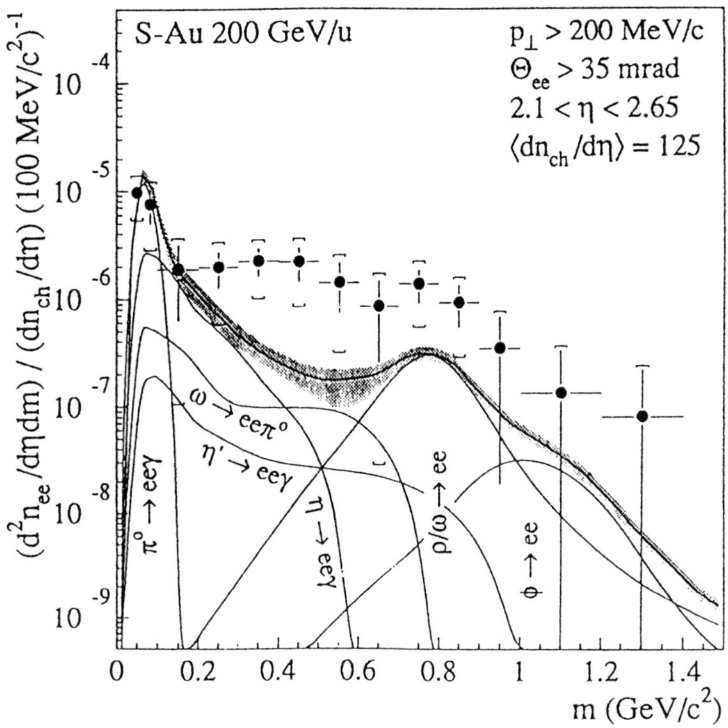

Figure 1: *Inclusive e^+e^- mass spectra in 200 GeV/u S-Au collisions (CERES, Ref.3).*

of an excess of vector mesons (ρ, ω, ϕ) over and above the normal yield in high energy pp-collisions, but also a substantial lowering of their masses, as the result of their propagation in a hot dense medium such as the QGP. And it is claimed[6], that simulations with relativistic transport models do reproduce the experimental findings. While we have no technical objection to the deductions and simulations of Li, Ko and Brown[6], we believe that the very notion of QGP, on which this approach is based, is very arduous and difficult, and one that is quite unlikely to occur, due to the peculiar features of particle formation in high energy nucleon-nucleon scattering [c]. It is for this reason that we have

[c] A space time analysis[7] of the particle formation in high energy inelastic events shows that most hadrons form well outside the interaction region. This surprising fact was first realized in the seventies, both theoretically and experimentally, when the fallacy of equating a nucleus

searched for a simpler, more realistic explanation of these interesting findings in a completely different direction, in the electromagnetic collective properties of a heavy nucleus, that are well known since a long time in the thoroughly studied phenomena related to the "giant resonances", with frequency (A is the atomic number) [8]

$$\omega(A) \simeq 78A^{-\frac{1}{3}}MeV \tag{1}$$

2 THE WEAK COHERENT E.M. FIELD OF THE GIANT RESONANCES.

It follows from QED [d] that around a nucleus, where a number of charged nucleons oscillate in phase with the frequency $\omega(A)$, there lingers a coherent (classical) electromagnetic field, which in the rest frame has the form (we use the units where $\hbar = c = 1$)

$$\vec{A}(\vec{x},t) = e^{-i\omega_A t}\frac{1}{(2\omega_A V_{CD})^{\frac{1}{2}}}\frac{8\pi}{3\sqrt{2}}\frac{\sin\omega_A r}{\omega_A r}\hat{e} + c.c. \tag{2}$$

where V_{CD} is the coherence domain (CD) of the mode with frequency ω_A:

$$V_{CD} = \frac{4\pi}{3}R_{CD}^3 \tag{3}$$

$$R_{CD} = \frac{\pi}{\omega_A} \tag{4}$$

In the case of S ($A_1 = 32$) and Au ($A_2 = 197$) we have $\omega_{A1} = \omega_1 = 23MeV$ and $\omega_{A2} = \omega_2 = 13MeV$, and $R_{CD1} = 27fm$, $R_{CD2} = 48fm$. The unit vector $\hat{e}$ obeys the transversality conditions :

$$\vec{x}\cdot\hat{e} = 0 \tag{5}$$

and

$$\vec{\nabla}\cdot\hat{e} \tag{6}$$

The physical origin and meaning of Eq (2) is that the 8π-independent modes of the e.m. field of frequency ω_A in the CD, through the interaction with the e.m. current generated by the giant resonance's fluctuations, get "aligned" in a coherent superposition, such as Eq.2, and in so doing they minimize the total

---

to a "very dense bubble chamber" was clearly exposed.

[d] In Ref.9 one can find a complete discussion, based on the fundamental QED field equations, of the existence and features of the solutions analyzed in this section.

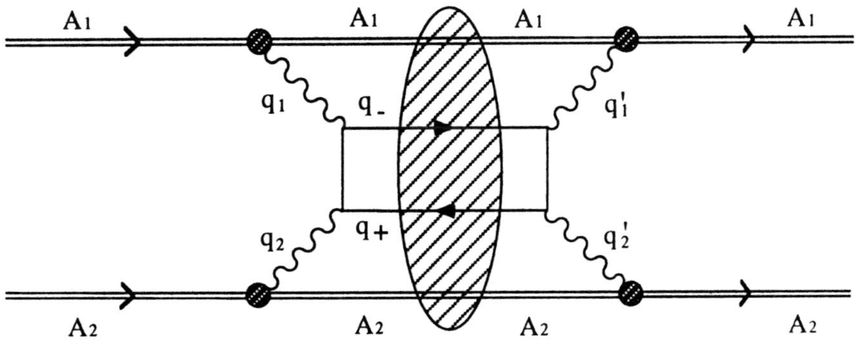

Figure 2: *Feynman graph for the complete process in the Heavy Ion collision.*

energy of the coupled matter-e.m. field system. As a result we can picture the nucleus A as surrounded by a "photon cloud" oscillating at the frequency of the giant resonance ω_A with a well defined amplitude (see Eq.(2)).
It should be clear that the interaction of the "photon clouds" of the Heavy Ions when they are brought to overlap at high energy can induce, in the strong electric fields that get created in the collision, the production of e^+e^- pairs of a well defined character, which we are now going to study.

3 THE e^+e^- PAIR PRODUCTION PROCESS.

The process we wish to calculate is depicted in Fig.2, where the Heavy Ions A_1 and A_2 surrounded by their "photon clouds" described by $A_\mu(q_1)$ and $A_\mu(q_2)$, the Fourier transforms of the boosted e.m. fields of Eq.(2), collide inelastically, while their "photon clouds" produce an e^+e^- pair through the well-known QED mechanism. From the factorizable structure of the diagram in Fig.2 we can write

$$dN_{(e^+e^-+hadrons)} = dN_{e^+e^-} \cdot dN_{hadrons} \tag{7}$$

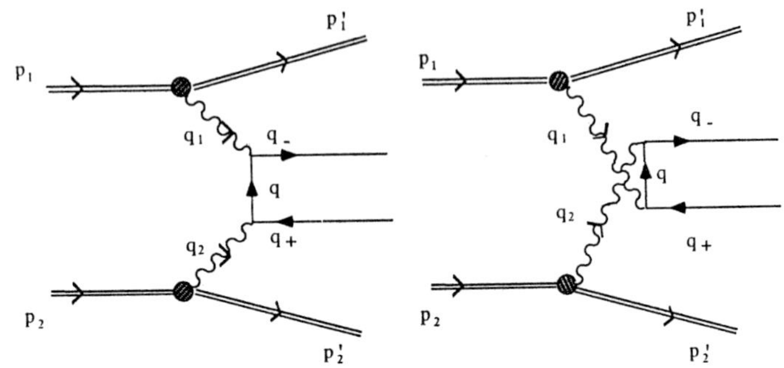

Figure 3: *Feynman diagram for the e^+e^- pair QED photoproduction mechanism.*

where the e^+e^- yield $dN_{e^+e^-}$ is calculated by squaring the amplitudes associated to the diagrams of Fig.3, which can be written as

$$S = \int \frac{d^4q_1}{(2\pi)^4}\frac{d^4q_2}{(2\pi)^4}(2\pi^4)\delta(q_1 + q_2 - q_- - q_+)A_{1\mu}(q_1)A_{2\nu}(q_2)T^{\mu\nu} \tag{8}$$

where $T^{\mu\nu}$ represents the well known leptonic part of the Feynman amplitude. A tedious, but straightforward calculation, to be reported elsewhere[10], through a well defined set of approximations allows us to write:

$$dN_{e^+e^-} = \int \frac{d^4q_1}{(2\pi)^4}\frac{d^4q_2}{(2\pi)^4}\frac{6\omega_1^2\omega_2^2}{\pi(2\pi)^4}|a(q_1)|^2\ |a(q_2)|^2 * 2M^2 * \sigma_{\gamma\gamma\to e^+e^-}(M) \tag{9}$$

where $a(q_1)$ and $a(q_2)$ are the 4-dimensional Fourier transforms of the boosted "photon clouds". The structure of Eq.(9) is particularly transparent, for it shows that the e^+e^- pair production is the result of the fusion of the photon fields "aligned" by the giant resonances of the colliding Heavy Ions. By inserting the appropriate Fourier transforms $a(q_1)$ and $a(q_2)$ into Eq.(9), we can straightforwardly compute[10] (η is the pseudorapidity variable) the normalized differential spectrum

$$\frac{\{\frac{dN_{e^+e^-}}{dM\,d\eta}\}}{\{\frac{dN_{ch}}{dM\,d\eta}\}} \tag{10}$$

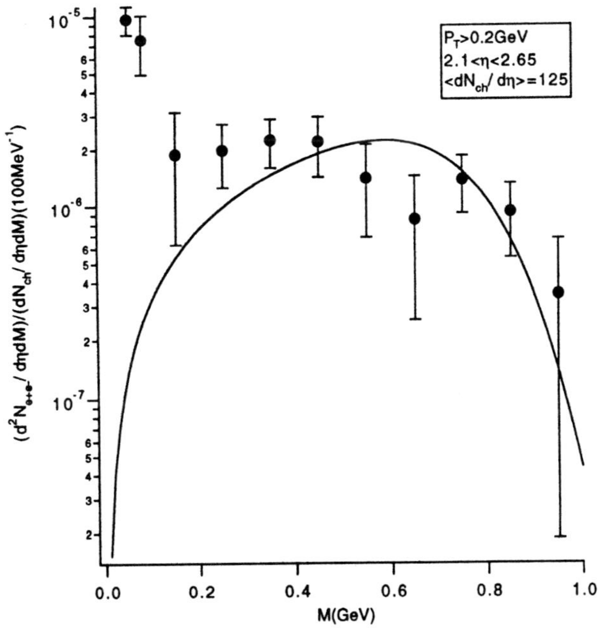

Figure 4: *Normalized differential spectrum of the photoproduced e^+e^- pairs calculated in our work and compared with the experimental data of Ref.3 .*

of the photoproduced e^+e^- pairs, which we compare with the experimental data in Fig.4 and Fig.5.
In Fig.6 we plot a more telling experimental quantity, namely:

$$\epsilon = \frac{\frac{dN_{e^+e^-}}{dM\,d\eta}|_{exp} - \frac{dN_{e^+e^-}}{dM\,d\eta}|_{known\ processes}}{\frac{dN_{e^+e^-}}{dM\,d\eta}|_{known\ processes}} \tag{11}$$

and compare it with our theoretical calculation:

$$\epsilon_{th} = \frac{\frac{dN_{e^+e^-}}{dM\,d\eta}|_{th}}{\frac{dN_{e^+e^-}}{dM\,d\eta}|_{known\ processes}} \tag{12}$$

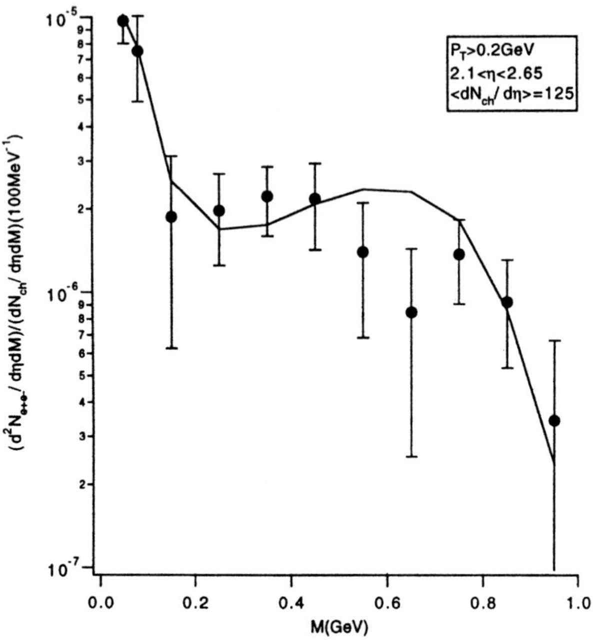

Figure 5: *Normalized differential spectrum of the entire e^+e^- pair production (QED + hadron sources) compared with the experimental data of Ref.3 .*

4 Conclusions and Outlook.

Figs.4, 5 and 6 summarize the achievement of our calculation, which, we wish to stress, has been obtained without any adjustable parameter and following a very simple logical path. In particular according to our results the remarkable excess of low-mass e^+e^- pairs discovered by the three CERN experiments [3,4,5] does not need for its explanation a very complicated, and rather doubtful, physical mechanism related to the formation of a hypothetical new state of hadronic matter, the QGP, whose evidence anywhere else (including the highly publicized J/Ψ suppression) seems, to say the least, dim. We believe that the simplicity and the quantitative character of our explanation recommend it as a good means to bring this unexpected phenomenon back to the realm of scientific rationality. And in so doing, to pinpoint the interesting role that even a weak, but coherent electromagnetic field associated to the well-known

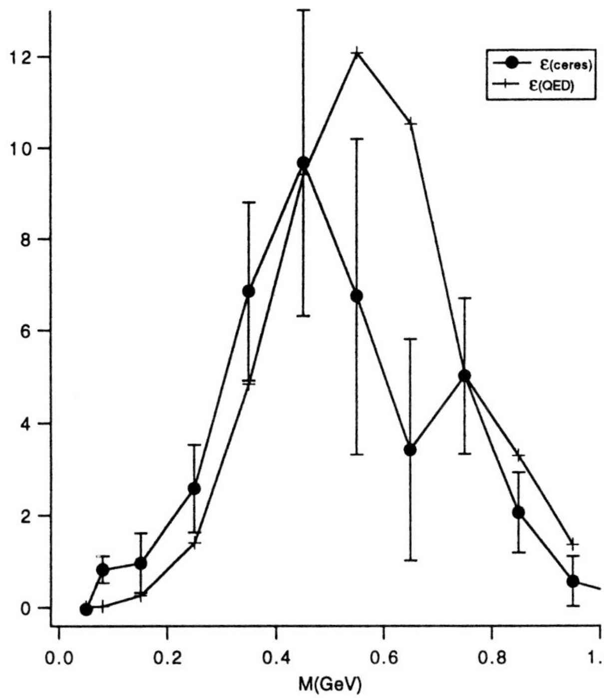

Figure 6: *Comparison of the experimental (CERES, Ref.3) e^+e^- pair relative excess of Eq.(11) with our theoretical relative photoproduction of Eq.(12) .*

giant resonances may play in the cataclysmic Heavy Ions' collisions.

Acknowledgments.

We thank Matteo Buzzacchi for help in calculations.

References

1. H. Satz, *Annu. Rev. Nucl. Part. Sci.* **35**, 245 (1985)
   H. Satz, *Colour deconfinement in hot and dense matter.* Talk given at CRIS '96.
2. A. Drees, Nuclear Physics **A630**, 449c (1998)

3. G. Agakichiev et al(CERES Collaboration), *Phys.Rev.Lett.* **75**, 1272 (1995)
4. T. Ullrich et al, *Nuclear Physics* **A610**, 317c (1996)
5. G.Agakichiev et al(CERES Collaboration), *Phys.Lett.* **B422**, 405 (1998)
6. G.Q. Li, C.M. Ko, G.E. Brown and H. Sorge, *Nuclear Physics* **A611**, 539 (1996)
7. A. Mueller, *in Proceeding of the 1981 ISABELLE Summer Workshop*, edited by H. Gordon (BNL, Upton, New York, 1982), p.636
   J.D. Bjorken, *Phys. Rev.* **D27**, 140 (1983)
8. F.E. Bertrand, *Nucl. Phys.* **A354**, 129c (1981)
9. G. Preparata, *QED Coherence in Matter*, World Scientific, Singapore 1995
10. G. Preparata et al, *QED Coherence, Giant Resonances and e^+e^- pairs in High Energy Nucleus-Nucleus Collisions*. In preparation.

FIFTY WAYS TO MEASURE SOURCE SIZE

SCOTT PRATT

Cyclotron Laboratory, Michigan State University, E. Lansing, MI 48824 (USA)

E-mail: pratt@nscl.msu.edu

In this talk I will review a variety of methods for measuring source size and lifetimes in heavy ion collisions. Various techniques based on correlations, coalescence and multiplicity fluctuations will be discussed. The phenomenogical powers of different observables will be compared and contrasted, and a short review of the status of the field will be presented.

LIGHT PARTICLE CORRELATIONS WITH HEAVY ION COLLISIONS: SPACE-TIME AMBIGUITIES

D. ARDOUIN
Laboratoire SUBATECH, University of Nantes, Ecole des Mines de Nantes, IN2P3, CNRS, France
E-mail: ardouin@sciences.univ-nantes.fr

We emphasize the importance of disentangling time and spatial effects for the quantitative interpretation of particle correlations in the domain of heavy ion collisions at intermediate energies. Several methods dealing with directional selections and complementary measurements of sizes using high granularity 4π detectors are presented. Unlike particle correlations are also shown to give access to the time ordering of emission between different particles. Thus particle correlations are providing detailed and invaluable information for a theoretical understanding of nucleus collision processes.

1 The H-B.T. effect and Nuclear Physics

The discovery of a two-particle interferometry effect in astronomy by Hanbury-Brown and Twiss [1] ("HBT effect") took place in 1954. The corresponding quantum effect was discovered in 1959 by Goldhaber *et al*[2] through the observation of an enhanced probability for identical pions emitted at small relative momenta. This was later interpreted as a result of Bose-Einstein statistics leading to a preference of bosons for a same quantum-state occupation. The real story of interferometry in nuclear physics only started later on with the theoretical papers of Kopylov and Podgoretsky [3] and Koonin [4] where formulations of correlation-functions were discussed in the frame of nuclear collisions together with the influence of final state interactions. Interferometry is interesting in nuclear physics provided it can help determining spatial and time extents of the emitting source(s). However, although significant results, regarding the dynamics of heavy-ion collisions, have been brought using this technique, still, many open questions remain, related to the ambiguities and difficulties to disentangle between space and time characteristics of emission. Distinctive features with the genuine HBT method can help to characterize these specific difficulties. In HBT, a space-time correlation is measured in the detector space which can be related after an averaging over time to the energy-momentum picture of the stellar source and then, using its dependence with the angular size to its transverse size. This can be understood by noticing that the two detected photons belong to overlapping energy wave-packets resulting from a superposition field. In other words, the intensity fluctuations are correlated as long as they correspond to two quanta belonging to the same quantum cell and representing the degree of coherence of the light. In nuclear physics correlations, the squared amplitude of the symmetrized two-particle wave function is averaged over the energy-momentum difference of the two detected particles. Through the resulting Fourier transform, the correlation contains a direct dependence on the (conjugate variable) spatial-size extension of the source.

Another distinctive feature of the nuclear physics domain is the dynamical evolution of the source resulting from the collision process: a large variety of successive emitting scenarios will take place while the source evolves. Then, additional space-velocity correlations are likely to appear which have to be taken out by appropriate detection and/or analysis selections. In addition, their interpretation will have to rely on some model descriptions: transport theories[5] using the time evolution of the one-body phase space Wigner density are presently able to simulate the global features of heavy ion collisions but still suffer from the lack of a correct description of the whole quantum nature of hadron systems based on two-body correlations.

For energetic reasons, most of the available measurements in heavy-ion collisions at intermediate energy have been made with baryons (mainly protons); thus, the original quantum effect is affected with additional final state interactions which can play a major role if space-time relative separations are small enough. Only photon-photon and low energy neutron-neutron correlations can avoid this situation and will be presented separately.

1.1 Photon-Photon correlations

More than 40 years after the discovery of the H.B.T. effect (see ref. [1]), very few investigations have been made in nuclear physics using the photon probe. Photons don't suffer the final state interactions or re-absorption and appear as an ideal probe. In the heavy-ion intermediate energy domain, it has been well established from different experimental features and B.U.U. calculations that the photon spectrum above the Giant Resonances mainly originates from the incoherent superposition of bremsstrahlung radiation emitted in first-chance individual n-p collisions. Only recently the difficulty to identify photons, among the hadronic and meson-decay background, could be overcome with the unique performances of new electromagnetic calorimeters like TAPS[6] or MEDEA[7]. Two experiments were performed at GANIL. In the TAPS collaboration[8], two systems were studied: the ^{86}Kr + Ni collision at 60 AMeV and Ta + Au at 40 AMeV. In a Q_{inv} representation and after a simulation of the π^0 decay, a rise observed below 50 MeV for the lighter system has been attributed to the expected interference (with $R_{rms-equiv.} = 15 fm$). Thus for the first time, the true HBT effect is revealed in nuclear physics. A new interesting interpretation of the oscillatory structure observed for the heavier system has been proposed by Schutz and Marques where they consider density oscillations of the di-nuclear system where bremsstrahlung photons originate from two successive compression phases. Another experiment [9] was carried out at GANIL using the MEDEA detector. The systems ^{36}Ar + Al, Sn, Au were studied at 95 AMeV. In order to extract the spatial and lifetime information, both relative momentum q_{rel} and q_o correlations have been studied leading to size and time determinations: $R_{rms} = 1.4fm, t = 2.0fm/c$ for the Al target; $R_{rms} = 3.0fm, t = 6.3fm/c$ for the Sn target; $R_{rms} = 3.8fm, t = 9.9fm/c$ for the Au target. These low values are found by the authors compatible with the first-chance collision picture for photon production, as indicated for instance by the comparison of the R_{rms} with the two-nuclei overlap zone as calculated from the geometrical estimation of the number of the participant nucleons.

1.2 Neutron correlations

The neutron-neutron correlations measurement is a very difficult experiment but a very attractive tool to exploit quantum-statistical interference effects. Indeed, the absence of Coulomb interaction plus the lowering of interactions between neutrons widely separately in an evaporative nuclear process were for instance exploited in the work of Dünnweber [10]. In the case of neutron correlations observed in the O + Mg collisions at low energy, the average distance between two successive neutrons is estimated around 100 fm, much larger than the singlet S-wave scattering length (~16 fm). The authors observed a dip in the correlation function at a time (or energy) difference corresponding to $E_{CM} = 0$ in the source rest-frame which they proposed to correspond to a quantum effect. From the observed width, and comparison with the quantum description of Kopylov an average lifetime has been deduced: $t = 6 \cdot 10^{-21} s$.

The study of the O + Ni $\rightarrow$ nn + X collision at 96 MeV by Gentner [11] allowed to observe a similar anti-correlation from which a lifetime was determined: $t = 1.2 \cdot 10^{-20} s$ in agreement with an evaporative simulation of the excited compound-nucleus. Conversely, positive correlations have been observed at higher energies [12,13] which reflect the increasing effects of final state interactions negligible at low energy as compared to Coulomb effects. These results are not reproducible by usual transport theories. Cross-talk rejection methods are of crucial importance for such correlation measurements. New ways of analysis have been more recently presented by A. Galonsky *et al* [14] for the Ar + Ho system at 25 AMeV, and J. Pluta *et al* [15] for the Ar + Au system at 60 AMeV, using a rejection method based on time-of-flight and energy kinematical conditions between two successive neutron layers. Kinetic energy selections as well as comparisons with statistical evaporation model allowed to observe the contribution of a pre-equilibrium (short lifetime) component and measure evaporative lifetimes. Thus the richness of neutron-neutron correlations to investigate space-time characteristics of heavy-ion collisions appears now very promising. One should also notice that they can provide unique investigation of the actual HBT effect with fermionic systems where coherent effects are absent.

The neutron-proton system appears as another interesting case for correlation studies [16,17,18] since there are no more antisymmetrization nor Coulomb effects. This also allows the study of the production of the d* singlet n-p state.

2 Proton-Proton correlations

2.1 Dynamical features of baryon correlations

Most of the available measurements in the intermediate energy domain have been made with baryons for energy consideration reasons; thus, the original quantum effect may be affected by additional final state interactions which can play a major role if space-time separations are small enough. Among determinant factors are the excitation energy which can drive the emission time, the observation C.M. angle which can trigger different origins among successive emission sequences. For energies of tens MeV per nucleon, i.e. around the Fermi energy, the energy relaxation is characterized by a competition between one and two-body potential effects. Two-body effects will result in fast pre-equilibrium emission of particles resulting in short timescale ($t < 50 fm/c$) characteristics; whereas for one-body dominated regimes, the complete or locally relaxed degrees of freedom will produce thermalized or pre-thermalized emission. For thermal excitation at temperatures $T < 5$ MeV, the emission times are estimated larger than a few hundreds fm/c. Here, the Coulomb interaction and Pauli principles should dominate again like in the previous low energy domain. An interesting challenge for two-particle correlations is then to follow subsequent evolution of the correlation patterns between these two extreme regimes in order to provide information on collision dynamics and characterize prevailing dissipative regimes. A crucial point to interpret the results is our ability to describe theoretically the time evolution of the source: the time dependence of the phase-space Wigner density has been studied, for heavy-ion collisions, in the frame of various one-body or N-body transport theories [5].

In the first experimental studies [19], a peak was indeed observed at q $\sim$ 20 MeV, attributed to the interplay between the short range attractive S-wave interaction, the long range Coulomb repulsion and the Pauli principle. Its height was related to the source dimension. But because of time averaging effects, one should be extremely careful in drawing quantitative conclusions. Non-equilibrium emission will, for instance, result in short time scales for particle emission causing the nuclear interaction to dominate the correlation pattern and reflect the spatial dimension of the source. Conversely, in the case of compound nuclei formation and equilibrium emission, the average distance between two successive emissions is larger: the Coulomb interaction and Pauli principle will dominate with a sensitivity to the emission rate and thus to the thermal energy of the emitting nucleus. This feature is very clearly observed in low energy experiments (see for example ref. [20]). In those cases, 3-body Coulomb trajectories can account for most of the correlation patterns and demonstrate the feasibility and the limitations of compound-nucleus lifetime determinations using light particle correlations.

For energies of tens MeV per nucleon, several cuts in particle momentum, impact parameter, geometry or relative directions of emission have been usually proposed in order to provide information on collision dynamics and characterize prevailing dissipative regimes. We refer the reader to the existing literature (see for instance ref. [21]) between 30 and 200 AMeV for detailed description. The results can be summarized by saying that, up to a very recent time, and in spite of clear qualitative distinctions, many ambiguities remain in evaluating quantitatively the average time durations (often set to fixed or negligible values) and their real influence in the observed correlation patterns. Most of these uncertainties originate from the failure of transport theories to describe all dynamical aspects associated with particle production: sequential decays (produced by unstable fragments), reliable description of two-body processes in nuclear matter (and their relation with the production preequilibrium emission). These observations stress the necessity of model-independent and reliable emission time measurements in this energy domain.

2.2 *Emission-Time determinations: directional selections and 4π detector methods*

It has been proposed by Pratt *et al*[22] that an elongated phase-space distribution, as predicted for long-lived emission sources can lead through phase space constraints to the emission time dependence of the correlation-function on the direction of the relative momentum. When constructing such correlations in the frame of the emitting source, a clear dependence has been observed for pp[23] and nn [24] measurements. From a contour-plot-of χ^2 as a function of R and τ quantities for longitudinal and transverse correlations, a set of (R, t) values has been extracted for the source characteristics. Although containing some model dependent ingredients, the use of these directional dependences opens promising possibilities of probing the space-time characteristics of heavy-ion collisions.

The availability of a new generation of 4π high granularity detectors, like INDRA at GANIL, offers very attractive alternative possibilities. Indeed, from an event-by-event reconstruction of the momentum ellipsoid, it is possible to determine directly the source characteristics. For instance, a separation of particles emitted by the forward (projectile-like) source can be made and related to the size of the emitting source after some density assumption. This has been performed for the Xe + Sn reaction at 50 AMeV [25] (Collaboration with D. Gourio and the INDRA collaboration). The measured correlations have been fitted using a quantum final state interaction model [26] with source characterizations (size, temperature) deduced from the above approach and the observed energy spectra. As a consequence and for the first time, only the time duration of the source remains as a parameter which can be determined by fitting the correlations. Figures 1 and 2 show the results obtained for protons and deuterons with three ranges of impact parameters resulting of a transverse energy selection. The calculations (with normal density assumptions) indicate two interesting features: first, a very little evolution with impact parameter in the case of protons; secondly very short times for all the cases. Values like 80 fm/c are smaller than (or of the order of) the estimated separation time of the two interacting nuclei. Even smaller values would be obtained with lower density assumptions associated with the creation of a more diluted system. Complementary observations of the proton correlations originating from a mid rapidity selected zone (instead of projectile region) have also indicated an angular anisotropy (with a minimum sitting at 90° cm) for peripheral collisions if high energy protons are selected (Fig. 3). This is reminiscent of the angular distributions known for free nucleon scattering. From this new type of exclusive measurements, it can thus be postulated that an important contribution of pre-equilibrium particles, likely resulting from N-N scattering, is found over a large rapidity domain. This is supported by recent transport calculations by Eudes *et al* [27] where an important mid rapidity emission, taking place at the early phase of interacting nuclei, is found for central collisions. This is also to be related to an (anti-)correlated observation of high energetic protons and energetic photons in the same domain of energy by Sapienza *et al* [28].

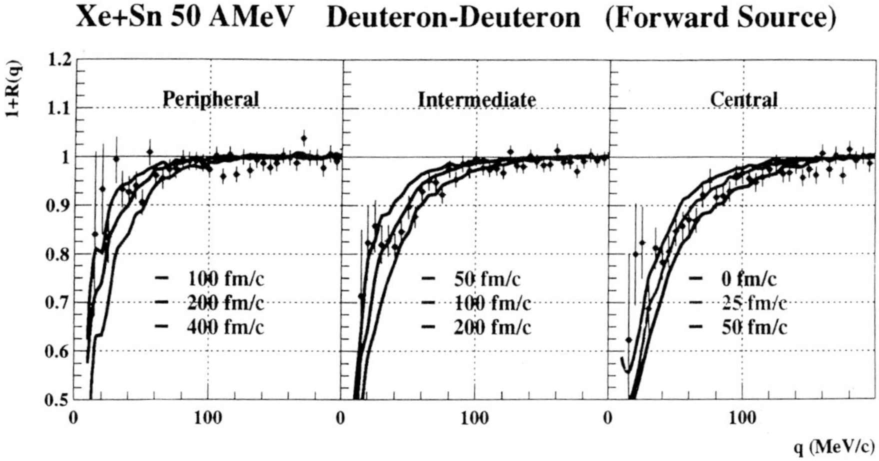

Figure 1: Proton-proton correlation as a function of relative momentum for the Xe + Sn reaction. The curves indicate calculations with indicated emission times. (Taken from ref.[25]).

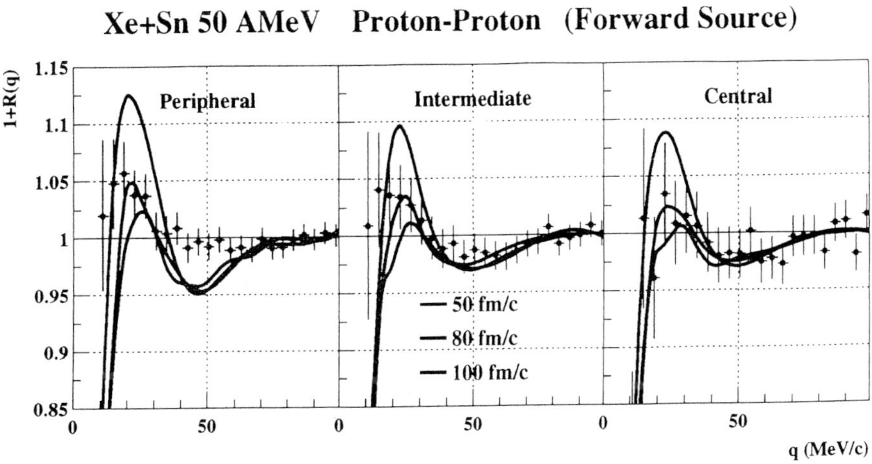

Figure 2: d-d correlations (same as Fig. 1)

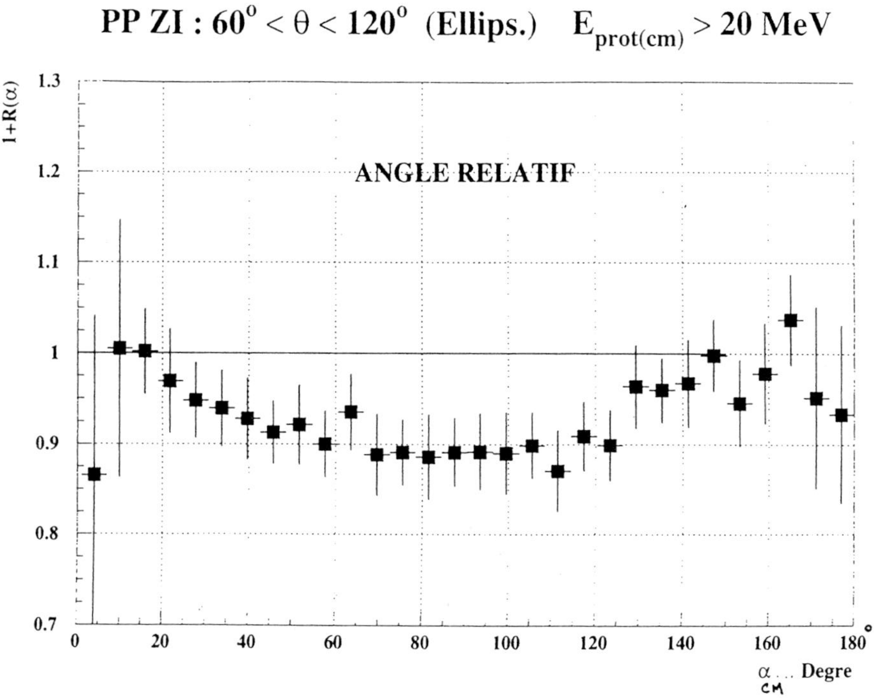

Figure 3: p-p correlation as a function of CM relative angle for high energetic protons in Xe+ Sn reaction. (Cf. ref.[25]).

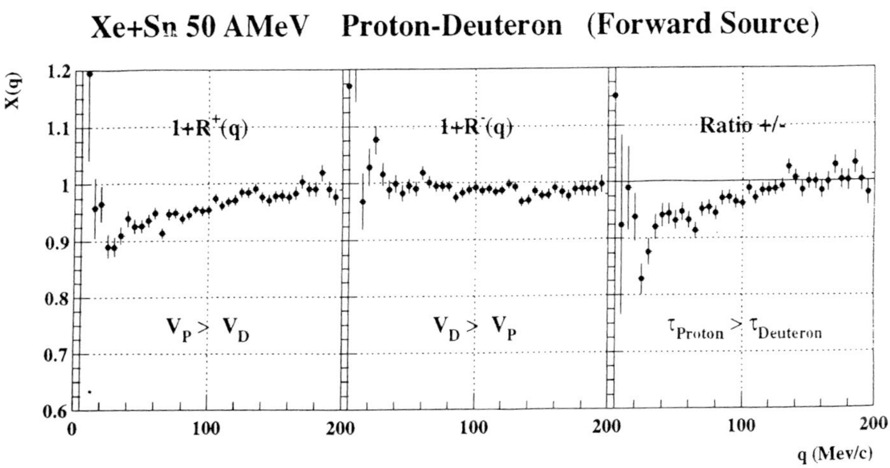

Figure 4: (p, d) predicted correlations for R+ ($\vec{k} \cdot \vec{v} > 0$) and R- ($\vec{k} \cdot \vec{v} < 0$) and their ratio indicating which particle is emitted earlier. (Taken from ref. [31]).

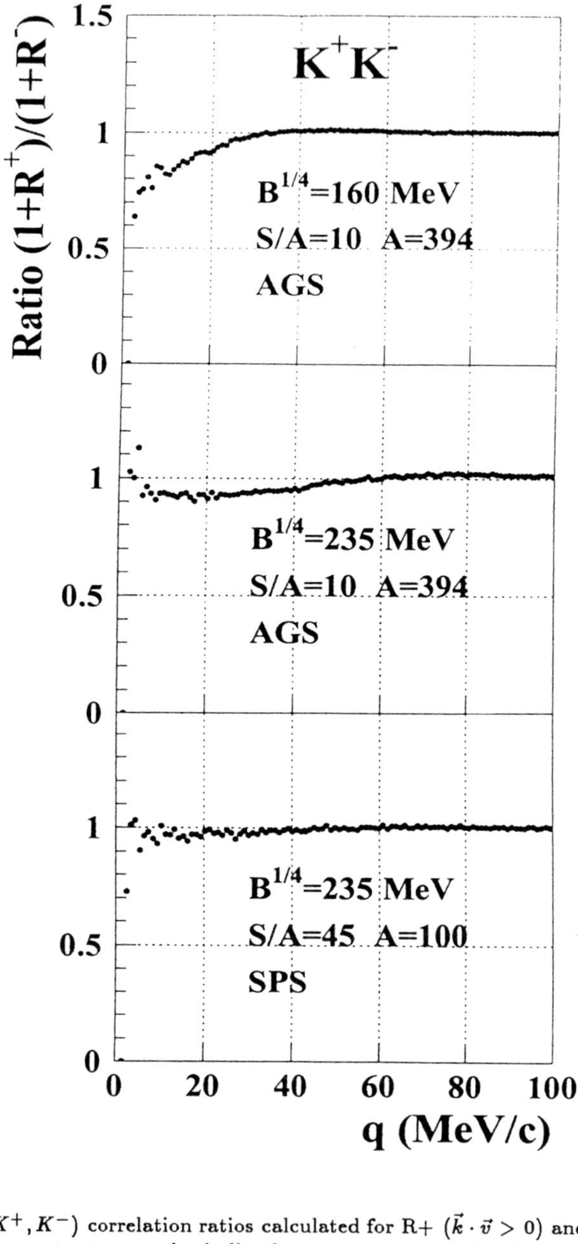

Figure 5: (K^+, K^-) correlation ratios calculated for R+ $(\vec{k} \cdot \vec{v} > 0)$ and R- $(\vec{k} \cdot \vec{v} < 0)$ for different bag constants scenarios indicating a way to search for the strangeness distillation process. (Taken from ref. [31]).

2.3 Time ordering of emission using unlike-particle correlations

It has been shown by Lednicky *et al*[29] that the occurrence of an anisotropy in the spatial-time distribution is reflected in the directional dependence of unlike-particle correlations and can be for instance used to measure the average delays (and their sign) between emissions of particles of different types. This sign can be determined by comparing correlation-functions R+ and R- where the sign of the scalar product $\vec{k} \cdot \vec{v}$ is fixed > 0 or < 0 (using their relative angle), respectively. Classically, it means that the interaction of the two particles in the case where the faster particle is emitted earlier will be different as compared to the case of its later emission. It has been shown that the Coulomb field of the residual nucleus slightly affects this result.

Such predictions have been compared to experimental (p,d) correlations measured at GANIL for various collisions and energies (see ref. [25] for example). As an example, in the case of Xe + Sn at E/A = 50 MeV, Fig. 4 shows that the ratio R+/R- is lower than unity: this evolution is attributed to an averaged deuteron emission earlier than the proton emission. The time difference is estimated to be $\sim$ 250 fm/c according to the comparison with Lednicky's calculations. This has to be understood as the result of averaging over a long time sequence between pre-equilibrium and thermal emission for protons whereas deuteron emission resulting mainly from hard nucleon-nucleon collisions is concentrated at a few tens of fm/c. This method can be also used for higher energy problems like the possible observation of strangelets in the frame of the distillation process following the creation of a quark gluon plasma as predicted by C. Greiner *et al*[30]: in this case, strange and anti-strange particles may not be produced at the same time in a baryon rich system under low bag constant scenarios. Such a prediction has been tested using (K^+, K^-) correlations[31] as demonstrated in Fig. 5 (collaboration with S. Soff, H. Stoecker, Inst. Theo. Phys. Frankfurt).

Thus for the first time, it has been demonstrated that correlation studies, combined with geometrical selections, provide unique tools to measure the time-ordering for particle emission.

3 Conclusions

Two particle correlations in heavy-ion collisions, have allowed in a unique way to investigate detailed aspects of the collisions process over a large variety of emission processes between equilibrated/thermalized mechanisms. Different kinematical and directionnal constraints have documented dynamical features of interest for the modelisation of the collisions. However, independent and complementary measurements of lifetimes using either directional selections or a new generation of dedicated high granularity detectors appear very useful to determine and disentangle quantitatively between spatial and time effects. Unlike particle correlations are also able to determine the emission time ordering of different particles. Pioneering works have been also made in neutron and photon interferometry where the actual HBT effect gives a direct and undistorted measurement of spatial and time properties of the source.

References

1. R. Hanbury-Brown and R.Q. Twiss, Phil. Mag. **45**, 663 (1954).
2. G. Goldhaber *et al*, Phys. Rev. **120**, 300 (1960).
3. G. Kopylov and Podgoretsky, Sov. J. N. Phys. **15**, 219 (1972); G. Kopylov and Podgoretsky, Sov. J. N. Phys. **18**, 336 (1974); G. Kopylov, Phys. Lett. B **50**, 472 (1974).
4. S.E. Koonin, Phys. Lett. B **70**, 43 (1977).
5. S. Pratt, Phys. Rev. Lett. **53**, 1219 (1984); H. Stoecker *et al*, Phys. Rep. **137**, 277 (1986); G. Bertsch *et al*, Phys. Rep. **160**, 189 (1988).
6. R. Novotny *et al*, IEEE Trans. Nucl. Sc. **38-2**, 379 (1991).
7. E. Migneco *et al*, Nucl. Instr. Meth. A **314**, 31 (1992).
8. F. Marques *et al*, Phys. Rev. Lett. **73**, 34 (1994).
9. A. Badala *et al*, Phys. Rev. Lett. **74**, 4779 (1995); A. Badala *et al*, Phys. Rev. C **55**, 2521 (1997).
10. W. Dünnweber *et al*, Phys. Rev. Lett. **65**, 297 (1990).
11. R. Gentner *et al*, Zeits. Phys. A **343**, 401 (1992).
12. B. Jakobsson *et al*, Phys. Rev. C **44**, 1238 (1991).
13. Y. Bayukov, Phys. Lett. B **189**, 291 (1987).
14. A. Galonsky, S. Gaff, in *Proceedings of CRIS' 98, Acicastello, Italy, June 8-12, 1998*, ed. S. Costa *et al* (World Scientific, Singapore, 1998).
15. J. Pluta *et al* (Demon Collaboration), Nucl. Inst. Meth., in press.
16. R. Kryger *et al*, Phys. Rev. C **46**, 1887 (1992).
17. M. Cronqvist *et al* (CHIC collaboration) Phys. Lett. B **317**, 505 (1993).
18. V. Bellini *et al*, in *Proceedings of CRIS' 98, Acicastello, Italy, June 8-12, 1998*, ed. S. Costa *et al* (World Scientific, Singapore, 1998).
19. W. Lynch *et al*, Phys. Rev. Lett. **51**, 1850 (1983); H. Gustafsson *et al*, Phys. Rev. Lett. **53**, 544 (1984); F. Zarbaksh *et al*, Phys. Rev. Lett. **46**, 1268 (1981); J. Pochodzalla, Phys. Rev. C **35**, 1695 (1987).
20. J. Alexander *et al*, Phys. Rev. C **48**, 2874 (1993).
21. D. Boal *et al*, Rev. Mod. Phys. **62**, 553 (1990); W. Bauer *et al*, Ann. Rev. Nucl. Sci. **42**, 77 (1992); D. Ardouin, Int. Journ. Mod. Phys. E **6**, 391 (1997).

22. S. Pratt *et al*, Phys. Rev. C **36**, 2390 (1987); G. Bertsch, Nucl. Phys. A **498**, 230 (1989).
23. M.A. Lisa *et al*, Phys. Rev. Lett. **71**, 2863 (1993); M.A. Lisa *et al*, Phys. Rev. C **49**, 2788 (1994).
24. N. Colonna *et al*, Phys. Rev. Lett. **75**, 4190 (1995).
25. D.Gourio (INDRA collaboration), to be published.
26. R. Lednicky *et al*, Sov. Journ. Nucl. Phys. **15**, 219 (1972); in *Proceedings of Corinne 90*, ed. D. Ardouin (World Scientific, Singapore, 1990).
27. P. Eudes *et al*, Phys. Rev. C **56**, 342 (1997).
28. P. Sapienza *et al*, Nucl. Phys. A **630**, 215 (1998).
29. R. Lednicky *et al*, Phys. Lett. B **373**, 30 (1996).
30. C. Greiner *et al*, Phys. Rev. D **44**, 3517 (1991).
31. S. Soff *et al*, Journ. Phys. G **23**, 789 (1997); D. Ardouin, to be published.

DETERMINATION OF FREEZEOUT CONDITIONS FOR MULTIFRAGMENTATION

W. G. Lynch

NSCL and the Department of Physics and Astronomy, Michigan State University, East Lansing, MI, 48824-1321. USA

Correlation measurements provide opportunities to examine the breakup conditions relevant to various modes of particle production. In the case of nuclear multifragmentation, correlation measurements provide a means to deduce whether the conditions attained may be relevant to low density phase transitions and to measure the relevant densities and temperatures. Correlation, however, may also indicate that cooling processes and other time dependent phenomena are important. This allows one to select reactions which either minimize or maximize the role played by such time dependent phenomena.

1 Introduction

During the latter stages of an energetic nuclear collision, the nuclear system can expand and fragment at subsaturation density into a number of light particles and intermediate mass fragments. Each of these particle species interact strongly with its nuclear environment as long as the system remains at non-negligible density. As the system continues to expand, however, there comes a (freezeout) time for each fragment where further interactions with the environment cease and the fragment evolves according to its own internal properties. Experimental observables largely reflect the fragment properties at this freezeout time. If the properties of the fragments and the fragmenting system are sufficient thermalized at freezeout, the fragmentation process may be described statistically.

Statistical theories, which replicate many of the features of fragmentation processes, frequently assume a common freezeout time for all particles and interpret the multifragmentation phenomenon as resulting from the proximity of a low density phase transition in nuclear matter[1,2]. Consistency with such approaches is greatest for peripheral collisions at high incident energies[3] , or for central collisions at $E/A \leq 50$ MeV[4,5]. Significant discrepancies with statistical theories have been reported for central collisions at higher incident energies[6,7] where strong collective expansion velocities have been observed.

Two particle correlation techniques can examine whether such equilibrium assumptions are valid and provide, for such cases, information about the breakup density and temperature[8]. Two particle correlation functions, in the limit of a common freezeout time, primarily determine the freezeout volume.

This can be illustrated simply in the thermodynamic limit wherein the correlation function $(1 + R)$ is given by:

$$1 + R = 1 + \frac{2\pi}{(2s_1 + 1)(2s_2 + 1)\, q^2 V} \sum_{\ell J} (2J + 1) \frac{d\delta_{\ell J}}{dq}, \tag{1}$$

where s_1 and s_2 are the spin of the two detected particles, q is the relative momentum, $\delta_{\ell j}$ is the phase shift of the two particle system for orbital and total angular momenta ℓ and J, respectively, and V is the freezeout volume. The freezeout density can subsequently be obtained when the deduced freezeout volume is combined with measurements of the mass of the fragmenting source[5].

Temperatures can be extracted from the relative populations of excited particle unstable nuclei[9]. In this thermodynamic limit, the decay spectrum for such unstable nuclei is given by:

$$Y(E^*) \propto const \cdot \sum_{\ell J} (2J + 1) \frac{d\delta_{\ell J}}{dE^*} \exp(-E^*/T), \tag{2}$$

where E^* is the excitation energy of the unstable nucleus and T is the freezeout temperature.

Adopting this general approach, the correlation functions and freezeout temperatures have been measured for a variety of systems and particle species. Many of these measurements focused upon particle pairs for which resonant channels dictated a strong energy dependence of the phase shifts. In such cases, the correlation functions display strong resonance peaks which explicitly reflect the interactions between the detected particles and the volume of the emitting system. The measurement of such structures require arrays with angular resolution of the order of 1° in the frame of emission.

2 Correlation Functions for Multifragmentation Processes

First generation multifragment detection arrays lack the angular resolution necessary for resonance dominated correlation functions. Correlation function investigations with such arrays have focused instead upon fragment-fragment correction functions[10,11]. Initial models for fragment-fragment correlations assumed the correlation function to be dictated by the mutual Coulomb final state interaction between the two detected fragments[10]. Later studies revealed the important role of multiparticle Coulomb interactions and the distortion of the correlation function due to the phase space constraints of total momentum conservation[12].

To illustrate some of these latter complications, Fig. 1 shows fragment-fragment correlations integrated over all fragment pairs emitted with $4 \leq$

$Z_{IMF} \leq 9$ for ^{86}Kr + ^{93}Nb at E/A=50 MeV[12]. Top and bottom panels show the results for fragments emitted at $25^o \leq \theta_{lab} \leq 50^o$ and $5^o \leq \theta_{lab} \leq 25^o$, respectively. The solid and open points show data for low- and high-fragment-energy thresholds, $E_{cm}/A \geq 3$ MeV and $E_{cm}/A \geq 7$ MeV, respectively. The dashed curves show the results of instantaneous breakup calculations assuming emission from a spherical volume of radius $R_S = 12$ fm and that the heaviest fragment (residue) is situated in the center of the breakup volume. The dotted curves show the corresponding calculations in which the heaviest fragment is placed randomly throughout the breakup volume. (Both correlation function simulations have been constrained to reproduce the fragment energy spectra and multiplicity distributions.) The differences between the two simulations are more pronounced for the lower energy gate. Calculated correlation functions which assume the heaviest fragment (residue) to lie at the center of the breakup volume display wider minima than those with the heaviest fragment randomly distributed; such a trend could also be obtained in the former calculation by assuming a smaller breakup volume. As the heaviest fragment is not one of the two fragments included in the correlation, this is definitely a many body correlation effect that makes the determination of the breakup volume dependent upon model assumptions concerning the placement of other fragments.

Despite these limitations, fragment fragment correlation functions have allowed clear comparisons with multifragmentation models that make specific predictions for the breakup configuration and timescale. For example, comparisons of fragment-fragment correlation functions measured for central Au+Au collisions at E/A=35 MeV to correlation functions calculated with the Statistical Multifragmentation Model[5] have allowed constraints of $\rho_o/6 \leq \rho_{freezeout} \leq \rho_o/3$ to be placed upon the freezeout density assumed within the Statistical Multifragmentation Model. Comparisons between fragment-fragment correlations functions measured for central collisions in other systems to sequential fragmentation models have also shown the upper limit on the fragmentation timescale assumed in such models to decrease with incident energy[13].

To reduce the complications of many body correlations, it would be interesting to measure fragment correlations with sufficient angular resolution to allow the construction of correlation functions that display resonant structures. As such structures are characteristic of the detected two particle system, they cannot be confused by multiparticle Coulomb interactions and some of the model dependences that complicate fragment-fragment correlations can be removed. To illustrate that resonant composite particle correlation functions can provide information that is equivalent to that provided by proton-proton correlations, Fig. 2 show systematics of Gaussian source radii extracted from

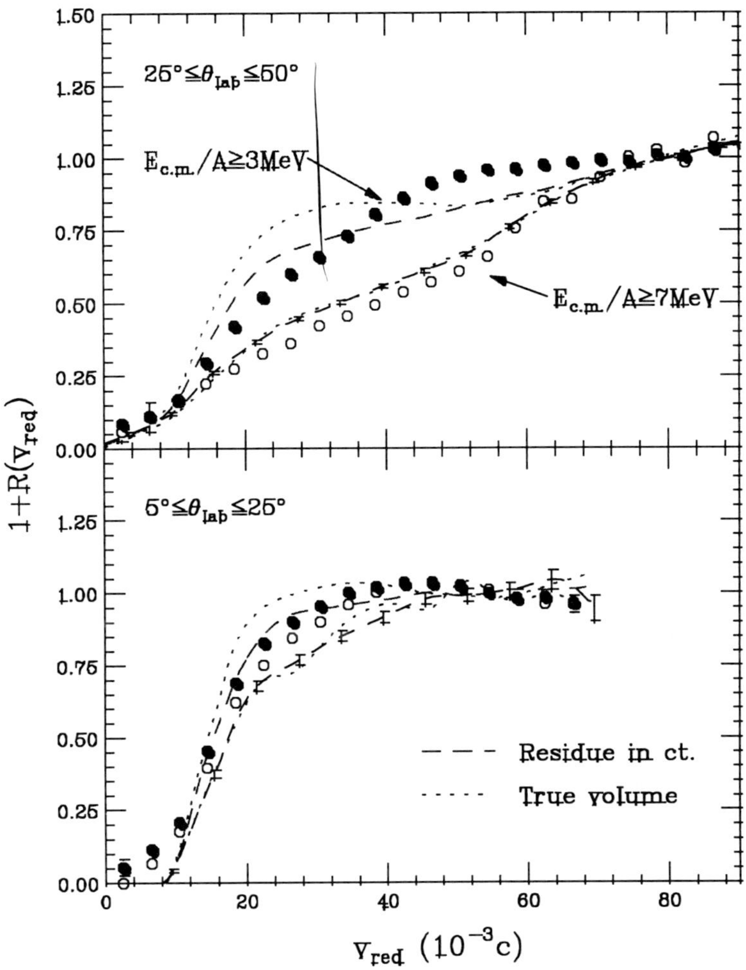

Figure 1: Measured and simulated fragment-fragment correlations for central Kr+Nb collisions at E/A=50 MeV. The various data points and calculated curves are described in the text.

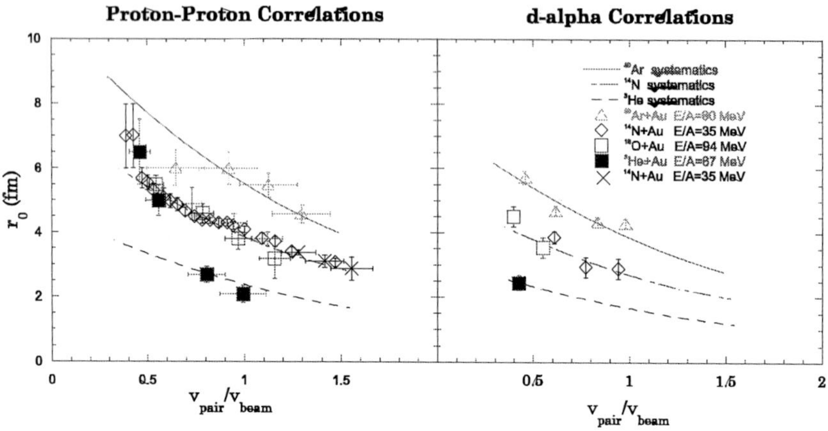

Figure 2: Systematics of light particle inclusive correlation functions. The various data points and curves are explained in the text.

inclusive two particle correlations plotted as a function of the ratio of the velocity of the two correlated particles divided by the beam velocity for a variety of asymmetric reactions[14]. The solid and open points in the left panel depict experimentally source radii extracted from two proton correlations. The solid line is an interpolation of the source radii extracted for ^{14}N projectiles. The dashed and dot-dashed lines are extrapolated from the solid line via the ratio of the projectile radius to the radius of ^{14}N. This suggests that the spatial extent of the emitting region for energetic protons is governed initially by the overlap of these relatively small projectiles with the heavy target and, therefore, scales with radius of the projectile. The scaling fails for less energetic protons, for which long time scale secondary decays play an important role in weakening the correlation function.

The right hand panel shows corresponding data for d-α correlation function which are obtained by reproducing the resonant peak in the d-α correlation function corresponding to the 3^+ state in ^{6}Li at $E^* = 2.19$ MeV. The solid dashed and dashed-dotted lines are taken from left hand panel, but renormalized by a multiplicative factor of 0.7. Clearly, p-p and d-α provide comparable information about the space-time evolution of the emitting source. Similar resonant correlation functions can be measured for heavier particle pairs[14]. However, the origin of the renormalization factor of 0.7 which relates the Gaussian radius parameters for the fits to the d-α correlation functions to those for the p-p correlation functions is not presently understood and should be investigated in order to put such composite correlation functions on a foundation that is as quantitative as that for p-p correlations.

3 Temperatures of Multifragmentation Processes

One of the general expectations for a system undergoing a first order phase transition is an enhanced heat capacity at temperatures where the phase transition occurs, reflecting the latent heat required to transform from one phase to the other. Calculations predict enhanced heat capacities for finite nuclear systems at temperatures of the order of 4-6 MeV[1,2], due to the transformation from the Fermi liquid which can be found in the interior of large nuclei in their ground and low excited states to a gas phase consisting of free nucleons and light clusters. Temperatures extracted from the isotopic abundances of helium and lithium fragments[15] produced by the fragmentation of Au projectiles[16] display a caloric curve consistent with the hypothesis of a mixed phase equilibrium. These results have stimulated a large body of experimental[4,16,17,19–22,32] and theoretical[24–31] work, both supportive and critical of the caloric and thermometric techniques employed.

To reduce the sensitivity to collective motion[32], temperatures for hot nuclear systems formed in nucleus-nucleus collisions have been extracted from the comparison of ratios of isotopic yields[4,16–22], T_{iso}, and of excited state populations[4,22,34–40], T_{ex}. For thermal distributions at low density and at chemical equilibrium, prior to the secondary decay of the excited fragments, the double ratios R_{iso} of the ground state yields of four suitably chosen isotopes are given by[15]:

$$R_{iso} = \frac{Y(A_1, Z_1)/Y(A_1+1, Z_1)}{Y(A_2, Z_2)/Y(A_2+1, Z_2)} = a \exp(B/T_{iso}) \tag{3}$$

where $Y(A_i, Z_i)$ is the yield for isotope with mass A_i and charge Z_i; a is a statistical factor determined by spin values and kinematics factors; $B = BE(A_1, Z_1) - BE(A_1+1, Z_1) - BE(A_2, Z_2) + BE(A_2+1, Z_2)$; and $BE(A_i, Z_i)$ is the binding energy of the ith nucleus. Such double yield ratios have the advantage of being insensitive to chemical potential terms which strongly influence the fragment isotopic distributions[15].

Similarly, the ratios R_{ij} of the yields of states i and j of a specific fragment, prior to the secondary decay of the excited fragments, are given (following Eq.

2) by :

$$R_{if} = \frac{Y_i}{Y_j} = \frac{(2J_i+1)}{(2J_j+1)} \exp\left(-\left[E_i^* - E_j^*\right]/T_{ex}\right) \tag{4}$$

where Y_i is the yield, E_i^* is the excitation energy, and J_i is the spin of the state i. Apparent temperatures obtained from Eqs.(3) or (4), often require corrections for the secondary decay of excited fragments. The exact magnitude of such corrections depends on the isotope or excited state ratio and the temperature[4,21,23,24]. These corrections can be minimized by choosing ratios characterized by large binding energy differences B in Eq. (3) or large $\Delta E^* = \left(E_i^* - E_j^*\right)$ in Eq. (4).

Consistent values of T_{iso} and T_{em} for large B and ΔE, respectively, have been obtained for central Au+Au collisions at incident energies $35MeV \leq E/A \leq 50MeV$[4,22], but these values diverge from each other as the incident energy is increased. To illustrate this effect, extracted values of T_{em} are plotted in Fig. 3 as solid circles for the ground and 16.7 MeV excited states of ^{5}Li, as solid diamonds for the ground and 20.1 MeV excited states of ^{4}He, and as solid squares for the 3.04 and 17.6 MeV excited states of ^{8}Be for central Kr+Nb collisions as a function of incident energy. The extracted temperatures are of the order of 4 - 4.5 MeV and show a weak variation with incident energy, a trend also observed for central Au+Au collisions[4,22].

Values of T_{iso} were also obtained via Eq. (3) at the four incident energies. Values for T_{iso}(C-Li) (open circles) obtained from (^{6,7}Li,^{11,12}C) isotope ratios vary little with incident energy, similar to the trends exhibited by the temperatures extracted from excited state populations. In contrast, values of T_{iso}(He-Li) (open diamonds) obtained from (^{6,7}Li, ^{3,4}He) isotopes increase monotonically with incident or excitation energy, consistent with trends recently reported for other systems[16,19,22]. Similarly increasing trends are extracted from other possible ratios based upon the large binding energy difference between ^{3}He and ^{4}He isotopes.

Some corrections to the measured apparent temperatures can be expected due to the secondary decay of heavier isotopes that feed the yields used in Eqs. (3) and (4)[4,23,24]. Secondary decay calculations, however, predict the secondary decay corrections to T_{ex}(^{5}Li), T_{ex}(^{4}He), T_{ex}(^{8}Be), and T_{iso} (He-Li) to be relatively small for apparent temperatures of the order of 4 - 4.5 MeV[4]; thus the predicted corrections to the observed values of T_{ex}(^{5}Li), T_{ex}(^{4}He) and T_{ex}(^{8}Be) are small. Moreover, the secondary decay corrections to T_{ex}(^{4}He) and T_{iso} (He-Li) should be nearly identical at all source temperatures because B and ΔE^* are comparable and because both are affected primarily by the feeding contributions to the ground state yield of ^{4}He[46]. The observed strong discrepancy between T_{ex}(^{4}He) and T_{iso} (He-Li) therefore cannot be reconciled

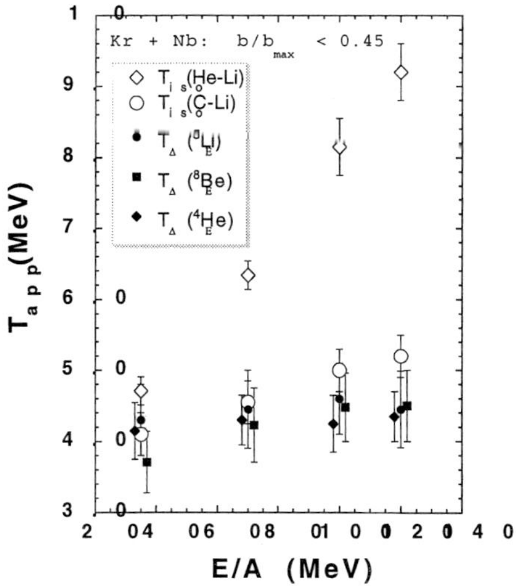

Figure 3: Comparison of excited state and isotope ratio thermometers. The various data points are discussed the text

by secondary decay calculations.

The much larger values for T_{iso} (He-Li) may reflect differences in the emission environments of ^{3}He as compared to the emission environments for fragments and alpha particles. Investigations of fragment-fragment correlations[47] and fragment charge distribution [47,48] have reported evidence for a hierarchy of timescales whereby light particle emission precedes fragment emission rendering the two processes out of equilibrium. Predictions of dynamic models for light particle emission and statistical emission rate approaches[43,45] also support this picture. Measurements of the energy dependence of the ^{4}He excited state thermometer show a dependence on the energy of the outgoing ^{4}He[49]. All of these observations point to the a time dependence of the temperature of the emitting source and a consequent breakdown of the assumption of a global freezeout time for these energetic multifragment breakups.

4 Summary

Present correlation measurements have provided insight into the breakup conditions pertinent to nuclear multifragmention. Measurements at lower incident energies are consistent with a common freezeout temperature and temperature and densities extracted for such low energy reactions are comparable to those assumed in statistical models that reproduce other multifragment observables. In contrast, temperatures extracted at higher incident energies are inconsistent and suggest that different particle species may be emitted at different times with different emission environments.

Clearly, additional work is needed to clarify this time dependence and to assess by which types of theoretical models can such measurements be most accurately interpreted. More precise correlation tools can be of great use in this task. Higher resolution correlationcorrelation functions which display resonant structures could be of great utility when arrays with greater angular resolution are constructed. Such correlations may eventually provide information for multifragmentation processes that are as quantitative as that currently provided by proton-proton correlations for other reactions.

Acknowledgments

This work was supported by the National Science Foundation under Grant Nos. PHY-95-28844 and PHY-93-14131.

References

1. J. P. Bondorf, et al., *Phys. Rep.* **257,** 133 (1995) and refs. therein.
2. D.H.E. Gross, *Rep. Prog. Phys.* **53**, 605 (1990) and refs. therein.
3. A. Schuttauf, et al., *Nucl. Phys. A* **607**, 457 (1996).
4. M. J. Huang, et al., *Phys. Rev. Lett.* **78**, 1648 (1997).
5. M. D'Agostino, et al., *Phys. Lett. B* **371,** 175 (1996).
6. W. Reisdorf, et al., *Nucl. Phys. A* **612**, 493 (1997)
7. C. Williams, et al., *Phys. Rev. C* **55**, R2132 (1997).
8. C. B. Chitwood, et al.., *Phys. Lett. B* **172**, :27 (1986).
9. W. Benenson, D.J. Morrissey, W. A. Friedman, *Ann. Rev. of Nucl. and Part. Sci.* **44**, 1994.
10. Y.D. Kim, Y.D., *Phys. Rev. Lett.* **67**, 14 (1991).
11. D.R. Bowman, et al., *Phys. Rev Lett.,* **70**, 3534 (1993).
12. R. Popescu, et al., *Phys. Rev. C 58*, **270** (1998).
13. E. Bauge, et al., *Phys. Rev. Lett.*, **70**, 3705 (1993).
14. F. Zhu, et al., *Phys. Rev. C* **44,** R582 (1991). F. Zhu et al., Ph.D. dissertation Michigan State University (1992).
15. S. Albergo et al., Nuovo Cimento 89, 1 (1985).
16. J. Pochodzalla et al., Phys. Rev. Lett. 75, 1040 (1995).

17. M.B. Tsang, W.G. Lynch, H. Xi, W.A. Friedman, Phys. Rev. Lett. 78, 3836 (1997). 7 is huang
18. R. Wada et al., Phys. Rev. C 55, 227 (1997).
19. Y-G. Ma et. al, Phys. Lett. B, B390, 41 (1997).
20. H.F. Xi et al., Z. fur Physik A 359 397 (1997).
21. H. Xi, W.G. Lynch, M.B. Tsang, W.A. Friedman, Phys. Rev. C 54 R2163, (1996).
22. V. Serfling et al., Phys. Rev. Lett. 80, 3928 (1998).
23. A. Kolomiets et al., Phys. Rev. C 54, R472 (1996).
24. F. Gulminelli and D. Durand, Nucl. Phys. A 615, 117 (1997).
25. J.B. Natowitz et al., Phys. Rev. C 52, R2322 (1995).
26. C. B. Das, L. Satpathy, Phys. Rev. C 57, R35 (1998).
27. P. F. Mastinu et al., Phys. Rev. C 57, 831 (1998).
28. Z. Majka et al., Phys. Rev. C55, 2991 (1997).
29. J. N. De, S. Das Gupta, S. Shlomo and S.. Sammadar, Phys. Rev. C 55, R1641 (1997);
30. S. J. Lee et al., Phys. Rev. C 56, 2621 (1997);
31. S. K. Samaddar, J. N. De, S. Shlomo, Phys. Rev. Lett. 79, 4962 (1997).
32. W.C. Hsi et al., Phys. Rev. Lett. 73, 3367 (1994).
33. S.C. Jeong et al., Phys. Rev. Lett. 72, 3468 (1994).
34. H.M. Xu et. al., Phys. Rev. C 40, 186 (1989).
35. T.K. Nayak et al., Phys. Rev. C 45, 132 (1992).
36. F. Zhu et al., Phys. Rev. C 52, 784 (1995); (Note: these are not reactions which display multifragmentation. Similar multifragmentation data does not currently exist.)
37. J. Pochodzalla et al., Phys. Rev. C35, 1695 (1987).
38. Z. Chen et al., Phys. Rev. C36, 2297 (1987).
39. C. Schwarz et. al. Phys. Rev. C 48, 676 (1993).
40. Xi Hongfei et al., Nucl. Phys. A552, 281 (1993).
41. R. Wada et al., Phys. Rev. C 55, 227 (1997). 26 is morr_rev
42. W.G. Friedman, Phys. Rev. C 42, 667 (1990).
43. R. Bougault et al., preprint, LPCC97-04 (1997).
44. H. Xi et al., Phys. Rev. C, Phys. Rev. C 57, R467 (1998).
45. G.J. Kunde et al., Phys. Lett. B416, 56 (1998).
46. H. Xi et al., MSU preprint MSUCL-1055 (1996); H. Xi, W.G. Lynch, M.B. Tsang, W.A. Friedman, and D. Durand, MSU preprint MSUCL-1107 (1998).
47. T. M. Hamilton, et al., Phys. Rev., C 53, 2273 (1996);
48. E. W. Cornell, et al., Phys. Rev. Lett. 77, 4508 (1996).
49. H.F. Xi, to be published (1998).

Formation of correlations at short time scales and consequences on interferometry methods

K. Morawetz
Fachbereich Physik, Universität Rostock, 18051 Rostock, Germany

H. S. Köhler
Physics Department, University of Arizona, Tucson, Arizona 85721

The formation of correlations due to collisions in an interacting nucleonic system is investigated shortly after a disturbance. Results from one-time kinetic equations are compared with the Kadanoff and Baym two-time equation with collisions included in second order Born approximation. A reasonable agreement is found for a proposed approximation of the memory effects by a finite duration of collisions. The formation of correlations and the build up time is calculated analytically for the high temperature and the low temperature limit. This translates into a time dependent increase of the effective temperature on time scales which interfere with standard fire ball scenarios of heavy ion collisions. The consequences of the formation of correlations on the two- particle interferometry are investigated and it is found that standard extracted lifetimes should be corrected downwards.

1 Introduction

To extract information about the space- time character of particle- emitting sources is one of currently much investigated tasks [1,2]. The size of stars are found to be measurable by coincident photons [3]. In laboratory this method is used to analyse heavy ion reactions [4]. The two-particle correlation function carries the information about nuclear interaction and correlations in the emitting source [5,6] and is given by the ratio of two - and single particle emission probability

$$R(P = p_1 + p_2, q = (p_1 - p_2)/2) + 1 = \frac{\Pi(p_1, p_2)}{\Pi(p_1)\Pi(p_2)} \tag{1}$$

which can alternatively be expressed by cross sections [7]. This two-particle correlation function can be written in terms of the the two-particle wave function $\varphi(q,r)$

$$R(P,q) + 1 = \int d^3r F_p(r) |\varphi(q,r)|^2 \tag{2}$$

while the average of the single particle Wigner functions of the source is given by [8,9]

$$F_p(r) = \frac{\int d^3R\, f(P/2, R + r/2, t^>)\, f(P/2, R - r/2, t^>)}{\left(\int d^3R\, f(P/2, R, t^>)\right)^2}. \tag{3}$$

Here $f(p, r, t>) = \int^{t^>} dt\, g(p, r - p(t^> - t)/m, t)$ is the phase-space Wigner distribution of particles of momenta p at position r at some time $t^>$ after emission with probability g. A rigorous derivation of this formula with the discussion of the necessary neglects has been given in [1] for the case of initially uncorrelated particles $|\varphi(p, r)|^2 \propto 1 \pm \cos(rq)$ corresponding to bosons/fermions. Eq. (3) is well suited for simulation of heavy ion reactions where the one-time distribution function $f(\mathbf{r}, \mathbf{p}, t)$ is then determined by solving appropriate kinetic equations (BUU) or simulating equation of motions (QMD). While most current codes rely on the quasiclassical Boltzmann equation including Pauliblocking effects, a quantum two-time theory for the time-evolution of real time Green's functions $G(\mathbf{r}, \mathbf{p}, t, t')$ has been developed using the Schwinger-Keldysh formalism already 30 years ago. The quantum image of the classical Boltzmann equation is usually referred to as the Kadanoff-Baym (KB) equations [10]. These equations have often been considered too complicated to solve numerically in the past. However, several numerical applications exist now [11].

These kinetic equations describe different relaxation stages. During the very fast first stage, correlations imposed by the initial preparation of the system are decaying [12,13]. These are contained in off-shell or dephasing processes described by two-time propagators. During this stage of relaxation the quasiparticle picture is established [14,15]. After this very fast process the second state develops during which the one-particle distribution relaxes towards the equilibrium value with a relaxation time. This is characterized by a nonlocal kinetic equation in agreement with the virial correction to the approached equation of state [16]. We will focus on the first stage which is related to the formation of correlations and will find a measurable effect on the two-particle correlation function. The extracted lifetimes are shown to be too high if one ignores this transient time effects. This can be considered to belong to the initial state correlations. While in [17,18] the initial state correlations are assumed to be of equilibrium type and discussed by density and temperature dependence we like to investigate here a nonequilibrium effect which consists in the fact that correlations need a specific time to be build up.

The formation of correlations is connected with an increase of the kinetic energy or equivalently the build up of correlation energy. This is due to rearrangement processes which let decay higher order correlation functions until only the one - particle distribution function relaxes.

2 Kinetic description

The time dependence of the kinetic energy as a one-particle observable will be investigated within the kinetic theory. This can only be accomplished if we employ a kinetic equation which leads to total energy conservation. It is immediately obvious that the ordinary Boltzmann equation cannot be appropriate for this purpose because the kinetic energy is in this case an invariant of the collision integral and constant in time. In contrast, we have to consider non-Markovian kinetic equations of Levinson type [19,20], which account for the formation of two particle correlations and which conserves the total energy [21] [22,23,24]

$$\frac{\partial}{\partial t} f_a(t) = \frac{2s_a}{\hbar^2} \sum_b s_b \int \frac{dp dq}{(2\pi\hbar)^6} V_D^2(q)$$

$$\times \int_{t_0}^{t} d\bar{t} \exp\left\{-\frac{t-\bar{t}}{\tau_g}\right\} \cos\left\{\frac{1}{\hbar}(t-\bar{t})\Delta_E\right\} \{f'_a f'_b \bar{f}_a \bar{f}_b - f_a f_b \bar{f}'_a \bar{f}'_b\}, \quad (4)$$

where $\Delta_E = \frac{k^2}{2m_a} + \frac{p^2}{2m_b} - \frac{(k-q)^2}{2m_a} - \frac{(p+q)^2}{2m_b}$ denotes the energy difference between initial and final states. The retardation of distributions, $f_a(k, \bar{t})$, $f'_a(k-q, \bar{t})$ etc., is balanced by the quasiparticle damping τ_g and $\bar{f} = 1-f$, the free particle dispersion $E = p^2/2m$ and the spin-isospin degeneracy s. The distribution functions are normalized to the density as $\int \frac{dp}{(2\pi\hbar)^3} f(p) = n/s$.

The Boltzmann collision integral is obtained from equation (4) if: (i) One neglects the time retardation in the distribution functions, i.e. the memory effects. This would lead to gradient contributions to the kinetic equation which can be shown to be responsible for the formation of high energy tails in the distribution function [25,26] and virial corrections [27,28,16]. This effect will be established on the second stage of relaxation. (ii) The finite initial time t_0 is set equal to $-\infty$ corresponding to what is usually referred to as the limit of complete collisions. This energy broadening or off-shell behavior in (4) is exclusively related to the spectral properties of the one-particle propagator and therefore determined by the relaxation of two-particle correlation.

Since we are studying the very short time region after the initial disturbance we can separate the one-particle and two-particle relaxation. On this time scale the memory in the distribution functions can be neglected $\bar{f}_a(\bar{t}) = f_a(0)$, $\exp\left\{-\frac{t-\bar{t}}{\tau_g}\right\} = 1$ but we will keep the spectral relaxation implicit in the off-shell cos-function of (4).

The eq. (4) can be integrated with respect to time and the resulting equation for $f(t)$ represents the deviation of Wigner's distribution from its initial value, $f_a(t) = f_a(0) + \delta f_a(t)$, and reads

$$\delta f_a(t) = 2\sum_b \int \frac{dpdq}{(2\pi\hbar)^6} V_{\rm D}^2(q) \frac{1-\cos\left\{\frac{1}{\hbar}t\Delta_E\right\}}{\Delta_E^2} \left\{f_a' f_b' \bar{f}_a \bar{f}_b - f_a f_b \bar{f}_a' \bar{f}_b'\right\}. \quad (5)$$

This formula shows how the two-particle and the single-particle concepts of the transient behavior meet in the kinetic equation. The right hand side describes how two particles correlate their motion to avoid the strong interaction regions. Since the process is very fast, the on-shell contribution to δf_a, proportional to t/τ, can be neglected in the assumed time domain and the δf has the pure off-shell character as can be seen from the off-shell factor $\Delta_E^{-2}\left(1-\cos\left\{\frac{1}{\hbar}t\Delta_E\right\}\right)$. The off-shell character of mutual two-particle correlation is thus reflected in the single particle Wigner's distribution.

The very fast formation of the off-shell contribution to Wigner's distribution has been found in numerical treatments of Green's functions [29,30]. Once formed, the off-shell contributions change in time with the characteristic time τ, i.e., following the relaxation (on-shell) processes in the system. Accordingly, the formation of the off-shell contribution signals that the system has reached the state the evolution of which can be described by the nonlocal Boltzmann equation [16], i.e., the transient time period has been accomplished.

From Wigner's distribution (5) one can readily evaluate the increase of the kinetic energy

$$E_{\rm corr}^{\rm static}(t) = -\sum_{ab} \int \frac{dkdpdq}{(2\pi\hbar)^9} V_{\rm D}^2(q) \frac{1-\cos\left\{\frac{1}{\hbar}t\Delta_E\right\}}{\Delta_E} f_a' f_b' \bar{f}_a \bar{f}_b. \quad (6)$$

This expression holds for general distributions f_a. We choose a model of two initially counter - flowing streams of nuclear matter and plot in figure 1 the time evolution of the kinetic energy and the correlation.

This build up of correlations is found to be independent of the form of the initial distribution. If for example we choose a (equilibrium) Fermi distribution as the initial distribution a build up of correlations will occur as well. This is due to the fact that the spatial correlations relate in momentum space to excitations, resulting in a distribution looking somewhat like a Fermi-distribution but with a temperature higher than that of the initial uncorrelated Fermi-distribution.[31]

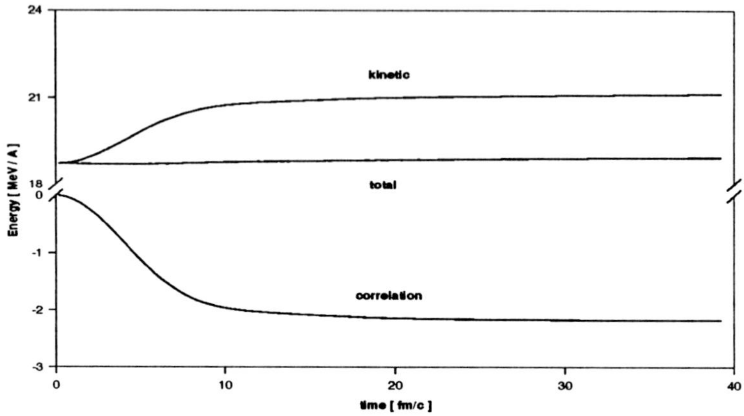

Figure 1: The time evolution of the correlation and kinetic energy for two initially counter - flowing streams of nuclear matter with density and temperature $n_1 = n_o/60 \quad T_1 = 0.5$ MeV and $n_2 = n_o/20 \quad T_2 = 0.1$ MeV moving with relative velocity of $1\hbar/fm$ which corresponds to a collision energy of 21 MeV/n. In this case the correlation time $\tau_c = 9 fm/c$. The equilibrated temperature is here (neglecting correlations) $T = 11.6$ MeV and a Gauß type of potential has been used.

3 The formation of correlations

We start now to calculate the observed increase of kinetic energy analytically. As a first example we shall consider a Yukawa-type potential of the form $V_{\rm Y}(r) = \frac{g_{ab}}{r} {\rm e}^{-\kappa r}$ where in nuclear physics applications g_{ab} is the coupling and κ the effective range of potential given by the inverse mass of interchanging mesons. As a second example we shall use a Gauß-type potential $V_G(r) = V_0 {\rm e}^{-(r/\eta)^2}$ which has been used in nuclear physics applications[32,31] with $\eta = 0.57 {\rm fm}^{-3}$ and $V_0 = -453$ MeV.

The high temperature limit of the time dependence of the correlation energy can be calculated analytically[33,15]. The low temperature value is of special interest, because it leads to a natural definition of the build up of correlations. One obtains

$$\frac{\partial}{\partial t} E_{\rm corr} = -\frac{1}{2} m^4 p_f \langle \frac{V^2}{\cos \frac{\theta}{2}} \rangle \tilde{I}_f \tag{7}$$

with standard notation of angular integrals[34] and finds the time dependence of the correlation energy[33]

$$
\begin{aligned}
E^{low}_{\rm corr}(t) - E^{low}_{\rm corr}(0) &= E^{low}_{\rm corr}(1+\frac{1}{3}(\frac{\epsilon_f+\epsilon_c}{\pi T})^2)^{-1} \\
\times &\left\{1-\frac{1}{x}\sin(x)+\left(\frac{\epsilon_f+\epsilon_c}{\pi T}\right)^2\left(\frac{1}{3}+\left[\frac{1}{x}\sin(x)\right]''\right)\right\}
\end{aligned}
\tag{8}
$$

with $x = 2\frac{\epsilon_f+\epsilon_c}{\hbar}t$ and the equilibrium correlation energy $E^{\rm low}_{\rm corr}$ for the Yukawa- or Gauß- potential reads[33]

$$
E_{\rm corr} = \frac{1}{6}s_1 s_2 \tilde{\epsilon}_f\left(T^2+\frac{1}{3}\left(\frac{4\tilde{\epsilon}_f}{\pi}\right)^2\right)\left(\frac{m}{\hbar^2}\right)^4\begin{cases}\frac{g^2}{2\kappa^3\pi^2}\left(\arctan\frac{1}{b_l}+\frac{b_l}{1+b_l^2}\right) & Yuk.\\ \frac{V_0^2\eta^5}{2\sqrt{2\pi}}{\rm erf}(\frac{p_f\eta\sqrt{2}}{\hbar}) & Gau\ss\end{cases}
\tag{9}
$$

where $\tilde{\epsilon}_f = \frac{\epsilon_f+\epsilon_c}{4}$ and $b_l = \frac{\hbar\kappa}{2p_f}$. The best choice for cut-off we found $\epsilon_c \approx \epsilon_f$. We see that the correlation energy is built up and oscillates around the equilibrium value damped with t^{-1} in time. We now define the build up time τ_c as the time where the correlation energy has reached its first maximum. This time is given by $\tau_c \approx \frac{\hbar}{\epsilon_F}$.

We assert here that the result (8) is valid for any binary interaction. We could use Born or T-matrix approximation and the same time dependence but different $E^{low}_{\rm corr}$ would result. This correlation time limits the validity of quasi-particle picture which is established at times greater than τ_c[15]. Incidentally, in the early 1950s the criterion $\hbar/k_BT < \tau$ was supposed to limit the validity of the Landau Fermi-liquid theory for metals[35]. Later it was shown by Landau that this criterion is irrelevant and he proposed the correct criterion $\tau > \hbar/\epsilon_F$.

In order to illustrate the weak temperature dependence of τ_c we plot in figure 2 (thick lines) results from the solution of the Kadanoff and Baym equations for a fixed chemical potential of 37.1 MeV and for three different temperatures. The figure shows the increase of the kinetic energy (equivalent to the decrease of correlation energy) with time. The KB results are compared with those from approximation (8). One sees that the agreement is good initially while correlations are built up. At low temperatures the oscillations discussed above are obvious in the approximate results while the KB calculations only show a slight overshoot at the lowest temperatures. We believe that the discrepancy is due to the damping that is neglected in (5).

We can give a Pade formula for the increase of kinetic energy which interpolates the results of K/B equation within the temperature range up to Fermi

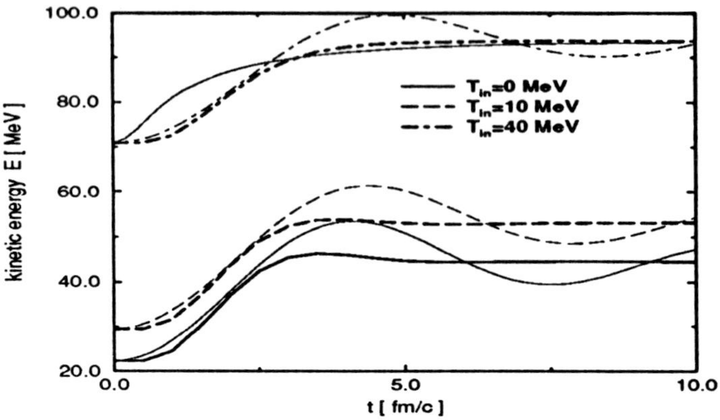

Figure 2: The formation of correlation plotted as an increase of the kinetic energy with time for temperatures $1, 10, 40$ MeV. The chemical potential is fixed to 37.1 MeV which corresponds to densities $0.16, 0.18, 0.35\text{fm}^{-3}$. The thick lines show results from KB calculations while the thin lines are approximate values via formula (8). The equilibrium correlation energy was chosen to be equal to the KB result. The oscillations are overestimated by the approximate formula. For $T = 40$ MeV we plotted also the high temperature approximate value [33] as thin solid line. The built up of correlations is too fast.

energy $\zeta_{\mathrm{K/B}}$ and the analytical result (8) signed $\zeta(t)$

$$\begin{aligned}
\zeta_{\mathrm{K/B}}(t) &= \frac{E_{\mathrm{corr}}(t)}{E_{\mathrm{corr}}(\infty)} = 1 - \frac{1.5225\ (1 - \cos(0.7337\,x) + \sin(0.6568\,x))}{\mathrm{e}^{0.5469\,x}\,x} \\
\zeta(t) &= 1 - \frac{\sin(x)}{x}
\end{aligned} \tag{10}$$

with $x = 4\epsilon_F t/\hbar$. The $\zeta_{\mathrm{K/B}}$ is a smooth curve through $\zeta(t)$. As we pointed out above, this result is universal for any binary interaction approximation. Therefore we can adopt this scaling as a universal scaling of the one-particle Wigner distribution. Provided we have solved any local Markovian kinetic equation like BUU, we can simply multiply the obtained Wigner function by the time dependent factor (10) in order to incorporate the transient time effect approximately. Since we fixed this factor to the observable kinetic energy, we expect to have correctly described the energy variables.

4 Application to interferometry

Now we employ the found scaling of short time effects to the formula (3) and adopt a space-time source parameterization of

$$g(p,r,t) = \rho_0 \Theta(r)\Theta(R_s - r)\Theta(t) p e^{-\frac{p^2}{2mT}} e^{-t/\tau} \zeta(t) \tag{11}$$

with the radius R_s and lifetime τ of the source and the scaling $\zeta(t)$ of (10). Without center of mass momentum P the correlation function becomes independent of lifetime and collapse to the standard interferrometry result [36]

$$R(0,q) + 1 = \frac{3}{4\pi R_s^3} \int_0^{2R_s} d^3r |\varphi(q,r)|^2 \left(1 - \frac{3}{4}\frac{r}{R_s} + \frac{1}{16}\left(\frac{r}{R_s}\right)^3\right). \tag{12}$$

Considering the momentum dependence and assuming initially free particles $|\varphi(p,r)|^2 \propto 1 \pm \cos(rq)$ we obtain a separable representation

$$R(P,q) + 1 = 1 \pm \Psi(q.R_s)\Pi(\zeta, q.p, \tau) \tag{13}$$

with

$$\Psi(x) = \frac{3}{2x^6}\left((-3 + 3x^2 - 2x^3 + 4x^4)\cos(2x) + (-6x - 6x^3 - 4x^4)\sin(2x)\right) \tag{14}$$

and

$$\Pi(\zeta, q.p/2m, \tau) = \frac{\int_0^\infty dt dt' e^{-(t+t')/\tau} \zeta(t)\zeta(t') \cos(q.p(t'-t)/2m)}{\left(\int_0^\infty dt e^{-t/\tau}\zeta(t)\right)^2}. \tag{15}$$

While the first factor Ψ gives the dependence of the correlation function on the source size, the second factor contains the effect of finite lifetimes. The result without transient time effects is that with increasing lifetime the correlation functions are diminished [9]. One gets for $\zeta = 1$ just

$$\Pi_0(1, q.p, \tau) = \frac{1}{1 + (q.p\tau/2m)^2}. \tag{16}$$

We can write the analytical result for (15) with $\zeta(t)$ from (10)

$$\Pi(a,c,\tau) = \left(4\left(\frac{-2\,a\,\tau}{1 + c^2\,\tau^2} + \arctan((a-c)\,\tau) + \arctan((a+c)\,\tau)\right)^2\right.$$

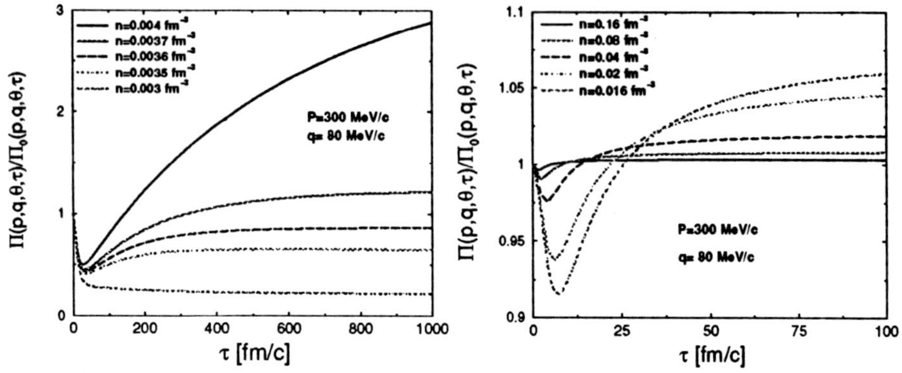

Figure 3: The ratio of correlation function Π of (15) with short time dynamics to Π_0 of (16) without short time dynamics versus lifetime of the source.

$$+ \left(\frac{-4\,a\,c\,\tau^2}{1+c^2\,\tau^2} - \log(1+(a-c)^2\,\tau^2) + \log(1+(a+c)^2\,\tau^2) \right)^2 \Bigg) \\ / \ (16\,(a\,\tau - \arctan(a\,\tau))^2) \qquad (17)$$

with $c = 4\epsilon_F/\hbar$ and $a = q.p/2m$. Figure 3 shows the influence of the transient time effect on the correlation function. We have plotted conveniently the ratio of reduced correlation functions Π of (15) with short time dynamics to Π_0 of (16) without short time dynamics.

We recognize that there is a suppression for smaller lifetimes. Since the effect of lifetime τ itself lowers the correlation function we can state that the extracted lifetimes from experiments are too high if short time or transient time effects are neglected. For higher lifetimes this effect is opposite. The extracted lifetimes from experiment should be too low. Interesting is the density dependence. While first with decreasing density the lowering/enhancement shifts to higher lifetimes, there is a critical density where we have only suppression. It is found to be around 0.0036fm^{-3}.

Due to the found universal scaling this effect is important in any recent simulation of BUU- type since the transient time effect of formation of correlations and therefore increase of kinetic energy has not been considered so far. For moleculardynamical methods this is in principle included but buried in artificial initial correlations due to the numerical set up. For a comparison between the kinetic and MD simulation see[15].

5 Summary

The gradient approximation of the kinetic equation in second order Born approximation is investigated. A finite duration approximation of the non Markovian collision integral is proposed which follows from time dependent Fermi's Golden Rule and which is in good agreement with the numerical solution of complete Kadanoff and Baym equation.

The build up time of correlations is investigated and it is found that the low temperature value is universal for any approximation at the binary collision level. It is shown that the formation time of correlations is nearly determined by the ratio of $\hbar$ to the transfer energy which can be considered as an analogue to the uncertainty principle.

A universal scaling is formulated which allows to map the Wigner function of Markovian simulation to a result including the transient time effect. As a consequence the influence to interferrometry methods is discussed. It is found that the extracted lifetimes from experiment should be too large for lower lifetimes and too small for higher values depending on the freeze out-density if transient time effects are neglected.

Acknowledgments

The authors like to thank P. Lipavský and V. Špička for interesting discussions and N. Kwong for many useful hints.

References

1. G. F. Bertsch, P. Danielewicz, and M. Herrmann. *Phys. Rev. C*, 49:442, 1994.
2. D. H. Boal, C. K. Gelbke, and B. K. Jennings. *Rev. Mod. Phys.*, 62:553, 1990.
3. R. Hanbury-Brown and R. Q. Twiss. *Nature (London)*, 178:1046, 1956.
4. J. Pochodzalla and et. al. *Phys. Rev. C*, 35:1695, 1987.
5. S. E. Koonin. *Phys. Lett. B*, 70:43, 1977.
6. W. G. Gong and et. al. *Phys. Rev. C*, 43:1804, 1991.
7. W. Bauer. *Nucl. Phys. A*, 471:604, 1987.
8. S. Pratt and M. B. Tsang. *Phys. Rev. C*, 36:2390, 1987.
9. W. G. Gong, W. Bauer, C. K. Gelbke, and S. Pratt. *Phys. Rev. C*, 43:781, 1991.
10. L. P. Kadanoff and G. Baym. *Quantum Statistical Mechanics*. Benjamin, New York, 1962.
11. H.S. Köhler. *Phys. Rev. E*, 53:3145, 1996.

12. N. N. Bogoljubov. *J. Phys. (USSR)*, 10:256, 1946. transl. in *Studies in Statistical Mechanics*, Vol. 1, editors D. de Boer and G. E. Uhlenbeck (North-Holland, Amsterdam 1962).
13. M. Bonitz and et. al. *J. Phys.: Condens. Matter*, 8:6057, 1996.
14. P. Lipavský, F. S. Khan, A. Kalvová, and J. W. Wilkins. *Phys. Rev. B*, 43(8):6650, 1991.
15. K. Morawetz, V. Špička, and P. Lipavský. Formation of correlations in plasma at short time scales. *Phys. Lett. A in press.*
16. V. Špička, P. Lipavský, and K. Morawetz. *Phys. Lett. A*, 240:160, 1998.
17. T. Alm. PhD thesis, Universität Rostock, 1992. report GSI-93-07,Darmstadt.
18. T. Alm, G. Röpke, and M. Schmidt. *Phys. Lett. B*, 301:170, 1993.
19. I. B. Levinson. *Fiz. Tverd. Tela Leningrad*, 6:2113, 1965.
20. I. B. Levinson. *Zh. Eksp. Teor. Fiz.*, 57(2):660, 1969. [Sov. Phys.–JETP **30**, 362 (1970)].
21. K. Morawetz. *Phys. Lett. A*, 199:241, 1995.
22. A. P. Jauho and J. W. Wilkins. *Phys. Rev. B*, 29(4):1919, 1984.
23. K. Morawetz, R. Walke, and G. Röpke. *Phys. Lett. A*, 190:96, 1994.
24. H. Haug and A. P. Jauho. Springer, Berlin Heidelberg, 1996.
25. K. Morawetz and G. Röpke. *Phys. Rev. E*, 51(5):4246, 1995.
26. V. Špička and P. Lipavský. *Phys. Rev. B*, 52(20):14615, 1995.
27. V. Špička, P. Lipavský, and K. Morawetz. *Phys. Rev. B*, 55(8):5084, 1997.
28. V. Špička, P. Lipavský, and K. Morawetz. *Phys. Rev. B*, 55(8):5095, 1997.
29. P. Danielewicz. *Ann. Phys. (NY)*, 152:305, 1984.
30. H. S. Köhler. *Phys. Rev. C*, 51:3232, 1995.
31. H.S. Köhler. *Phys. Rev. C*, 51:3232, 1995.
32. P. Danielewicz. *Ann. Phys. (NY)*, 152:239, 1984.
33. K. Morawetz and H. S. Koehler. *Phys. Rev. C*, 1997. sub.
34. H. Smith and H. Hojgaard-Jensen. Clarendon Press, Oxford, 1989.
35. R. Peierls. *Quantum Theory of Solids.* Oxford University Press, London, 1955.
36. B. K. Jennings, D. H. Boal, and J. C. Shillcock. *Phys. Rev. C*, 33:1303, 1986.

SEPARATION BETWEEN SOURCES OF PIONS AND PROTONS IN CENTRAL AU+AU COLLISIONS AT THE AGS (E877).

D. MIŚKOWIEC
Gesellschaft für Schwerionenforschung mbH
Planckstr. 1, D-64291 Darmstadt, Germany
E-mail: D.Miskowiec@gsi.de

Two-particle pion-pion, proton-proton, and pion-proton correlations were measured in central Au+Au collisions at 11 GeV/c per nucleon. The data were analyzed using one- and three-dimensional correlation functions. The pion-proton correlation functions exhibit asymmetry which indicates that the sources of protons and pions are separated in the direction along the accelerator beam.

1 Introduction

The two-particle correlation function, defined as the ratio between the two-particle density and the product of single particle densities

$$C(\mathbf{p}_1, \mathbf{p}_2) = \frac{\sigma \dfrac{d^6\sigma}{d^3p_1 d^3p_2}}{\dfrac{d^3\sigma}{d^3p_1}\dfrac{d^3\sigma}{d^3p_2}} , \tag{1}$$

contains important information about the space-time-momentum distribution at the time of last interaction (freeze-out) of particles produced in a nuclear collision. [1] The correlations of identical pions, dominated by the Bose-Einstein statistics and by mutual Coulomb interaction, exhibit a peak at $\mathbf{q} := \mathbf{p}_2 - \mathbf{p}_1 = 0$, the width of which is inversely proportional to the source size. Analysis of C as a function of different components of the momentum difference yields the source dimensions in different directions. The two-proton correlation is a product of the Fermi statistics as well as the mutual Coulomb and strong interactions. This correlation shows a peak at the relative momentum $q \approx$ 40 MeV/c, the amplitude of which is roughly inversely proportional to the volume of the proton source. The unlike-pions and pion-proton correlations are generated mostly by the Coulomb interaction and thus have an exponential-like shape at $\mathbf{q} = 0$. Similarly as for identical pions, even if with somewhat lower sensitivity, the width of this peak depends on the average r.m.s. distance between the particles at freeze-out.

2 Experiment and data analysis

I present here results of the correlation analysis of particles produced in central Au+Au collisions at 11 GeV/c per nucleon. Five millions central (4% of the geometrical cross section) collision events were recorded in Fall 1994 by the E877 Collaboration at the AGS. The experimental setup, presented in Figure 1, consisted of three calorimeters, used to determine the centrality and the orientation of the reaction plane and to generate the centrality trigger, and a forward spectrometer for measuring momenta of pions and protons. The momentum resolution of the spectrometer was $\Delta p/p \approx 3\%$. The acceptance allowed simultaneous measurement of positive and negative particles (Figure 2).

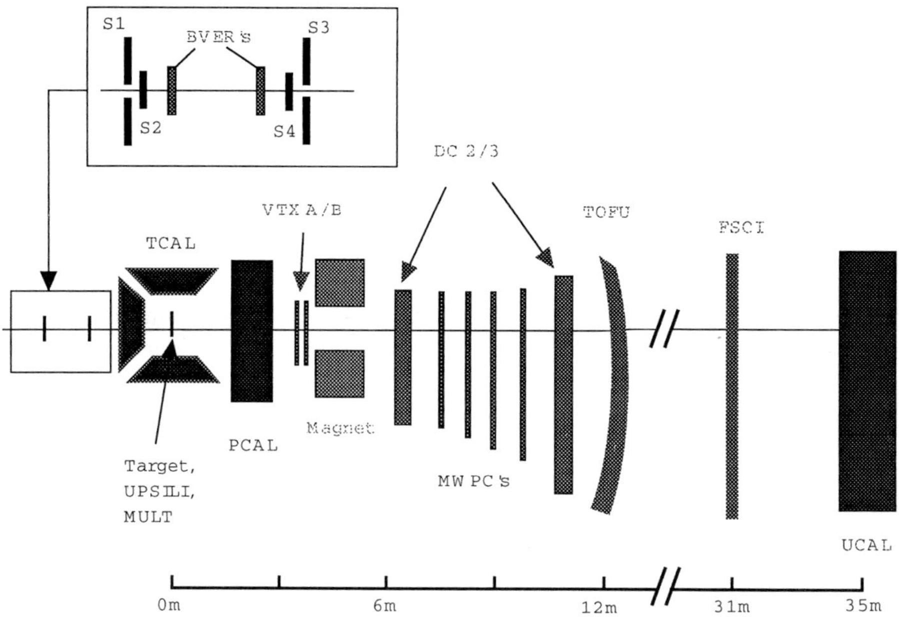

Figure 1: The E877 setup at the AGS. Three calorimeters, TCAL, PCAL, and UCAL, were used to determine the collision centrality and the orientation of the reaction plane. The particles used in the correlation analysis were measured in the forward spectrometer.

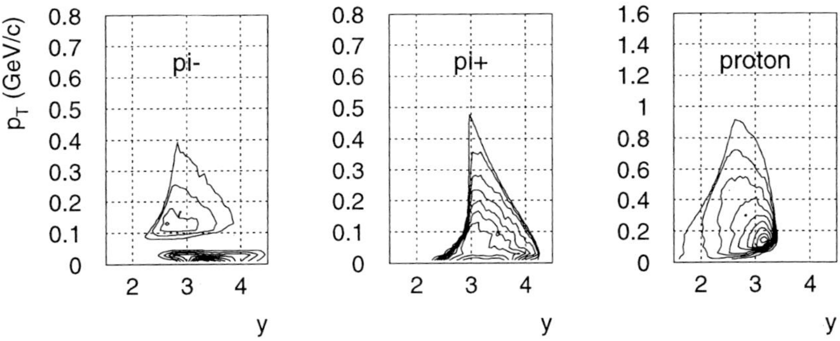

Figure 2: Distribution of analyzed particles in transverse momentum and rapidity. The target and the beam rapidities were 0 and 3.1, respectively. The hole in the acceptance for negative pions was caused by the accelerator beam going through the detectors.

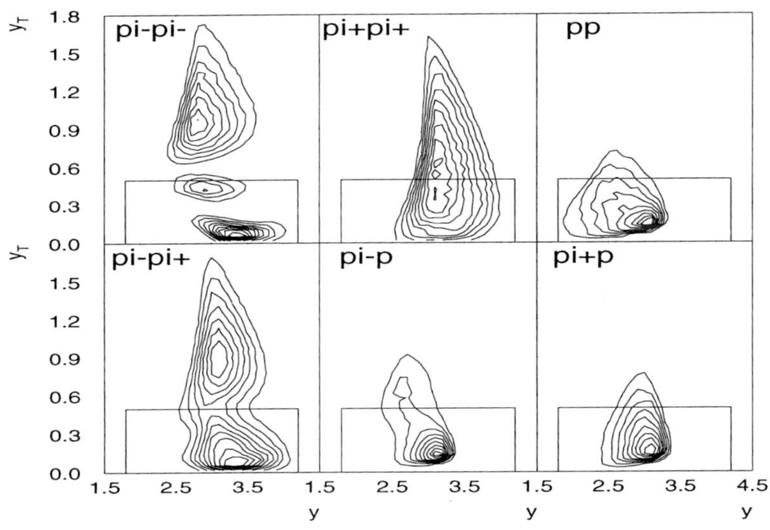

Figure 3: Distribution of accepted pairs. Pair rapidity y is defined as $\frac{1}{2}\log((E+P_z)/(E-P_z))$, with E and P_z being the pair energy and longitudinal momentum in the LAB frame. The pair transverse rapidity $y_{\rm T}$ is defined as $\frac{1}{2}\log((E^*+P_{\rm T})/(E^*-P_{\rm T}))$ where $E^* = \sqrt{E^2-P_z^2}$ is the pair energy in the longitudinally comoving system (LCMS), and $P_{\rm T}$ is the pair transverse momentum. The rectangle represents the analysis cut.

The numerators of the correlation functions were generated using pairs of identified particles detected in the forward spectrometer. The distributions of the analyzed particles and particle pairs are shown in Figures 2 and 3, respectively. The denominators were obtained by event mixing. We mixed only events with similar centralities and similar orientations of the reaction plane. The two track resolution problems were avoided by rejecting from the numerator and the denominator all track pairs which did not have enough separation in the tracking detectors. The correlation functions were normalized to unity at the flat part between $q = 100$ MeV/c and $q = 400$ MeV/c. No Coulomb correction was applied. The relative momentum $\mathbf{q}$ was evaluated in the pair c.m.s. which is the natural choice when analyzing correlation functions of particles with different masses. The q_x, q_y, and q_z components are defined as follows: $\hat{\mathbf{z}}$ is the accelerator beam axis, $\hat{\mathbf{x}}$ is in the reaction plane, and $\hat{\mathbf{y}}$ is perpendicular to the beam and to the reaction plane. The reaction plane is determined from the azimuthal distribution of the transverse energy. The sign of x is defined such that the transverse flow of forward protons is in the positive-x direction.

3 Results

The obtained one-dimensional correlation functions are presented in Figure 4. The correlation functions of identical pions are dominated by the positive Bose-Einstein correlation, slightly suppressed by the repulsive Coulomb interaction. The bumb in the two-proton correlation is caused by the attractive strong interaction, and the minimum at low relative momenta is caused by the Coulomb interaction and the Fermi statistics. The non-identical particle correlations are mainly due to Coulomb interaction. The little peak in the π^-–proton correlation function is caused by Λ^0 decays. The position and the width of this peak was used as a check of the momentum calibration and of the knowledge of the momentum resolution.

The line represents the result of a calculation described in the next section. The calculation overestimates both pion-proton correlation functions while all other correlations are reasonably reproduced.

The projections of the three-dimensional correlation functions are shown in Figure 5. The projections were done while restricting the other two components of $\mathbf{q}$ to the range from -50 MeV/c to 50 MeV/c. The line, again, represents the calculations. Inspection of Figure 5 allows to trace back the origin of the disagreement between the data and the calculation in the pion-proton correlations to the asymmetric shape of the peak in the data, not reproduced by the calculation.

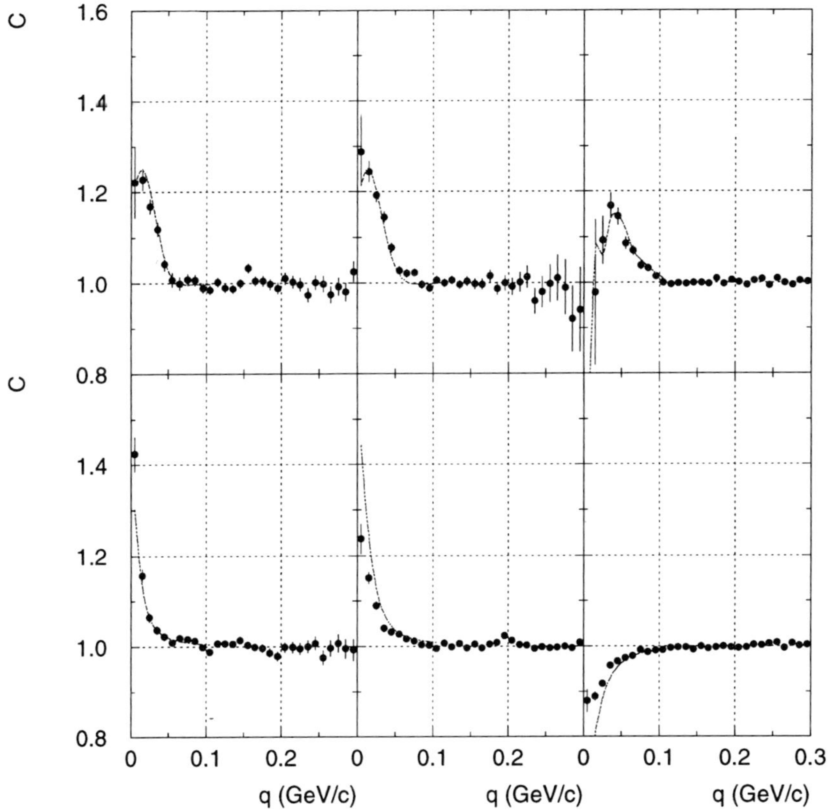

Figure 4: One-dimensional correlation functions vs. relative momentum in the pair c.m.s.

As observed by Lednicky *et al* [2], an asymmetric peak in the correlation function between non-identical particles is a sign of spatial or temporal separation between the sources of the two particle species. (The mechanism of this effect was explained by B. Erazmus in her talk during this conference.) Since the asymmetry in our data is in the q_z direction of both pion-proton correlations, we conclude that there is a finite separation between the sources of pions and protons in the beam direction. The sign of the asymmetry indicates

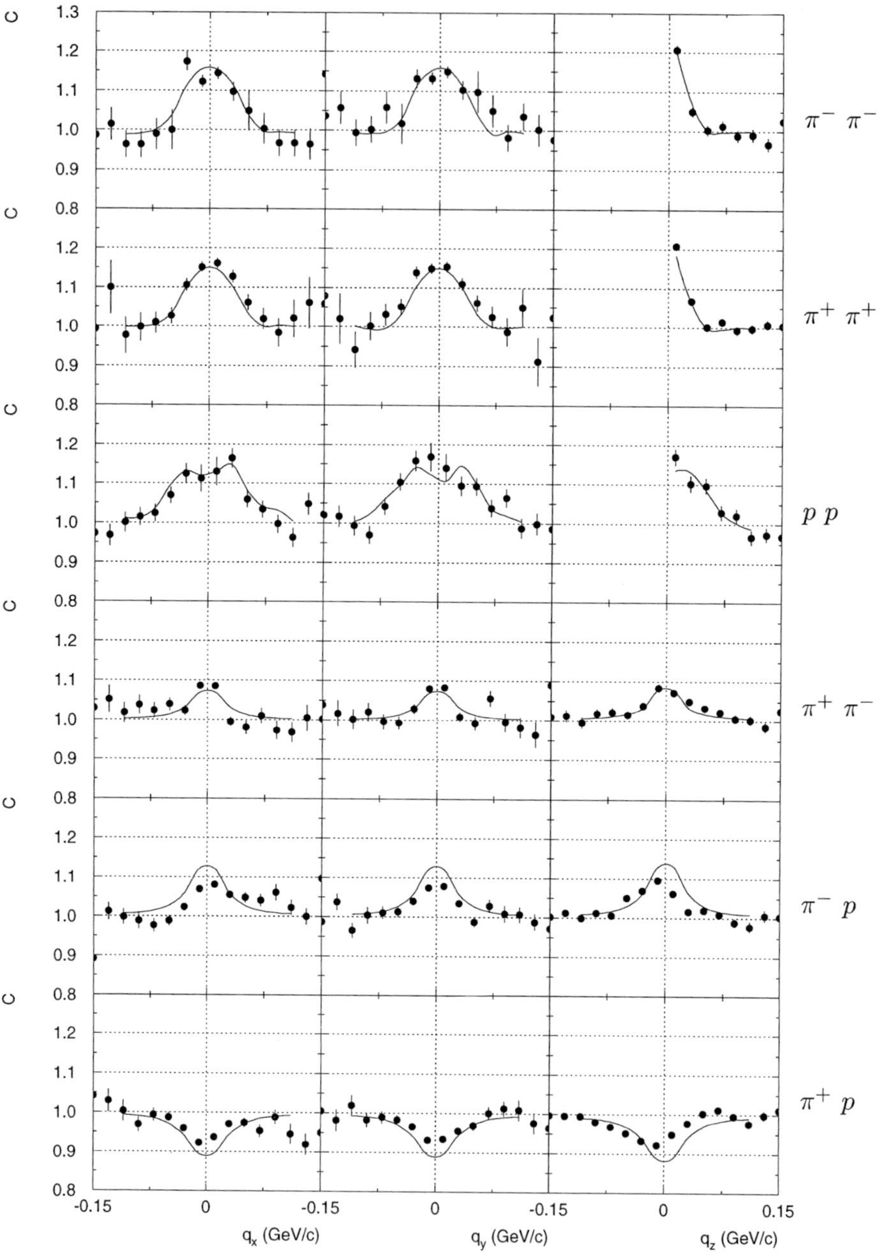

Figure 5: Projections of three-dimensional correlation functions.

that at freeze-out protons are more forward than pions with similar (forward) rapidities. In order to estimate the size of this displacement we performed the calculation described below.

4 Calculation

We generated pairs of particles. Each particle was described by four position and four momentum coordinates. The position coordinates were generated according to a double-gaussian distribution for pions (core-halo model [4]) and a single-gaussian for protons:

$$g^{\text{pion}}(\mathbf{r}) = f\ \exp\left(-\frac{x^2+y^2}{2R_\text{T}^2} - \frac{z^2}{2R_\text{z}^2}\right) + (1-f)\ \exp\left(-\frac{r^2}{2R_\text{halo}^2}\right) \quad (2)$$

$$g^{\text{proton}}(\mathbf{r}) = \exp\left(-\frac{x^2+y^2}{2R_\text{T}^2} - \frac{z^2}{2R_\text{z}^2}\right) \quad (3)$$

The parameter f represents the fraction of pions coming from the core, i.e. not from long-lived resonances. It is related to the correlation strength λ by

$$f^2 = \lambda\ . \quad (4)$$

The momenta were sampled from the experimental momentum distributions. The Lednicky code [5,6] was used to calculate a weight for every pair. The code accounts for the (anti)symmetrization effect as well as for the Coulomb and strong interactions. Subsequently, the momenta were smeared by the experimental momentum resolution and the two-track separation cut was applied. Finally, every pair was processed through the same analysis as the experimental data, except that the Lednicky's weight and unity were used to increment the histograms for the numerator and the denumerator of the correlation function, respectively (no event mixing).

The parameters of the gaussian position distributions were adjusted to fit the data. The three-dimensional proton-proton correlation function allows to determine the transverse and longitudinal proton source size, even if the coupling between the direction of $\mathbf{q}$ and $\mathbf{r}$ is not as clean as it is for pions. The best agreement was obtained for $R_\text{T}^\text{proton} = 4.75$ fm and $R_\text{z}^\text{proton} = 3.25$ fm. The $\pi^-\pi^-$, $\pi^+\pi^+$, and $\pi^+\pi^-$ correlations could be simultaneously reproduced with $f = 0.72$, $R_\text{T}^\text{pion} = 5.0$ fm, $R_\text{z}^\text{pion} = 7.8$ fm, and $R_\text{halo}^\text{pion} = 30$ fm. The correlation functions calculated using these parameters are represented by lines in Figures 4 and 5. The pion-proton correlations, calculated using the same parameter values and assuming that the pion and proton sources are centered at the same point, disagree with the data.

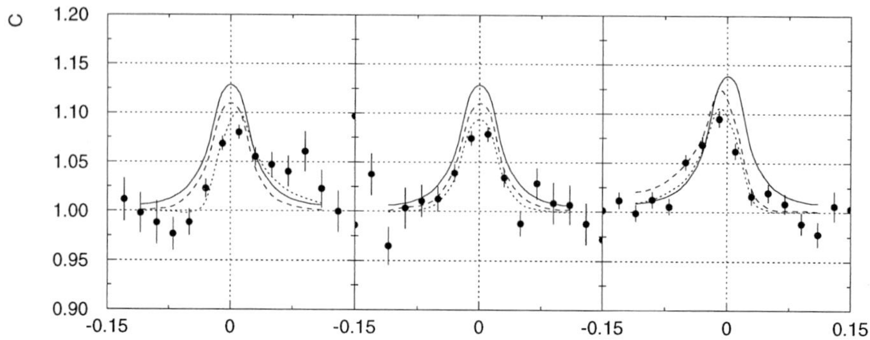

Figure 6: Projections of three-dimensional $\pi^- p$ correlation functions. The solid, dashed, and dotted lines represent calculations with ($\Delta x = 0$ fm, $\Delta z = 0$ fm), ($\Delta x = 0$ fm, $\Delta z = 10$ fm), and ($\Delta x = -10$ fm, $\Delta z = 10$ fm), respectively. $\mathbf{\Delta r}$ is defined as the difference between the positions of the sources of protons and pions: $\mathbf{\Delta r} = \langle \mathbf{r}^{\mathrm{proton}} \rangle - \langle \mathbf{r}^{\mathrm{pion}} \rangle$.

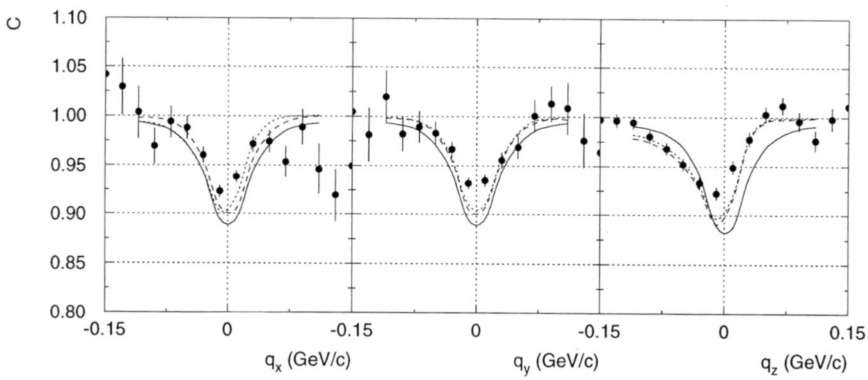

Figure 7: Projections of three-dimensional $\pi^+ p$ correlation functions. The solid, dashed, and dotted lines represent calculations with ($\Delta x = 0$ fm, $\Delta z = 0$ fm), ($\Delta x = 0$ fm, $\Delta z = 10$ fm), and ($\Delta x = 5$ fm, $\Delta z = 10$ fm), respectively.

The discrepancy can be removed by assuming a finite displacement between the sources of pions and protons. Calculations using different values of the displacement are compared to the data in Figures 6 and 7. A reasonable agreement was achieved with a 10 fm separation in z and -10 fm or +5 fm, depending on the charge of the pions, in x.

5 Conclusion

The corresponding source distributions are shown in Figure 8. The proton source is located 10 fm more forward than the sources of pions. The π^- source is, in addition, shifted in the reaction plane by 10 fm in the direction of positive x, i.e. in the direction of the flow of the forward protons. The π^+ source is shifted by 5 fm towards negative x.

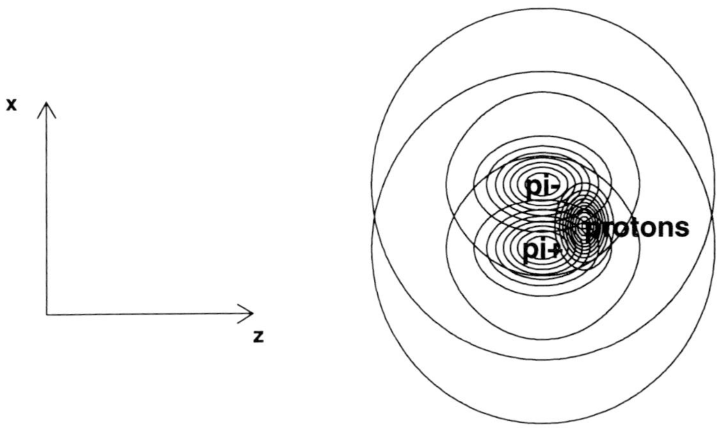

Figure 8: The π^-, π^+, and proton source distributions yielding the best fit to the experimental correlation functions.

A similar displacement of the pion and proton sources was observed in the RQMD generated events. [7] In hydrodynamical models the displacement in z is a natural consequence of longitudinal expansion combined with the mass difference between pions and protons. [8,9] However, the magnitude of the observed displacement seems large compared to the size of the system and the expected duration of the emission. Indeed, even assuming that the pion source stays at the origin (in the nucleus-nucleus c.m. system), and the proton source moves with beam rapidity, it would take $\tau_f = 10$ fm/c proper time to reach the separation of 10 fm. During this time the system would expand to the total length of

$$L = 2\sinh(\frac{y_\mathrm{B}}{2})\ \tau_f \approx 50\ \mathrm{fm}\ . \tag{5}$$

Since it hard to imagine that the pion source does not move at all, this value can be considered as the lower limit.

In summary, asymmetry in non-identical particle correlation functions was observed for the first time. The asymmetry indicates a displacement between the sources of pions and protons. The magnitude of this displacement is related to the total duration of the reaction.

References

1. D. H. Boal, C-K. Gelbke, and B. K. Jennings, *Rev. Mod. Phys.*62553 1990.
2. R. Lednicky *et al*, *Phys. Lett.* B **373**, 30 (1996).
3. B. Erazmus, this conference.
4. T. Csörgő, B. Lörstad, and J. Zimányi, *Z. Phys.* C **71**, 491 (1996).
5. R.Lednicky, V.L.Lyuboshitz, Sov.J.Nucl.Phys. 35 (1982) 770.
6. R.Lednicky *et al*, SUBATECH-94-22, Nantes).
7. R. Lednicky, S. Panitkin, and S. Voloshin, *Phys. Rev. Lett.* **79**, 4766 (1997).
8. U. Heinz, private communication.
9. T. Csörgő, private communication.

TEMPERATURE AND SIZE OF EXPANDING NUCLEI IN AU+AU COLLISIONS

G. IMMÉ, N. GIUDICE, G.RACITI, G. RICCOBENE, F.P. ROMANO, A. SAIJA, C. SFIENTI, G.VERDE

Dipartimento di Fisica - Universitá di Catania and I.N.F.N. - Laboratorio Nazionale del Sud and Sezione di Catania
I-95100 Catania, Italy

M.BEGEMANN-BLAICH, S. FRITZ, C. GROß, U. KLEINEVOß, W.D. KUNZE, U. LYNEN, M. MAHI, W.F.J. MÜLLER, B. OCKER, T.ODEH, M. SCHNITTKER, C. SCHWARZ, V. SERFLING, W. TRAUTMANN, A. WÖRNER

Gesellschaft für Schwerionenforschung mbH, Planckstrasse 1
D-64291 Darmstadt, Germany

W. SEIDEL

Forschungszentrum Rossendorf ,
D-01314 Dresden, Germany

J. POCHODZALLA, A.SCHÜTTAUF

Max Planck Institut Heidelberg ,Saupfercheckweg 1
D-69177 Heidelberg, Germany

S. GAFF, G.J. KUNDE, T. MÖHLENKAMP, H. XI

National Superconducting Cyclotron Laboratory Michigan State University ,
East Lancing,MI 48824, USA

R. BASSINI, I. IORI

I.N.F.N. Sezione di Milano and Dipartimento di Fisica, Universitá di Milano,
I-20133 Milano, Italy

A. TRZCINSKI, B. ZWIEGLINSKI

Soltan Institute for Nuclear Studies ,
00-681 Warsaw,Hoza69, Poland

The differences, increasing with the beam energy, of temperatures obtained from isotopic yields and from populations of particle- unstable resonances have stimulated the interest in the time-space evolution of the emitting source. Fragments from the interaction zone in central Au+Au collisions at incident energies between 100 and 200 AMeV were studied. In order to explore the statistical and dynamical evolution of an hot expanding source, we measured correlation functions of light charged particles. In particular, from the correlation functions of d-alpha and p-alpha pairs, large and constant radius parameters are obtained, while, instead, from proton pairs they are dependent on the beam energy. The deduced nuclear densities, smaller than the saturation density, reveal source expansion, that can determine cooling of the decaying system. It is intriguing that the constant low values of density and temperatures are both related with alphas and heavy clusters that seem to sample the cooling source at the latest stages of the expansion.

In order to understand the thermodynamic properties of strongly interacting systems under extreme conditions of pressure and temperature, it is interesting to measure simultaneously freeze-out temperature and density of the system. In particular temperatures can be determined through two different thermometers: the isotopic yield double ratios (*Isotope Temperature* [1]), as already used to extract the well known *nuclear caloric curve* [2] and population ratios of excited states of emitted fragments (*Emission Temperature* [3,4]). Complementary information on the density of the system at freeze-out can be provided by particle interferometry of light charged particles.

To this aim we studied central collisions between symmetric systems at intermediate energies, where, in the geometric participant spectator scenario, the overlap region was investigated. In this case the available center of mass energy is shared between random (thermal) energy and collective expansion motion. During the collision, it is assumed that the participant region is strongly compressed, afterwards the system expands and the compression energy is transferred to collective radial flow.

In particular we studied Au + Au reactions at 100, 150 and 200 AMeV [5,6], with beams provided by the heavy-ion synchrotron SIS at the GSI laboratory. The experimental set-up (shown in Fig.1) mainly consisted of three, high granularity, hodoscopes of Si-CsI(Tl) telescopes (in total 216), that, in closely packed geometries, were placed on opposite sides with respect to the beam axis. Their angular resolution and granularity were optimized in order to permit the detection of correlated particles. The hodoscopes were placed in the forward hemisphere in order to cover the angular region ($24^o \leq \theta_{lab} \leq 58^o$) close to 90° in the center of mass system, in order to minimize contributions from target like and projectile like sources and to obtain a good coverage at mid-rapidity. Additional detectors were employed in order to probe the charged-particle multiplicity for impact parameter selection. The angular range $6^o \leq \theta_{lab} \leq 20^o$ was

Figure 1: Experimental setup at the ALADIN spectrometer for the measurements with Au beams of 100-200 MeV/u.

covered by an azimuthally symmetric array of 36 CaF_2- plastic phoswich detectors. Within the angular range 30° to 55°, a solid angle of 0.7 sr was covered by an array of 48 elements of Si-strip detectors. For more experimental details see ref. [5].

In Fig. 2 isotope (T_{HeLi}) and emission temperatures from ^{5}Li decay are compared[5]. In particular for central collisions the differences between the two thermometers show up very pronounced. With increasing beam energy there is a growing disagreement between the two temperatures, and this discrepancy increases up to 200 AMeV incident energy, where the isotopic temperatures are higher by a factor of two compared to the temperatures from excited states. The dynamical evolution of the fragment formation [7] and emission [8,9] has recently been investigated. Although in ref. [7] is suggested that the asymptotic fragments can be identified during the early stages of the process, a time delay between the emission of p, d, t and ^{3}He on one side and alpha and larger clusters on the other side, is predicted in the Friedmann "Expanding Emitted Source" model[8].

In order to take into account the collective radial flow [10] in the cooling of the expanding system formed in central collisions, in Fig. 3 let us consider an adiabatic expansion of an ideal gas. Starting within a given volume with initial temperature deduced from the total available CM energy, subtracted of the radial flow, we follow (lines in Fig.3) the temporal evolution of the cooling system, expanding with radial flow velocity from the systematics of ref.[11]. The comparison with the experimental temperatures indicates that, while the T_{HeLi} isotope temperature reflects the one at the first stages of the

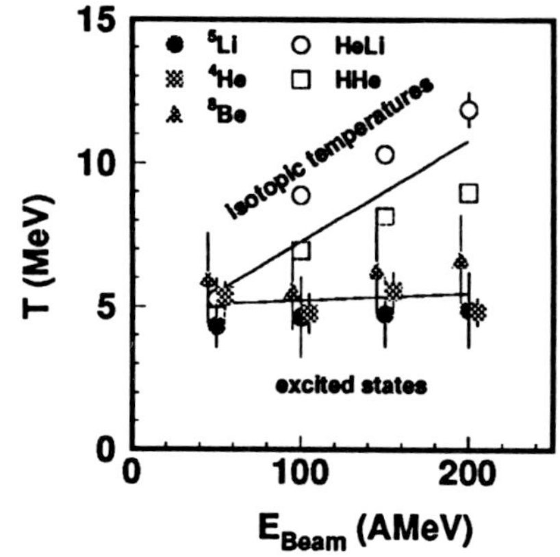

Figure 2: Comparison between isotope and emission temperatures for central collisions.

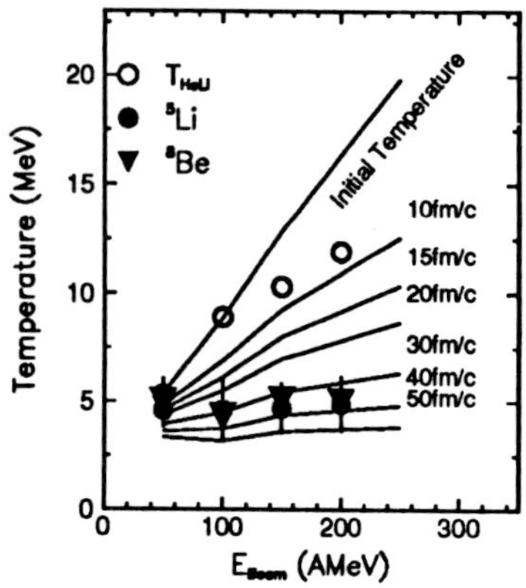

Figure 3: Adiabatic expansion of an ideal gas. Points refer to experimental values of Fig.2.

expanding system, the cluster (^{5}Li, ^{8}Be) temperatures recall the system in the later stages of the adiabatic process. This rough comparison opens the question whether the different emission time between particles and cluster [8] and the longest time (lower density) required by the excited fragments to get their properties, are the reason of the difference in the measured temperatures.

In order to shed light on this question the different freeze-out conditions, in particular the temporal evolution of the expanding system, can be probed by particle - particle correlations. In fact the expanding source scenario also influences the p-p correlation function, from which information about the space-time behavior of the source may be obtained [12]. Then, following the common approach used in particle-particle HBT interferometry [13,14], we construct the experimental correlation functions from measured coincident proton events.

The correlation functions for proton pairs from reactions at 100, 150 and 200 AMeV incident energies [15,16] are reported in Fig.4. The uncorrelated background was generated by event mixing and normalized to the coincident yield at large relative momenta q$\geq$ 150 MeV/c. In particular we have to note that the peak at q=20 MeV/c, corresponding to the final state interaction between two protons, becomes more pronounced with increasing bombarding energy. For a given impact parameter value, due to the constant size of the overlap system, one expects naively equal emission radii for any beam energy. However, the correlation function rises with incident energy, as shown in Fig.4,

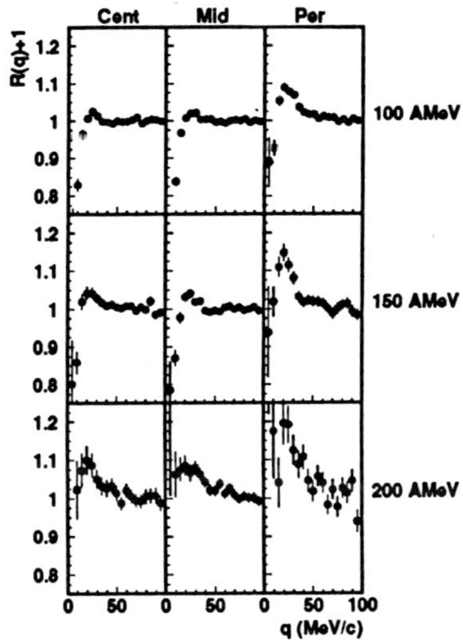

Figure 4: p-p correlations for different impact parameter with incident energies of 100,150 and 200 AMeV.

both for peripheral and central collisions. In order to select the same source as in the temperature measurements, we'll limit our analysis to the central collisions.

In order to deduce reliable values for the radius parameter from the experimental correlations several contributions must be taken into account. From Quantum Statistical Model (QSM) [17,15] calculations, with parameters selected to describe measured yields, the fractions of protons emitted from long lived resonances were estimated to be 0.21, 0.29 and 0.31 for 100, 150 and 200 AMeV, respectively. Source sizes were deduced by comparing the experimental correlation functions with predictions of the Koonin-Pratt formalism [18] corrected for the emission of long lived resonances. A Monte Carlo generated expanding emission source with zero lifetime and Maxwellian shapes for the energy spectra served as input for this theoretical description.

The occurence of the collective expansion, that represents a significant fraction (about 40%) of the energy available in the center of mass frame, as confirmed by our kinetic energy spectra analysis [6], was taken into account. In particular, in the collective motion, superimposed to the radial flow, the coulomb energy was included [15], due to the coulomb interaction of the two protons with the remainder core. The large charge of the emitter, in fact, strongly

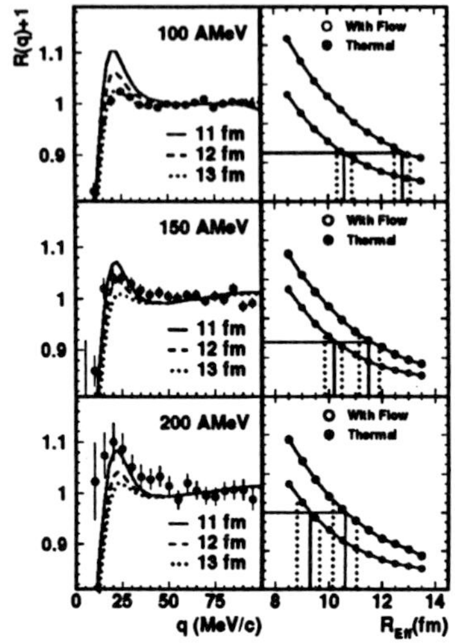

Figure 5: Central p-p correlation functions. Influence of radial flow. In the right part peak height is reported vs radius.

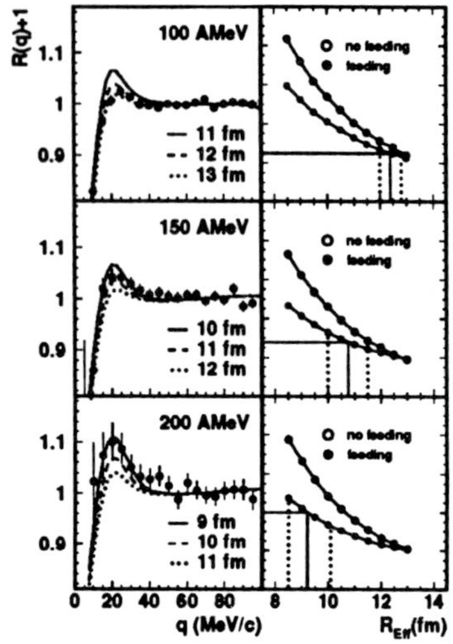

Figure 6: Central p-p correlation functions. Influence of the sequential feeding decay.

influences the proton trajectories, producing a weakening and a smearing of the correlation function. The simulated protons were filtered through the detection set-up acceptance.

In Fig.s 5, 6 experimental p-p correlation functions for central impact parameter and for the three incident energies are compared with calculations for different radii, and with the correction due to the flow contribution, (Fig. 5), and, considering in addition contribution from sequential feeding, (Fig. 6). In the right part of the figures the dependence of the height of the peak at q=20 MeV/c of the correlation function on the source radius is drawn. If we compare the extracted radii values with those deduced from a purely thermal source, we find a deviation of 10-20 % when flow contribution is taken into account. A contribution of the same order is obtained by including the sequential feeding effect. This deviation becomes 30% in the emission density. The extracted radii, from 12.4 fm at E_{inc} = 100 AMeV to 9.2 fm at 200 AMeV, evidence that the space-time extent of the emitter decreases as the bombarding energy increases. Notice that, however, these values are larger than the radius of the combined system of two Au nuclei at ground state density, further confirming an expansion of the system.

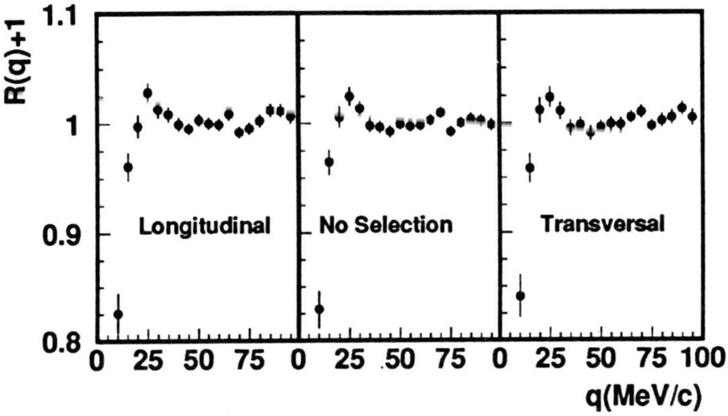

Figure 7: Transversal and longitudinal correlation function for E_{inc} =100 AMeV.

A possible scenario could be that, for the high incident energy, the protons come from a source with higher density, due to the strong compression in the first stage of the reaction, with respect to the protons in the low incident energy case, that seem to be emitted only after a certain expansion of the system and therefore from a more diluted source. That the two - particle correlations at small relative momenta show sensitivity to the expansion dynamics could be confirmed by investigation on the emission time. However, the big size of the system is decisive for the sensitivity lacking in the comparison between transversal and longitudinal correlation functions that result very similar each other, as shown in Fig.7. Nevertheless, due to the expansion of the system, we can obtain information on the emission time from the density, that we can deduce from the extracted radii, knowing the expansion rate. Break-up densities, compared with a systematic study for different systems, performed from the Diogene collaboration at the Saturne accelerator [19], show (Fig.8) a general trend of a rise with increasing beam energy, that is substantially more pronounced for the gold system, indicating strong dependence on the system size.

In order to get more information on the time-space extent of the emission source we considered also correlation functions for different pairs of particles, p-alpha and d-alpha. As reported in Figs. 9, 10 they exhibit a different behavior than observed for p-p correlations. The peak due to the scattering in the 3^+ state of ^{6}Li (for d-alpha) and the ground state $3/2^-$ (for p-alpha)

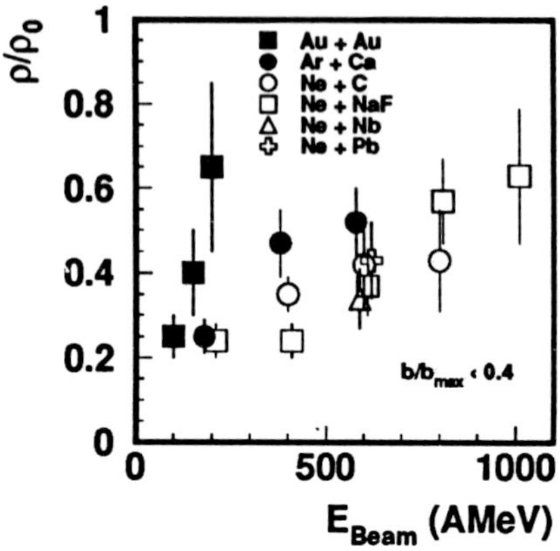

Figure 8: Systematics of the freeze-out densisties for different systems and incident energies [19].

remain unchanged for all incident energies. By performing similar analysis as the one for the p-p correlation functions, including the already named effects, we extracted radii also from d-alpha and p-alpha correlation functions.

In Fig. 11 we report the extracted radii versus the incident energy. The radii extracted from p-p correlation drop with increasing beam energy, in contrast to the d-alpha and p-alpha correlation functions that refer to constant radii. These differences could be related to the lower density reached by the expanding system at the freeze-out time, when alpha correlations are sampling the emitting source. This scenario agrees well with the Friedmann prediction [8] that the emission time for alpha and larger clusters is delayed to respect to p, d, t, ^{3}He.

It is interesting to correlate this information with the discrepancies (Fig. 2) between the freeze-out temperatures deduced by double ratios of isotopes and from ratios of unbound excited states. The different energy dependence of the two thermometers can be explained by different freeze out conditions (i.e. time). In the initial stage of the reaction the density will be high due to the compression in the collision, and fragments will not yet have their identity. Below a certain density fragments will be formed. When the fragments have reached their final identity (chemical equilibrium is achieved) the isotopic ratios temperatures reflect this condition. A further step in time order is the formation of excited states, that will be formed later in time as compared to stable fragments. Therefore the lower temperatures extracted from the cluster

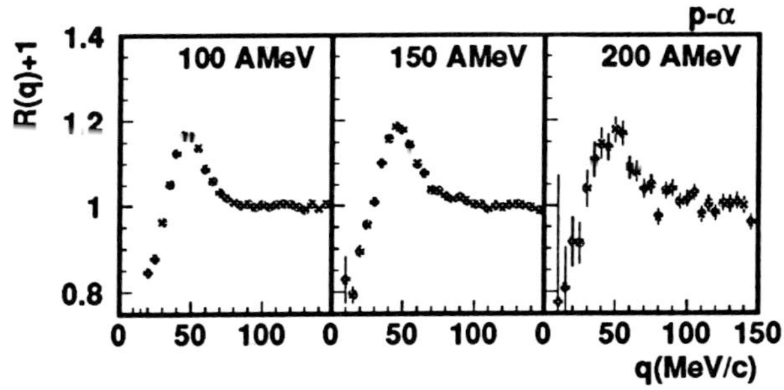

Figure 9: p-α correlation functions for different incident energies.

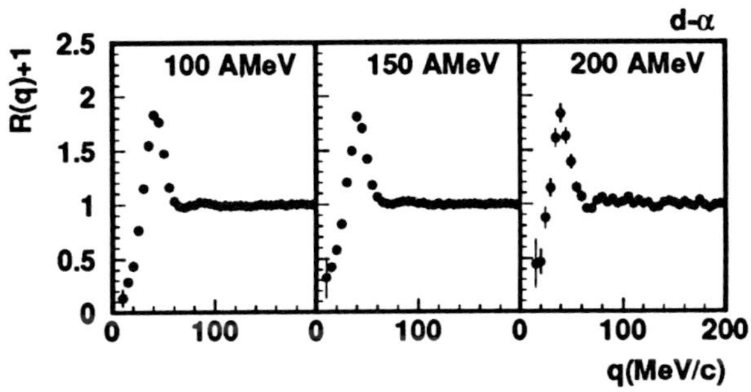

Figure 10: d-α correlation functions for different incident energies.

excited states might be explained by increasing cooling of the decaying system during its expansion. The combined analysis of temperature and size parameters of an expanding source seems to open the possibility to a quantitative understanding of the evolution of the multifragmentation process.

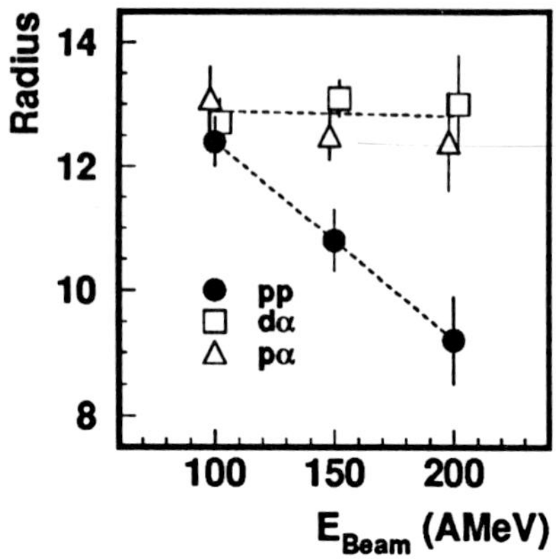

Figure 11: Radii extracted from p-p, p-α and d-α correlation functions.

References

1. S. Albergo et al., *Nuovo Cimento* **89**, 1 (1985).
2. J. Pochodzalla et al., *Phys. Rev. Lett.* **75**, 1040 (1995).
3. J. Pochodzalla et al., *Phys. Rev.* C **35**, 1695 (1987).
4. C. Schwarz et al, *Phys. Rev.* C **49**, 3316 (1994).
5. V. Serfling, *Phys. Rev. Lett.* **80**, 3928 (1998).
6. H. Xi et al., *Z. Phys.* A **359**, 397 (1997).
7. C.O. Dorso and J.Aichelin, *Phys. Lett.* B **315**, 197 (1995).
8. W.A. Friedmann, *Phys. Rev.* C **42**, 667 (1990).
9. G. Papp and W. Norenberg, *XII International Workshop on Gross Properties of Nuclei and Nuclear Excitations* Hirschegg1994,p.87.
10. W.C. Hsi et al., *Phys. Rev. Lett.* **73**, 3367 (1994).
11. M.A. Lisa et al., *Phys. Rev. Lett.* **75**, 2662 (1995).
12. H. Xi *GSI Scientific Report* 37(1995).
13. D.H. Boal et al., *Rev. Mod. Phys.* **62**, 553 (1990).
14. Y.D. Kim et al., *Phys. Rev. Lett.* **67**, 14 (1991).
15. B. Ocker, *PhD Thesis, Frankfurt University* 1998.
16. C. Schwarz et al., *Proceedings of the 35th International Winter Meeting on Nuclear Physics, Bormio, Italy* (Jan. 1997).
17. D. Hahn and H.Stöcker, *Nucl. Phys.* A **476**, 718 (1988).
18. S.E. Koonin, *Phys. Lett.* B **70**, 43 (1977).
19. P. Charmensat, *PhD Thesis, Universite Blaise Pascal - Clermont II* 1989.

PROTON-PROTON CORRELATIONS IN CENTRAL HEAVY-ION COLLISIONS AT SIS ENERGIES AND THE SPACE-TIME EXTENT OF THE EMISSION SOURCE

R. KOTTE
Forschungszentrum Rossendorf, PF 510119, D-01314 Dresden, Germany
E-mail: kotte@fz-rossendorf.de

AND THE FOPI COLLABORATION

Small-angle correlations of proton pairs produced in central collisions of ^{58}Ni+^{58}Ni at a beam energy of 1.93 A·GeV are investigated with the outer Plastic Wall/Helitron detector combination of the FOPI detector system at GSI Darmstadt. Simultaneous comparison of longitudinal and transverse correlation functions with the predictions of the Koonin-Pratt formalism allows to unravel the space-time ambiguity of the source extension. The determination of the source rest frame is found essential for the identification of this effect. For the present system, the selection of central events comprising about 8% of the total integrated cross section allows for the isolation of a compact source in momentum space which is centered around the c.m. velocity of the colliding nuclei. Taking into account the strong collective expansion of the participant zone, which introduces a reduction of the extracted source radius of more than 30%, r.m.s. radius and lifetime parameters of $R_{rms} = (4.2 \pm 1.2)$ fm and $t_{rms} = (11^{+7}_{-5})$ fm/c are extracted, respectively. In contrast, the analysis of the angle-integrated correlation function mixes the spatial and temporal size of the source and gives an upper limit $R_{rms} = (7.0 \pm 0.8)$ fm of the source radius. Additionally, recent results of correlation functions of proton pairs produced in central collisions of the systems $^{96}_{44}$Ru($^{96}_{40}$Zr)+$^{96}_{44}$Ru($^{96}_{40}$Zr) at 400, 1530 and 1690 A·MeV are presented.

1 Introduction

Two-proton correlation functions at small relative momenta can probe the space-time extent of the reaction zone created in heavy-ion reactions. This is due to the fact that the magnitude of nuclear and Coulomb final-state interactions as well as antisymmetrization effects, depend on the spatial and temporal separation of the two protons[1−12]. Usually, the correlation functions are evaluated as a function of the relative momentum $|\mathbf{q}| = \frac{1}{2}|\mathbf{p}_1 - \mathbf{p}_2|$. The attractive s-wave nuclear interaction leads to a maximum in the correlation function at $q \sim 20$ MeV/c, whereas Coulomb repulsion and antisymmetrization cause a minimum at q=0 [1]. Most analyses give only upper limits for the spatial extent of the source due to an ambiguous interplay of radius and lifetime of the source. Thus, model calculations simulating large sources with short lifetimes can give similar correlation functions as model calculations simulating small sources with long lifetimes [1,5]. However, this ambiguity may be avoided by applying cuts on the angle between the vectors of the relative momentum $\mathbf{q}$ and the total pair momentum $\mathbf{P} = \mathbf{p}_1 + \mathbf{p}_2$ [1,3,5,9].

For a source of finite lifetime τ, the phase space distribution of the emitted protons is elongated in the direction of $\mathbf{P}$ by $\langle v \rangle \tau$, where $\langle v \rangle$ is the emission velocity of the pair. For such distributions, the Pauli suppression due to the antisymmetrization of the wave functions is smaller in the elongated (longitudinal, $\mathbf{q} \| \mathbf{P}$) direction as compared to the not elongated (transverse, $\mathbf{q} \perp \mathbf{P}$) direction. Reliable impact parameter sampling is needed for the selection of well-defined sources, where certain longitudinal and transverse cuts can be applied in the source rest frame. Otherwise, the small effect of the directional dependence (about 5-10% of the height of the the correlation peak [11]) is expected to be washed out.

2 The Experiment

The experiment has been performed at the heavy-ion synchrotron SIS at GSI Darmstadt. A ^{58}Ni target of 1% interaction thickness has been irradiated by ^{58}Ni ions of 1.93A·GeV beam energy. The data are taken with the outer Plastic Wall/Helitron combination (which covers polar angles from 8.5 to 26.5 degrees) of the FOPI detector system [12], where the Plastic Wall delivers the velocity β and the nuclear charge Z of the particles and the Helitron allows to measure the curvature (which is a measure of the momentum over charge (p/Z)) of the particle track in the field of a large superconducting solenoid. Since the momentum resolution of the Helitron is rather moderate, this detector component serves for particle identification only. The corresponding mass is determined via $m = (p/Z)_{Hel}/(\beta\gamma/Z)_{PlaWa}$. From the velocity, which is measured with a resolution of about 1%, other kinetic quantities like the proton momenta are deduced.

3 Event Classification

Central events comprising about 8% of the total integrated cross section are selected with the online central trigger, which demands large charged particle multiplicities to be measured with the outer wall. Simulations with the IQMD model [13] predict an averaged impact parameter of 2 fm for this multiplicity cut.

In a previous investigation of central Au+Au collisions between 100 and 400 MeV per nucleon beam energy it was found that the correlation function of pairs of intermediate mass fragments (IMF) is strongly affected by the collective directed sideward flow of nuclear matter [14,15]. The directed sideflow causes an enhancement of correlations at small relative momenta. This enhancement results from mixing of differently azimuthally oriented events; it vanishes if the events are rotated into a unique reaction plane [16]. Therefore, the technique of event rotation is applied to the data in order to exclude such dynamical correlations.

4 The Correlation Function

Let $Y_{12}(\mathbf{p}_1, \mathbf{p}_2)$ be the coincidence yield of proton pairs. Then the two-particle correlation function is defined as

$$1 + R(\mathbf{p}_1, \mathbf{p}_2) = \mathcal{N} \frac{\sum_{events,pairs} Y_{12}(\mathbf{p}_1, \mathbf{p}_2)}{\sum_{events,pairs} Y_{12,mix}(\mathbf{p}_1, \mathbf{p}_2)}. \tag{1}$$

The sum runs over all events fulfilling the above mentioned global selection criteria and over all pairs satisfying certain conditions given below. Event mixing, denoted by the subscript "mix", means to take particle #1 and particle #2 from different events. We only mix events found within the same event class. $\mathcal{N}$ is a normalization factor fixed by the requirement to have the same number of true and mixed pairs. The correlation function (1) is than projected onto the relative momentum

$$q = |\mathbf{q}| = \frac{1}{2}|\mathbf{p}_1^{cm} - \mathbf{p}_2^{cm}|. \tag{2}$$

Besides the above described global event characteristics we use gate conditions on the angle α between $\mathbf{q}$ and the c.m. sum momentum of the proton pair $\mathbf{P}_{12}^{cm} = \mathbf{p}_1^{cm} + \mathbf{p}_2^{cm}$ and on the normalized pair momentum $p_{12}^o = |\mathbf{P}_{12}^{cm}|/(2 \cdot (p/A)_{proj}^{cm})$.

5 Results

Fig. 1 shows the two-proton correlation function and the relative momentum distribution of proton pairs from collisions selected with the central trigger condition. The dashed line in the lower part represents the distribution of the uncorrelated background which is produced with the event mixing technique. The statistical errors of all the correlation functions presented here are governed by those of the coincidence yield, since the mixed yield is generated with two orders of magnitude higher statistics.

Obviously, the correlation peak is governed by the contribution of very few proton pairs in the relative momentum range 10 MeV/c $< q <$ 50 MeV/c which show up above the background distribution. The corresponding yield is less than 1 per thousand. For large relative momenta the mixed yield is perfectly on

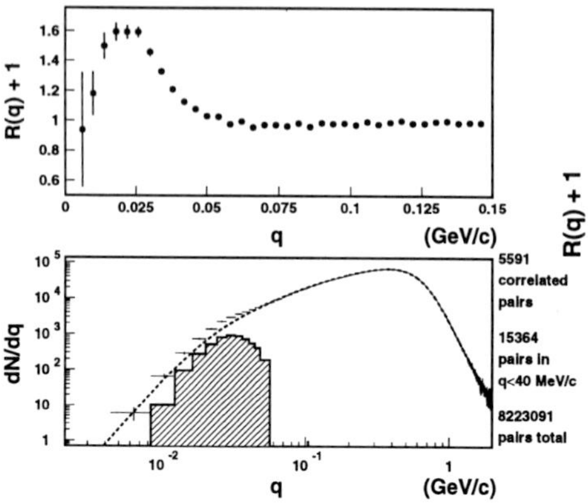

Figure 1: The experimental correlation function (upper part) and the corresponding relative momentum yield (lower part) of proton pairs from central trigger events. The dashed line represents the uncorrelated background which is constructed via event mixing. The difference of the true and mixed yield given as hatched histogram delivers the distribution of the correlated pairs.

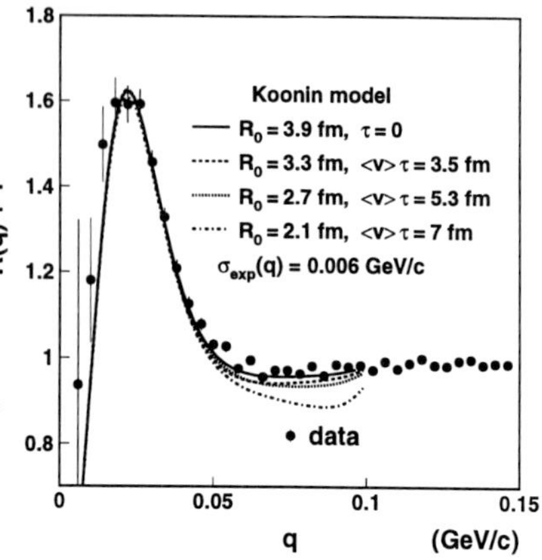

Figure 2: Correlation functions of proton pairs from Ni+Ni collisions at $1.93A{\cdot}$GeV selected with the the central trigger condition (dots). The different lines are results of Koonin model simulations with different combinations of Gaussian radius and lifetime. The model results are folded with an experimental resolution function of Gaussian shape with dispersion $\sigma_{exp}(q) = 6$ MeV/c.

top of the coincidence yield. Consequently, a normalization procedure where the constant $\mathcal{N}$ in Eq. (1) is determined by the requirement that $R(q)$ vanishes for large q (cf. refs.[8,9,10,11]) would give the same correlation function.

As in previous fragment-fragment correlation analyses [14,15] an enhanced coincidence yield at very small relative angles is observed, which is due to double counting caused mainly by scattering in the scintillator strips. This surplus yield is eliminated by excluding, around a given hit, positions within a rectangular segment of $|\phi_1 - \phi_2| < 4^o$ and $|\theta_1 - \theta_2| < 2^o$. The relative-momentum distribution of the excluded proton pairs allows for a direct experimental determination of the q resolution. For these dispersions a typical value of $\sigma_{exp}(q) \simeq (6 \pm 1)$ MeV/c is estimated. In order to keep the influence of the exclusion of small angles onto the correlation function as small as possible, the same procedure is applied to the uncorrelated background. Additionally, the bias of the correlation function at very small relative momenta due to the exclusion procedure is tested with GEANT simulations using the IQMD model

[13] as event generator. It has been found that for $q > 12$ MeV/c the correlation peak, which is a measure of the source size, is almost unbiased.

In Fig. 2 the angle-integrated two-proton correlation function for central trigger events is compared with the predictions of the Koonin-Pratt formalism [1] with different combinations of the Gaussian parameters of radius R_o and lifetime (times proton pair velocity) $\langle v\rangle\tau$. The figure clearly demonstrates the space-time ambiguity of the source extension carried by the angle-integrated correlation function. The experimental correlation peak is equally well described by the model simulating either a large source ($R_{rms} = \sqrt{\frac{3}{2}}R_o = 4.8$ fm) of zero lifetime or a smaller source of finite lifetime.

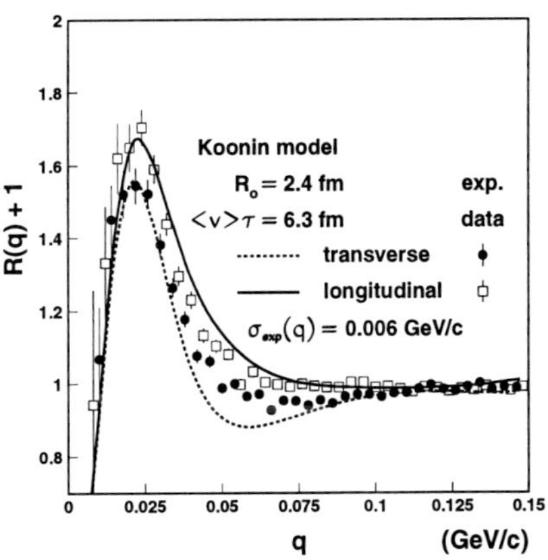

Figure 3: Transverse (dots) and longitudinal (squares) correlation functions of proton pairs from Ni+Ni collisions at 1.93A·GeV selected with the central trigger condition. Dashed and full lines are the corresponding results of Koonin model calculations using the optimum Gaussian source parameters of radius and lifetime $R_o = 2.4$ fm and $\langle v\rangle\tau = 6.3$ fm, respectively, as delivered by the χ^2 minimum of Fig. 4.

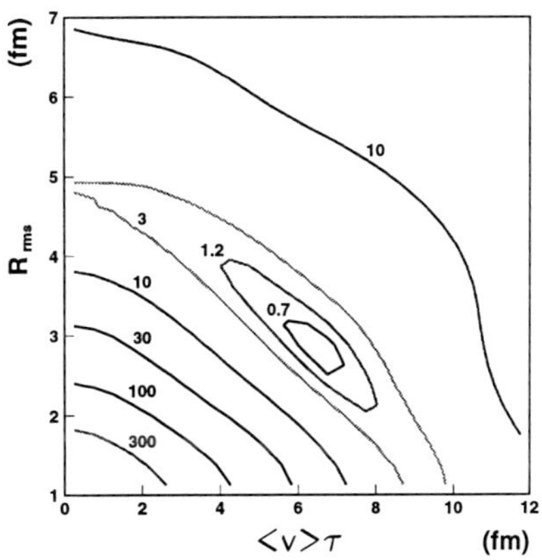

Figure 4: The distribution of χ^2 per degree of freedom as determined by simultaneously comparing measured longitudinal and transverse correlation functions (over the range 12 MeV/c $< q <$ 36 MeV/c) with the correlation functions predicted by the Koonin model. The quantity is given in dependence of the r.m.s. radius R_{rms} and Gaussian lifetime parameter $\langle v\rangle\tau$.

The space-time ambiguity of the angle-integrated correlation function can be reduced by analysing two-proton correlations with cuts on the angle $\alpha = cos^{-1}(\mathbf{P}\cdot\mathbf{q}/(Pq))$ between $\mathbf{P}$ and $\mathbf{q}$. As mentioned in Section 1, the two-proton correlation function carries a directional sensitivity due to the Pauli suppression in the non-elongated direction [1,5]. Since the pair momentum P depends on the rest frame of the source, but the relative momentum q does not, the angle α depends on the rest frame of the emitting source. Thus, a careful selection of a well-defined emission source is required. In the present analysis, we investigate correlation functions of proton pairs from central trigger events. The pair momentum is defined in the rest frame of the central source, $\mathbf{P} \equiv \mathbf{P}_{12}^{cm} = \mathbf{p}_1^{cm} + \mathbf{p}_2^{cm}$. Experimental transverse and longitudinal correlation functions are generated here by selecting proton pairs with angles $|cos(\alpha)| < 0.5$ and $|cos(\alpha)| > 0.5$, respectively. These cuts are applied both to the coincidence yield and to the uncorrelated background. The normalization constant $\mathcal{N}$ in Eq. (1) is determined individually for each α cut. The longitudinal and transverse correlation functions are presented in Fig. 3.

The wanted effect is clearly visible: We find a suppression of the transverse correlations with respect to the longitudinal ones of about 10% which is consistent with model predictions for the emission from a source of finite lifetime [1,3,5]. Koonin model calculations of longitudinal and transverse p-p correlations are performed for a wide range of Gaussian source parameters R_o and $\langle v\rangle\tau$. The results of the model are folded with the experimental resolution of $\sigma_{exp}(q) = 6$ MeV/c. In order to find the optimum set of parameters, for each set the agreement between experimental and simulated transverse and longitudinal correlation functions is tested by evaluating the χ^2 per degree of freedom in the peak region of $q = 12 - 36$ MeV/c. Fig. 4 shows, as contour plot, the distribution of this quantity as function of R_{rms} and $\langle v\rangle\tau$.

The best agreement of experimental data and model calculations is reached for Gaussian source parameters of $R_o = (2.35 \pm 0.65)$ fm and $\langle v\rangle\tau = (6.3 \pm 1.7)$ fm. The corresponding r.m.s. quantities are $R_{rms} = (2.9 \pm 0.8)$ fm and $\langle v\rangle t_{rms} = \langle v\rangle\tau/\sqrt{2} = (4.5 \pm 1.2)$ fm. The results of the Koonin model with the optimum parameter set are depicted in Fig. 3. It should be mentioned that for zero lifetime in Fig. 4 one would predict a r.m.s. radius of $R_{rms} = 4.9$ fm in nice agreement with the results of the previous analysis of the angle-integrated correlation function (cf. Fig. 2).

For the determination of the mean lifetime of the emission process one has to take into account the emission velocity of the proton pairs. For pairs of protons from central trigger events which have q values in the region of the correlation peak and which are detected within the Plastic Wall/Helitron acceptance this velocity exhibits a distribution with mean value $\langle v\rangle = 0.42$ c and dispersion $\sigma_v = 0.1$ c. Considering these quantities a r.m.s. lifetime $t_{rms} = (11^{+7}_{-5})$ fm/c can be derived.

6 Preliminary Results of Experimental p-p Correlations for the Systems ${}^{96}_{44}$Ru(${}^{96}_{40}$Zr)+${}^{96}_{44}$Ru(${}^{96}_{40}$Zr) at 400, 1530 and 1690 A·MeV

The FOPI collaboration has continued experiments of almost complete measurements of central and semi-central heavy-ion collision. For systematic investigations of small-angle p-p correlations with respect to system size and projectile energy data sets of the following systems are available:

- ^{40}Ca+^{40}Ca at 400, 600, 800, 1000, 1500, 1930 A·MeV and Ca+Au at 1500 A·MeV (in course of calibration)

- ^{58}Ni+^{58}Ni at 1930 A·MeV (published in ref. [19])

- all projectile-target combinations of ${}^{96}_{44}$Ru(${}^{96}_{40}$Zr)+${}^{96}_{44}$Ru(${}^{96}_{40}$Zr) at 400 and 1530 A·MeV, ${}^{96}_{44}$Ru+${}^{96}_{44}$Ru(${}^{96}_{40}$Zr) at 1690 A·MeV (in course of analysis)

- Au+Au at 400, 600, 800, 1000, 1200, 1490 A·MeV (in course of calibr.)

In Fig. 5 the available angle-integrated correlation functions of proton pairs in central collision are summarized for all Ru(Zr) combinations and compared with the corresponding result of the Ni+Ni system. Obviously, the data group into three classes of correlation functions. There is one group with a low correlation peak at 1.13 for all Ru(Zr)+Ru(Zr) combinations at 400 A·MeV, another one at 1.3 for the same collision systems at the high beam energies, and a third one at 1.6 for the Ni+Ni system. The observation of the highest peak for the Ni+Ni system is quite natural since the system has the smallest participant number in the central source. Thus, one would expect a smaller source in coordinate space and consequently a higher peak than for the Ru(Zr)+Ru(Zr) system at (comparable) high beam energies. However, the strong increase of the correlation peak with increasing energy for the Ru(Zr)+Ru(Zr) systems has to be attributed to different time scales. A faster collective expansion in central nucleus-nucleus collisions at higher beam energies would reduce the influence of the time component on the angle-integrated correlation. Following the procedure described above (cf. Figs. 3, 4), the space-time ambiguity of the source extension is resolved by introducing directional cuts on the correlation function. A simultaneous fit of the transverse and longitudinal correlation functions of the Ru(Zr)+Ru(Zr) system at 1.53 A·GeV (Fig. 6) with the predictions of the Koonin-Pratt formalism delivers an optimum set of r.m.s. radius and lifetime parameters of $R_{rms} = (4.9 \pm 0.5)$ fm and $t_{rms} = (10^{+5}_{-4})$ fm/c, respectively. (Here, the proton pair velocity distribution with mean value of 0.36 c and dispersion of 0.1 c has been taken into account.)

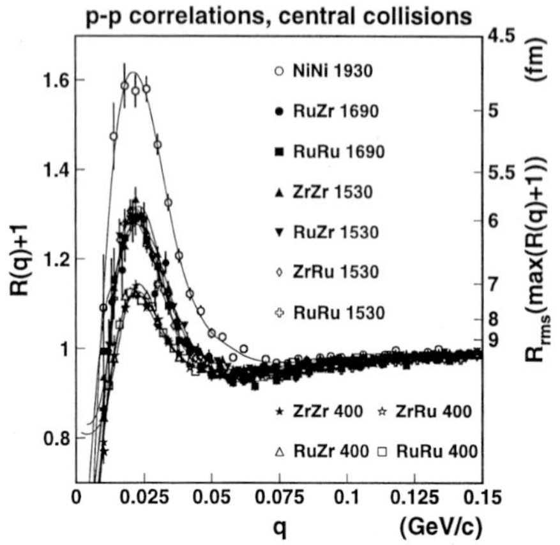

Figure 5: Angle-integrated p-p correlation functions for central events comprising an integrated cross section of about 200 - 300 mb. The projectile and target nuclei and the corresponding beam energies are indicated in the symbol list (e.g. "RuZr1530" for a 1530 A·MeV $^{96}_{44}$Ru projectile on a $^{96}_{40}$Zr target). The curves are smooth lines through the data points in order to guide the eye. The right-hand ordinate gives the r.m.s. radius $R_{rms} = \sqrt{\frac{3}{2}}R_o$ for which Koonin model simulations with zero lifetime (folded with the experimental q resolution of 6 MeV/c) would deliver the same peak height as the experimental correlation function.

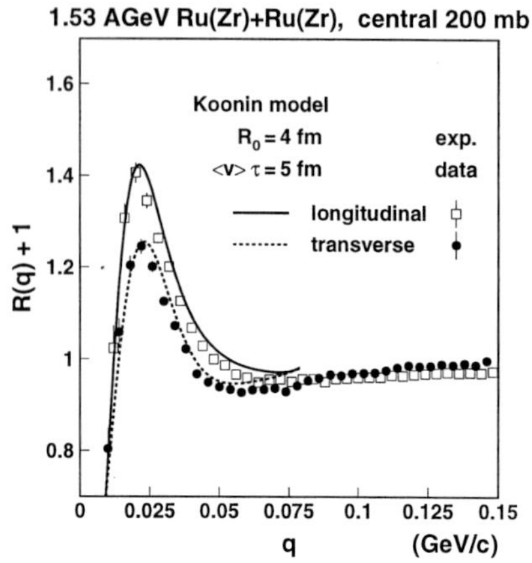

Figure 6: The angle-dependent p-p correlation functions (transverse: dots, longitudinal: squares) for central collisions of Ru(Zr)+Ru(Zr) at 1.53 A·GeV beam energy (all projectile-target combinations). Dashed and full lines are the corresponding results of Koonin model calculations using the optimum Gaussian source parameters of radius R_o = 4 fm and lifetime $\langle v\rangle\tau.$ = 5 fm as delivered by a χ^2 minimization procedure in analogy to Fig. 4. The results of the model are folded with experimental q resolution parameters of 7 MeV/c and 4 MeV/c for transverse and longitudinal correlations, respectively.

7 Discussion

Since central nucleus-nucleus collisions are characterized by a collective expansion of nuclear matter after the compression phase the question arises how does this correlation of coordinate and momentum space influence the extracted source radii. For this reason, within the Koonin model we have modified the single-proton phase space emission function [1]

$$g(\mathbf{p}, \mathbf{r}, t) \propto \exp\left(-\frac{p^2}{2mT} - \frac{r^2}{R_o^2} - \frac{t^2}{\tau^2}\right) \tag{3}$$

by introducing a linear velocity profile which accounts for the radial expansion

$$\exp\left(-\frac{p^2}{2mT}\right) \longrightarrow \exp\left(-\frac{(\mathbf{p}-\alpha\mathbf{r})^2}{2mT}\right). \tag{4}$$

After a lengthy but straightforward transformation one gets a relation containing a reduced radius

$$R^* = \frac{R_o}{\sqrt{1+\frac{2}{3}\frac{e_{flow}}{T}}} \tag{5}$$

instead of the true source radius R_o. Here the parameter α is connected with the (non-relativistic) radial flow energy

$$e_{flow} = \frac{\langle \alpha^2 r^2 \rangle}{2m} = \frac{3}{4}\frac{\alpha^2 R_o^2}{m} = \frac{mc^2\beta_r^2}{2} \tag{6}$$

and with the radial blast velocity β_r whereas T represents the temperature.

For a quantitative estimate of the apparent reduction of the source radius we have analysed c.m. kinetic energy spectra of p, d, and t particles produced in central collisions and measured by the Plastic Wall/Helitron combination within a c.m. angular cone of $55^o < \Theta_{cm} < 70^o$. A simultaneous fit of the shape of the three spectra with the Siemens-Rasmussen formula [18] of an expanding spherical shell has been performed. From this analysis we get for the temperature and the radial blast velocity $T = (66 \pm 6)$ MeV and $\beta_r = 0.48 \pm 0.02$, respectively [19]. (For the Ru(Zr)+Ru(Zr) system at 1.53 A·GeV a fit to the p, d, t, ^3He, and α spectra within $55^o < \Theta_{cm} < 65^o$ delivers the very similar results $T = (71 \pm 5)$ MeV and $\beta_r = 0.48 \pm 0.02$.)

Taking into account these quantities the correction factor in Eq. (5) amounts to 1.45 ± 0.06 and gives a true r.m.s. radius of $R_{rms} = (4.2 \pm 1.2)$ fm. The corresponding sharp sphere radius $R_{ss} = \sqrt{5/3}R_{rms}$ allows to determine the breakup volume $V = A/\rho = 4\pi R_{ss}^3/3 = 670$ fm^3 of the source. For an estimated participant multiplicity of the central source of $A = 80$ this volume gives a breakup density of $\rho = 0.70\rho_o$. In contrast, the radius which is deduced from the angle-integrated correlation function after correcting for collective expansion amounts to $R_{rms} = (7.0 \pm 0.8)$ fm. It would give a 4.5 times larger volume than the value delivered by the correlation function with directional cuts and a correspondingly smaller breakup density of $\rho = 0.15\rho_o$.

The estimated characteristic source emission time of is the shortest one determined with the method of directional cuts on the two-proton correlation function known so far. It does well compare with predictions of calculations we have performed with the BUU transport model [6,17] for central (b=2 fm) Ni+Ni collisions at 2 A·GeV. The simulation delivers a spread ($\pm 1\sigma$) of about

10-15 fm/c for the distribution of the time instants when the nucleons undergo the last collision. Furthermore, the BUU model predicts about 3 fm/c for the time from the begin of the density increase until a maximum value of $\rho \simeq 1.8\rho_o$ is reached followed by a time of about 10 fm/c for the expansion to a density of $\rho \simeq \rho_o/4$. Thus, proton pairs contributing to the peak in the correlation function are supposed to be emitted over almost the whole radial expansion time (or even over the whole reaction time). It is certainly appropriate to assume that - in the 1-2 A·GeV energy region - proton emission is not governed by sequential decay chains but by a sudden explosion/expansion process. Dealing with such short time scales the notation "lifetime" almost looses its meaning.

8 Acknowledgement

This work was supported by the German BMBF under contract No. 06DR666.

References

1. S.E. Koonin, *Phys. Lett.* B **70**, 43 (1977).
2. S. Pratt, *Phys. Rev. Lett.* **53**, 1219 (1984).
3. S. Pratt and M.B. Tsang, *Phys. Rev.* C **36**, 2390 (1987).
4. W.G. Gong *et al.*, *Phys. Rev. Lett.* **65**, 2114 (1990).
5. W.G. Gong et al., *Phys. Rev.* C **43**, 781 (1991).
6. W. Bauer, C.K. Gelbke and S. Pratt, *Ann. Rev. Nucl. Part. Sci.* **42**, 77, (1992); W. Bauer, *Nucl. Phys.* A **545**, 369c (1992).
7. G.J. Kunde *et al.*, *Phys. Rev. Lett.* **70**, 2545 (1993).
8. M.A. Lisa *et al.*, *Phys. Rev. Lett.* **70**, 3709 (1993).
9. M.A. Lisa *et al.*, *Phys. Rev. Lett.* **71**, 2863 (1993).
10. M.A. Lisa *et al.*, *Phys. Rev.* C **49**, 2788 (1994).
11. D.O. Handzy *et al.*, *Phys. Rev.* C **50**, 858 (1994).
12. A. Gobbi *et al.* (FOPI collaboration), *Nucl. Instrum. Methods* A **324**, 156 (1993), J. Ritman, *Nucl. Phys.* (Proc. Suppl.) B **44**, 708 (1995).
13. S.A. Bass, C. Hartnack, H. Stöcker and W. Greiner, *Phys. Rev.* C **51**, 3343 (1995).
14. B. Kämpfer et al. (FOPI collaboration), *Phys. Rev.* C **48**, R955 (1993).
15. R. Kotte *et al.* (FOPI collaboration), *Phys. Rev.* C **51**, 2686 (1995).
16. P. Danielewicz and G. Odyniec, *Phys. Lett.* B **157**, 146 (1985).
17. Gy. Wolf, W. Cassing, U. Mosel, *Nucl. Phys.* A **552**, 549 (1993), *Phys. Lett.* B **271**, 43 (1991).
18. P.J. Siemens and J.O. Rasmussen, *Phys. Rev. Lett.* **42**, 880 (1979).
19. R. Kotte *et al.* (FOPI collaboration), *Z. Phys.* A **359**, 47 (1997).

TRANSVERSE MOMENTUM DEPENDENT TWO-PION BOSE-EINSTEIN CORRELATIONS IN AU+AU COLLISIONS AT 11.6 A·GeV/c

J.H. Lee
Physics Department, Brookhaven National Laboratory
Upton, NY 11973, USA
E-mail: jhlee@bnl.gov

(for the E866 Collaboration)

Bose-Einstein correlations of $\pi^+\pi^+$ and $\pi^-\pi^-$ pairs collected by the BNL-E866 Forward Spectrometer in 11.6 A·GeV/c Au + Au collisions have been measured. The data were analyzed using three-dimensional correlation functions parameterized by the Yano-Koonin-Potgoretskii and Bertsch-Pratt formalism to study transverse momentum dependent source parameters. Rapid decreases of longitudinal source radii and slower decreases in the transverse parameters with increasing transverse momentum were observed, which suggests a strong longitudinal and some transverse expansion. A freeze-out time τ_0 was derived as 4.5 – 5 fm/c, under the assumption of the freeze-out temperature T = 130 MeV, and the duration of emission was found to be $\delta\tau \approx 2 - 4$ fm/c.

1 Introduction

Bose-Einstein Correlation measurements are used to obtain the information on the space-time structure of the particle-emitting source in heavy ion collisions[1]. But the dynamical nature of collisions[2–6] complicated by the evolution of the system in the finite source lifetime limits the Bose-Einstein Correlation measurements only to the extent of space-time "region of homogeneity" from which bosons of similar momenta are emitted[6]. Correlation studies as a function of transverse momentum enable one to access more complete picture of the dynamical space-time structure of the source[6,7]. Recently much attention has been paid to the experimental measurements of transverse momentum dependent Bose-Einstein correlations, mainly at the CERN SPS energy[8,9,10]. In this contribution, we present results of the transverse momentum dependent source size measurements in Au+Au collisions at the AGS energy.

2 Experiment

The beam of 11.6 A·GeV/c Au ions from the BNL Tandem-AGS complex was normally incident upon a Au target (1.5% or 3% of an interaction length), and the data were taken using the E866 Forward Spectrometer[11]. The Forward

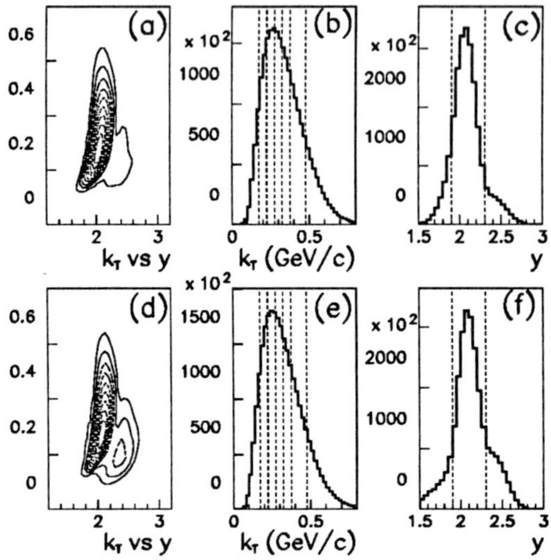

Figure 1: The distribution of the average transverse momentum (K_T) vs. the rapidity (y) and their projections for the $\pi^+\pi^+$ [(a,b,c)] and $\pi^-\pi^-$ [(d,e,f)] pairs. The dotted lines in the figures (b) and (e) represent the k_T bins used in the analysis, and the dotted lines in the figures (c) and (f) show the rapidity regions selected.

Spectrometer was designed for handling the high multiplicity environment. It covers the forward rapidity region from 6 degrees to 28 degrees to the beam by rotating the spectrometer. It has a small aperture of about 5 msr, and consists of two dipole magnets, two tracking stations, and a Time-of-Flight (ToF) wall. The first magnet (M1) sweeps away low momentum particles, including delta rays, in order to reduce the hit multiplicity on the first tracking station which follows the M1. Each tracking station consists of a time projection chamber (TPC) sandwiched between a pair of projective drift chambers. Accurate and efficient track reconstruction is provided by pattern recognition in the two TPC's (TPC1 and TPC2) and fitting the tracks found in the TPC's with additional projected drift chamber hits to straight lines. If the two independently reconstructed straight tracks match inside the second magnet (M2), the momentum associated with the matched trajectory is obtained from the bending angle between the two tracks. After the momentum of the track is determined, the trajectory is projected toward the target through M1 to ensure the tracks originated from the target. Particle identification up to 4 GeV/c for $\pi^\pm$ is obtained by the ToF hodoscope behind the second tracking station with 75 $p$sec timing resolution.

The data were taken in 1995. The total numbers of analyzed $\pi^+\pi^+$, $\pi^-\pi^-$, and $\pi^+\pi^-$ pairs are 1.5×10^6, 1.5×10^6, and 2.5×10^6, respectively. A large fraction of the data comes from a single experimental setup: the Forward Spectrometer at 10 degrees, with magnetic fields of M1 and M2 −0.2 Tesla and +0.4 Tesla, respectively. This minimizes systematic uncertainties contributed by different experimental running conditions. The 30% (≈ 1.65 barn) of the "minimum-biased" Au+Au cross section events corresponding to the most central collision gated by the New Multiplicity Array [12] were selected for the analysis. Fig. 1 shows the rapidity, y, and the average transverse momentum, K_T, distributions for $\pi^+\pi^+$ and $\pi^-\pi^-$ pairs. The K_T used for the analysis is from 0.175 GeV/c to 0.475 GeV/c in 5 bins, corresponding $< K_T >$ = 0.2, 0.25, 0.3, 0.35, and 0.425, respectively, as displayed in Fig. 1. The rapidity range of pairs was limited to $1.9 \leq y_{\pi\pi} \leq 2.3$ for this analysis. The mid-rapidity is $y_{\rm CMS} = 1.6$.

3 Correlation Function Parameterization

The correlation function C_2 is the ratio of the two-particle probability to the product of the single-particle probabilities for particles with momenta $\boldsymbol{p}_i$. Experimentally it is given by the normalized ratio of the number of correlated (signal) pairs S to the number of uncorrelated (background) pairs B:

$$C_2(\boldsymbol{q}, \boldsymbol{K}) = \frac{P(\boldsymbol{p}_1, \boldsymbol{p}_2)}{P(\boldsymbol{p}_1)P(\boldsymbol{p}_2)} = \mathcal{N} \cdot \frac{S}{B \cdot w_{Coulomb}}\,, \tag{1}$$

where $(q_0, \boldsymbol{q}) = (E_1 - E_2, \boldsymbol{p}_1 - \boldsymbol{p}_2)$ and $\boldsymbol{K} = \frac{\boldsymbol{p}_1 + \boldsymbol{p}_2}{2}$ denote the relative and average momentum of the pair, respectively.

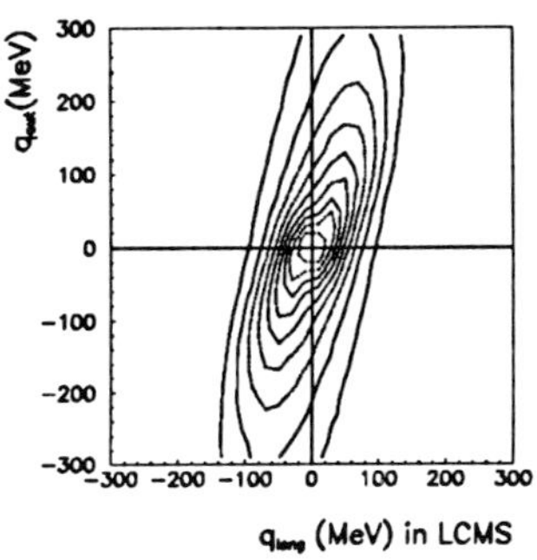

Figure 2: Contour plot of the correlation of $q_{out} - q_{long}$ for $\pi^-\pi^-$. The $\pi^+\pi^+$ plot looks similar.

The signal pairs are from the same events and the background pairs are obtained by mixing different events with the same experimental condition. The background pairs were constructed with 20 times more statistics than the number of correlated pairs, and weighted with the Coulomb factor $w_{Coulomb}$ to remove the correlation created by the Coulomb effect in the correlated pairs. To compare with other published results[13,14], the one-dimensional correlation functions for $\pi^+\pi^+$ and $\pi^-\pi^-$ were parameterized by,

$$C_2(Q_{inv}) = \mathcal{N}(1 + \lambda e^{-R_{inv}^2 Q_{inv}^2}), \tag{2}$$

where the Lorentz invariant variable Q^2_{inv} is defined as $q_0^2 - \boldsymbol{q}^2$.

For the three-dimensional parameterization, the data were fit with the Yano-Koonin-Potgoretskii (YKP) [6,15,16] scheme which is a function of the space-time components $(R_\perp, R_\parallel, R_0)$ and the average longitudinal velocity v:

$$C_2(\boldsymbol{q}, \boldsymbol{K}) = \mathcal{N}[1 + \lambda e^{-R_\perp^2(\boldsymbol{K})q_\perp^2 - R_\parallel^2(\boldsymbol{K})(q_\parallel^2 - q_0^2) - (R_0^2(\boldsymbol{K}) + R_\parallel^2(\boldsymbol{K}))(q \cdot U(\boldsymbol{K}))^2}]. \quad (3)$$

Here $q_\perp = \sqrt{q_{side}^2 + q_{out}^2}$ and $q_\parallel$ is the longitudinal component. The four-velocity U only has a longitudinal spatial component:

$$U(\boldsymbol{K}) = \gamma(\boldsymbol{K})(1, 0, 0, v(\boldsymbol{K})), \text{ with } \gamma = \frac{1}{\sqrt{1 - v^2}}. \quad (4)$$

The data have been also parameterized with the Bertsch-Pratt (BP) Cartesian decomposition [2,17]:

$$C_2(\boldsymbol{q}, \boldsymbol{K}) = \mathcal{N}[1 + \lambda e^{-R_{long}^2(\boldsymbol{K})q_{long}^2 - R_{side}^2(\boldsymbol{K})q_{side}^2 - R_{out}^2(\boldsymbol{K})q_{out}^2 - 2R_{out,long}^2(\boldsymbol{K})q_{out}q_{long}}] \quad (5)$$

where λ is a chaoticity parameter, with λ=0 corresponding to a totally coherent and λ=1 corresponding to a totally chaotic source. The longitudinal component of $\boldsymbol{q}$, q_{long} is parallel to the beam, q_{side} is perpendicular both to the beam and to the average momentum, and q_{out} is orthogonal to q_{side} and q_{long}. The source radius parameters R_{long}, R_{side}, and R_{out} correspond to three dimensional radius parameters in the $(q_{long}, q_{side}, q_{out})$ space. The "out-longitudinal" cross term [18] $R^2_{out,long}$, which can be either positive or negative, is included in the function to accommodate the rotation of the q_{out}-q_{long} plane as seen in Fig. 2. The correlation functions, Eq. 3 and Eq. 5, are evaluated in the Longitudinally Co-Moving System (LCMS), which is the longitudinally boosted frame where the average pair momentum $\boldsymbol{K}$ has no longitudinal component, $\beta_L = 0$.

4 Analysis

The $\boldsymbol{q}$ resolution functions calculated from the Monte-Carlo simulations of the detector system using GEANT [19] and the data reconstruction chain are shown in Fig. 3. The resolution factors contribute to an underestimation of the source parameters. The systematic uncertainties of the fitted radius parameter are estimated to be less than 10%. No correction for this effect has been applied to the data.

A minimum separation of 1 cm between two tracks at the middle planes of TPC1 (2.4 m from the target) and TPC2 (3.8 m from the target) in both the x and y directions is applied in order to exclude regions of inefficient track reconstruction. The same condition for the minimum track separation is applied in the selection of mixed events to prevent introducing any bias from the cut.

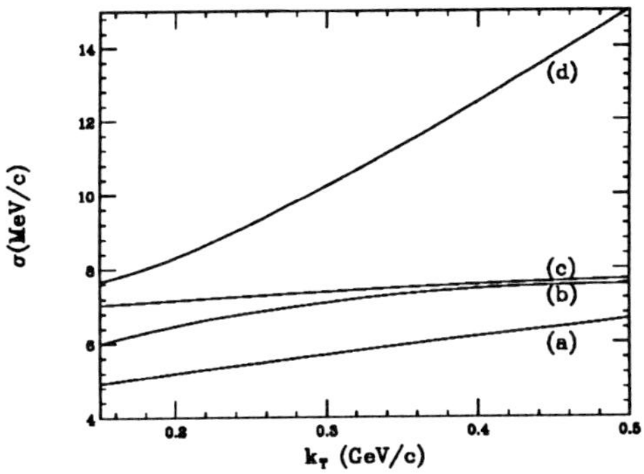

Figure 3: Resolution functions of the q components, obtained by averaging over the other q components, are shown as a function of K_T for Q_{inv} [(a)], q_{long} [(b)] in LCMS, q_{side} [(c)], and q_{out} [(d)].

The final state Coulomb interaction between two pions is corrected for by an iterative method which takes into account the pion source size [14] and the momentum resolution. Fig. 4 shows Coulomb correction factors for two pions with and without the momentum resolution smearing. For the comparison, the Gamov function which assumes a point-like source is also displayed in the figure. To test the quality of the Coulomb corrections, the correction functions are applied to $\pi^+\pi^-$ correlations which is dominated by the mutual Coulomb interactions for small q. It is demonstrated in Fig. 5 that the $\pi^+\pi^-$ correction function becomes close to unity after applying the Coulomb correction function calculated for a finite source and folded with the experimental momentum resolutions. No correction is applied for Coulomb effects between the pair and the remaining particles.

Fitting the data to the correlation functions was done by minimizing the logarithm of the likelihood function:

$$-\ln \mathcal{L} = \sum_i^n [C_i B_i - S_i \ln(C_i B_i) + \ln(S_i!)], \tag{6}$$

where C_i is the correlation function, and S_i is the number of signal counts in the i-th three dimensional $\boldsymbol{q}$ bin, and B_i is the number of background events weighted by a Coulomb factor in the bin. The minimization is done using MINUIT [20]. The projections of the $\pi^+\pi^+$ correlations and of the fits for the YKP and BP parameterizations, obtained by averaging over the two other components of $-50 < \boldsymbol{q} < 50$ MeV/c weighted by the number of background pairs, are plotted in Figs. 6 and 7. The fits reproduce the data for all bins in K_T, and for both parameterizations.

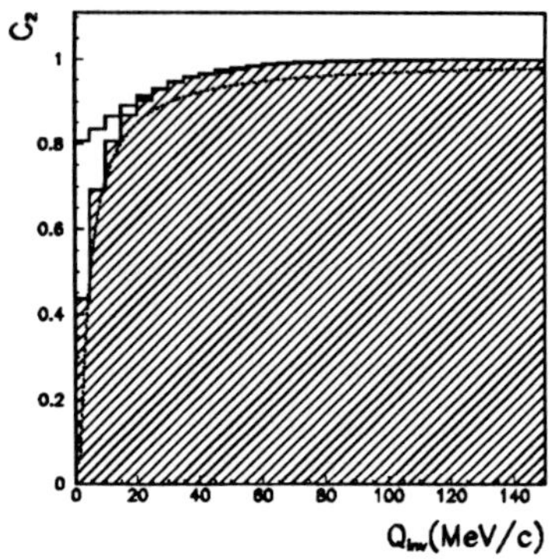

Figure 4: Coulomb correction functions without (shaded distribution) and with (unshaded) momentum resolution. The Gamov function is shown as a dotted curve.

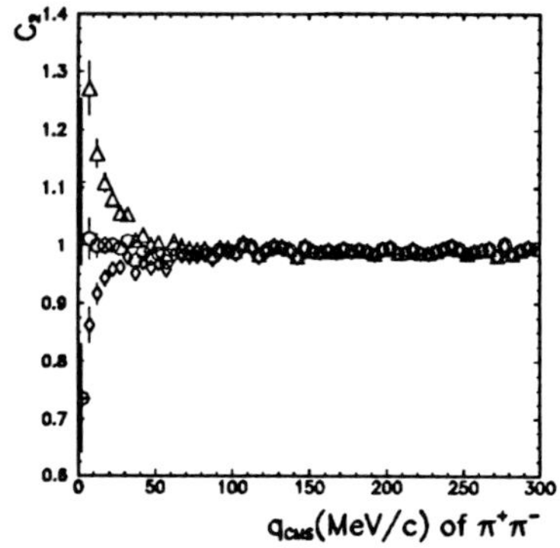

Figure 5: $\pi^+\pi^-$ correlation functions C_2 as a function of $|\boldsymbol{q}|$ (in CMS system): The raw correlation ($\triangle$), correlations corrected with the Coulomb function ($\diamond$) and corrected for Coulomb and momentum resolution ($\bigcirc$) are shown.

5 Results

The fitted values from the one dimensional parameterization of Eq. 2 are R_{inv} = 5.69 ± 0.09, λ = 0.47 ± 0.01 for $\pi^+\pi^+$ and R_{inv} = 6.12 ± 0.09, λ = 0.48 ± 0.01 for $\pi^-\pi^-$. The difference in the size between $\pi^+\pi^+$ and $\pi^-\pi^-$ sources may be attributed to the Coulomb interaction between each of the pions and the rest of the system. With the simplified assumption [21] that the Coulomb field of the rest of the system can be described by a central Coulomb potential with an effective charge Z_{eff}, the difference in the source size results in Z_{eff} $\approx$ 40.

Multi-dimensional radius parameters for the five K_T bins are shown in Fig. 9 for the YKP parameterization, and for the BP parameterization in Fig. 10. The fitted regions for each points are indicated by lines in the horizontal direction. Fitted λ parameters for both parameterizations are shown in Fig. 11. The longitudinal component $R_{\parallel}$ (and R_{long}) features a clear K_T dependence. Since the slope of the decrease is expected to grow with the expansion rate of the source [3,7,22,23], it can be interpreted as an indication of a longitudinal expansion of the source before freeze-out. To quantify this systematic decrease, $R_{\parallel}$ values are fit with $R_{\parallel} \propto 1/m_T^{\alpha}$, where $m_T = \sqrt{K_T^2 + m_\pi^2}$, since the approximate analytic "m_T scaling" ($R_{\parallel} \propto 1/\sqrt{m_T}$) from the hydrodynamical models [3,22] has been commonly used as a basis for experimental

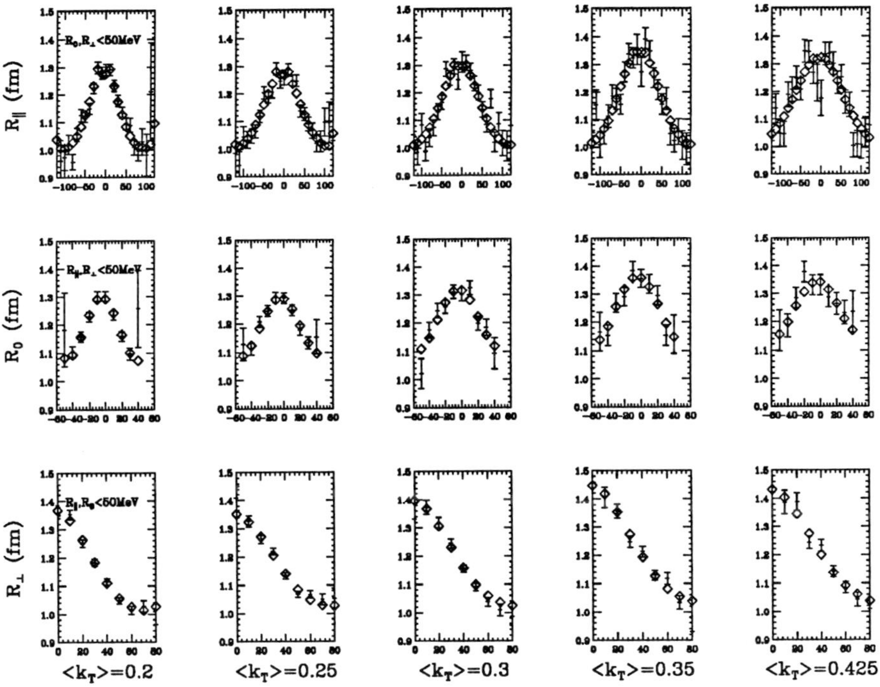

Figure 6: Projections of the $\pi^+\pi^+$ correlation function and the three dimensional fit using the YKP parameterization. The first, second, and third rows correspond for $R_{\|}$, R_0, and $R_\perp$ and the each column represents bins in K_T. The data are shown with statistical error bars and the projections of the correlation functions are displayed with symbols ($\diamond$).

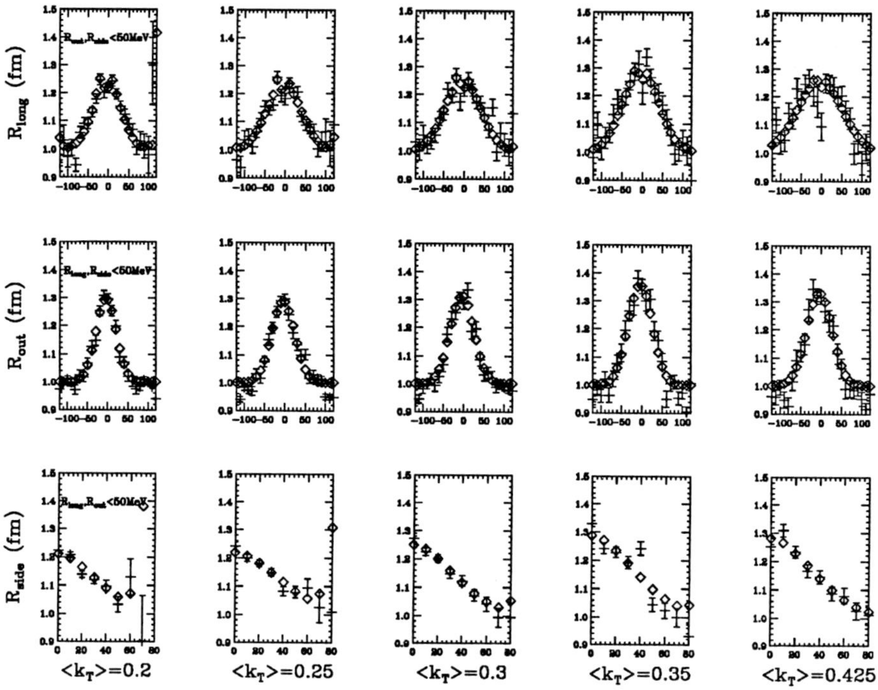

Figure 7: Projections of the $\pi^+\pi^+$ correlation function and the three dimensional fit using the BP parameterization. The first, second, and third rows correspond for R_{long}, R_{out}, and R_{side} and the each column represents bins in K_T. The data are shown with statistical error bars and the projections of the correlation functions are displayed with symbols (◇).

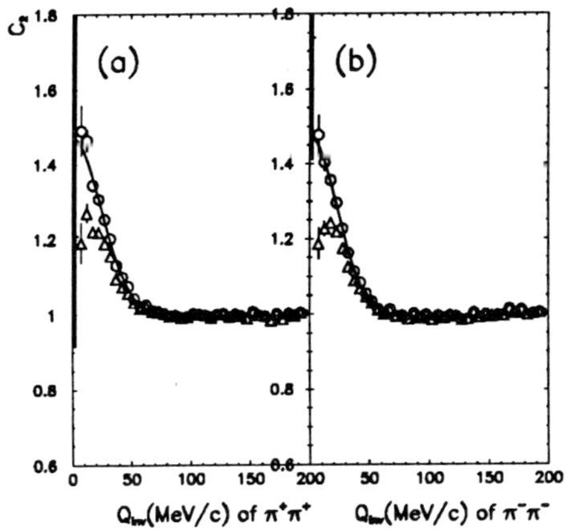

Figure 8: One-dimensional $\pi^+\pi^+$ [(a)] and $\pi^-\pi^-$ [(b)] correlation function $C_2(Q_{inv})$. The raw correlation function is shown with triangles ($\triangle$) and the Coulomb function corrected one is shown with circles ($\bigcirc$).

and theoretical characterization of the dependence of the source radii on the transverse momentum. Fitted exponent α values for $R_{\|}$ in Fig. 9 are 0.75 ± 0.04 (with $\chi^2 = 1.4$) for $\pi^+\pi^+$, and 0.57 ± 0.09 ($\chi^2 = 0.29$) for $\pi^-\pi^-$, which indicates that the K_T-dependence of $R_{\|}$ is similar with or stronger than the $1/\sqrt{m_T}$ behavior.

The values of $R_\perp$ show a weaker and less clear K_T-dependency compared to $R_{\|}$. Fitting $R_\perp$ to $1/m_T^\alpha$ for a quantitative comparison with $R_{\|}$ results in $\alpha = 0.26 \pm 0.07$ ($\chi^2 = 0.7$) for $\pi^+\pi^+$, and 0.15 ± 0.08 ($\chi^2 = 2.0$) for $\pi^-\pi^-$. This can support the interpretation that the source undergoes some transverse expansion [23] as well as longitudinal expansion. The geometrical source parameter value $R_\perp \approx 4.5$ fm at small K_T corresponds to r.m.s. radius of $r_{\text{r.m.s.}} = \sqrt{3}R_\perp \approx 7.8$ fm to be compared with an r.m.s. radius of Au nuclei measured by electron scattering[24] of 5.33 fm. The $r_{\text{r.m.s.}}$ values from the correlation measurement give larger radii than the Au r.m.s. radius. The difference can be understood as a transverse expansion contribution [6] to the geometrical radius $R_\perp$. The source sizes differences between $\pi^+\pi^+$ and $\pi^-\pi^-$ are only apparent for the $R_\perp$ dimension, which qualitatively agrees with nuclear Coulomb field effects [25]. The $R_\perp$ values show a reasonable overall agreement with R_{side} within systematic uncertainties. The rapidity of YKP frame ($y_{YKP} = \frac{1}{2}\ln\frac{1+v}{1-v}$ ≈ -0.3) appears to be slower than the LCMS, which can be understood as an effect of the thermal smearing of the pion velocities in the source[23].

One can derive an estimate of the approximate average freeze-out time τ_0 by assuming a purely longitudinally expanding system without transverse flow from[3,22]:

$$R_{||} = \tau_0 \sqrt{\frac{T}{m_T}}, \tag{7}$$

where T is a freeze-out temperature. By assuming[26] $T = 130$ MeV, the freeze-out time is $\tau_0 \approx 5$ fm/c at $< K_T > = 200$ MeV/c, and $\tau_0 \approx 4.5$ fm/c at $< K_T > = 400$ MeV/c.

An estimate of the phase of freeze-out, i.e., duration of the particle emission, $\delta\tau$ can be provided by the temporal parameter R_0, or can be obtained[5] from the difference of R^2_{out} and R^2_{side} from the BP parameterization,

$$\delta\tau = \frac{1}{\beta_\perp}\sqrt{R^2_{out} - R^2_{side}}, \tag{8}$$

where $\beta_\perp$ is the transverse velocity of the pair. The durations of emission given by both parameterizations are $\delta\tau \approx 2 - 4$ fm/c showing $\delta\tau \lesssim R_\perp$. That excludes the possibility of a long-lived state created from a phase transition[27] at the AGS energy. It is shown that the $\delta\tau$ values for $\pi^+\pi^+$ are larger than for $\pi^-\pi^-$, and they increase at small K_T for both $\pi^+\pi^+$ and $\pi^-\pi^-$. Those systematic features of the effective lifetime are consistent with both parameterizations within the systematic uncertainties as shown in Fig. 12.

6 Summary

Preliminary results on the transverse momentum dependent source size measurement for $\pi^+\pi^+$ and $\pi^-\pi^-$ in Au+Au collisions using the E866 Forward Spectrometer at the AGS have been presented. Multi-dimensional Bose-Einstein correlation functions parameterized by the Yano-Koonin-Potgoretskii method show a rapid decrease of $R_{||}$ and a slower and less clear decrease of $R_\perp$ as the transverse momentum increases. This may indicate a strong longitudinal and a moderate transverse expansion before freeze-out. An approximate freeze-out time was estimated as $\tau_0 \approx 4.5 - 5$ fm/c with the duration of emission $\delta\tau \approx 2 - 4$ fm/c. These results agrees a Bertsch-Pratt parameterization within the systematic uncertainties. The data indicate systematic differences in the source parameters between $\pi^+\pi^+$ and $\pi^-\pi^-$. The π^- source has a larger transverse size, smaller K_T dependence, and smaller duration of emission. This may be explained by the Coulomb interaction between pions and nucleons, but to clarify the effects, further theoretical and experimental investigation are needed.

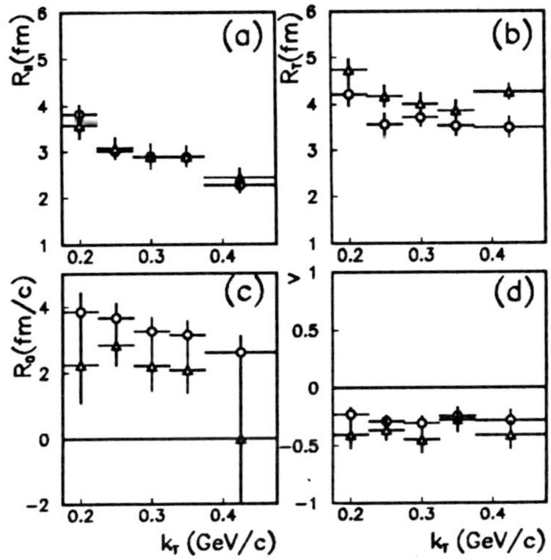

Figure 9: Fitted radius parameters for the YKP parameterization for $R_{\parallel}$ [(a)], $R_{\perp}$ [(b)], R_0 [(c)], and v [(d)]. Circles (◯) shown for $\pi^+\pi^+$ and triangles (△) for $\pi^-\pi^-$. The error bars are statistical only.

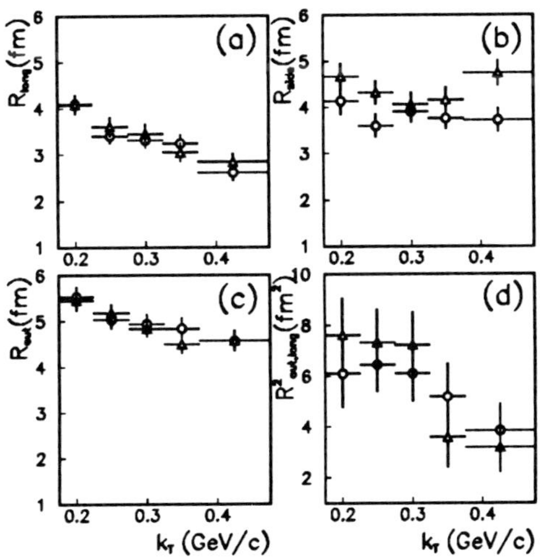

Figure 10: Fitted radius parameters for the BP parameterization for R_{long} [(a)], R_{side} [(b)], R_{out} [(c)], and $R_{out,long}$ [(d)]. Circles (◯) shown for $\pi^+\pi^+$ and triangles (△) for $\pi^-\pi^-$. The error bars are statistical only.

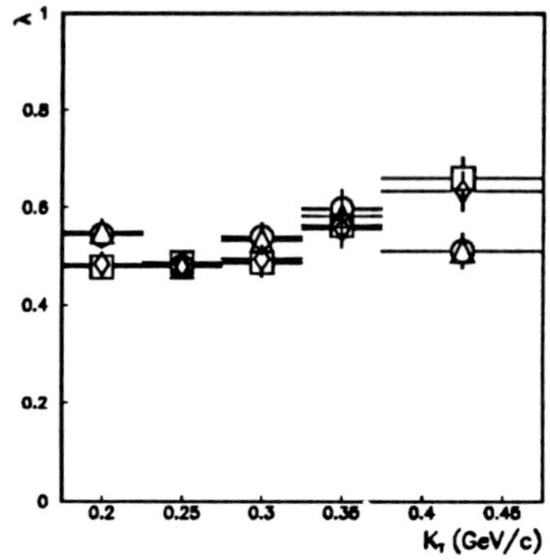

Figure 11: λ parameters for the YKP parameterization ($\triangle$: $\pi^+\pi^+$, $\diamond$: $\pi^-\pi^-$) and the BP parameterization ($\bigcirc$: $\pi^+\pi^+$, $\square$: $\pi^-\pi^-$).

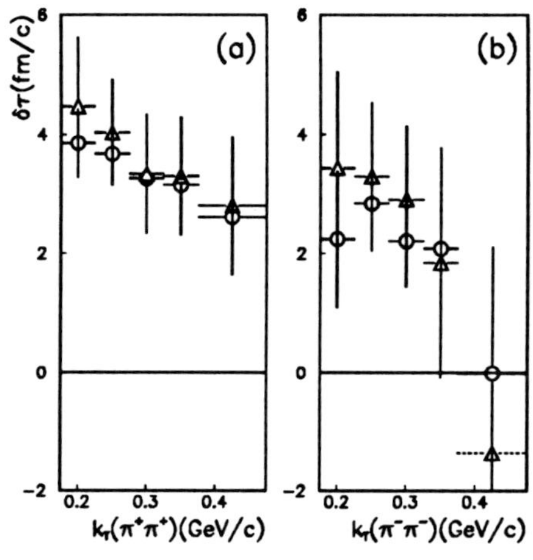

Figure 12: $\delta\tau$ calculated from the BP parameterization for $\pi^+\pi^+$ [(a)] and $\pi^-\pi^-$ [(b)] as functions of K_T are shown with triangles ($\triangle$). For the last point with dotted a line in the (b) panel, where $R_{side} > R_{out}$, an absolute value has taken in the square root in Eq. 8. For a comparison, R_0 values from YKP parameterization are also shown in the figures with circles ($\bigcirc$).

Acknowledgments

This work was supported by the U.S. Department of Energy under contracts with BNL (DE-AC02-98CH10886), Columbia University (DE-FG02-86-ER40281), LLNL (W-7405-ENG-48), MIT (DE-AC02-76ER03069), UC Riverside (DE-FG03-86ER40271), and by NASA (NGR-05-003-513), under contract with the University of California, and by Ministry of Education and KOSEF in Korea, and by the Ministry of Education, Science, and Culture of Japan.

References

1. See D.H. Boal, C.G. Gelbke, and B. Jennings, *Rev. Mod. Phys.* **62**, 553 (1990) for a review.
2. S. Pratt, *Phys. Rev.* D **33**, 1314 (1986).
3. A.N. Makhlin and Yu.M. Sinyukov, *Z. Phys.* C **39**, 69 (1988).
4. T. Csörgő and B. Lörstad, *Phys. Rev.* C **54**, 1390 (1996).
5. H. Heiselberg, *Phys. Lett.* B **379**, 27 (1996).
6. U. Heinz, contribution to this volume.
7. U.A. Wiedemann, P. Scotto, and U. Heinz, *Phys. Rev.* C **53**, 918 (1996).
8. J. Sullivan, contribution to this volume.
9. R. Ganz, contribution to this volume.
10. H. Appelshäuser *et al.*, *Eur. Phy. J.* C **2**, 661 (1998).
11. H. Sako, Ph.D. Thesis, University of Tokyo, INS-J-186 (1997).
12. L. Ahle, Ph.D. Thesis, MIT (1997).
13. J. Barrett *et al.*, *Phys. Rev. Lett.* **78**, 2916 (1997).
14. M. Baker *et al.*, *Nucl. Phys.* A **610**, 213c (1996).
15. F. Yano and S. Koonin, *Phys. Lett.* B **78**, 556 (1978).
16. S. Chapman, J.R. Nix and U. Heinz, *Phys. Rev.* C **52**, 2694 (1995).
17. G. Bertsch, *Nucl. Phys.* A **498**, 173c (1989).
18. S. Chapman, P. Scotto and U. Heinz, *Nucl. Phys.* B **52**, 2694 (1995).
19. CERN Computing Division, CERN Program Library Long Writeup W5013.
20. F. James, CERN Program Library Long Writeup D506.
21. G. Baym and P. Braun-Munzinger, *Nucl. Phys.* A **610**, 286c (1996).
22. J. Bolz *et al.*, *Phys. Lett.* B **300**, 404 (1993).
23. Wu Y.-F. *et al.*, *Eur. Phy. J.* C **1**, 599 (1998).
24. M.A. Preston and R.K. Bhaduri, *Structure of the Nucleus*, p.99, Addison-Wesley, Massachusetts (1975).
25. H.W. Barz, contribution to this volume.
26. J. Stachel, *Nucl. Phys.* A **610**, 509c (1996).
27. D.H. Rischke and M. Gyulassy, *Nucl. Phys.* A **608**, 479 (1996).

PROTON-PROTON INTERFEROMETRY IN ^{40}Ar + ^{197}Au, ^{12}C COLLISIONS AT 95 A MEV.

V. Bellini[1,3],C. Connelli[1], R.Potenza[2,3], G. Riera,[3]
V. Russo[2], M.L. Sperduto[1,3], C.M. Sutera[2] and M. Urrata[3]

[1] *INFN/LNS, Via S. Sofia 44, 95123 Catania, Italy*

[2] *INFN, Sezione di Catania, Corso Italia 57, 95129 Catania, Italy*

[3] *Dipartimento di Fisica, Universitá di Catania, Corso Italia 57, 95129 Catania, Italy*

Abstract

We measured two-proton correlation functions for the reactions ^{40}Ar +^{197} Au,^{12} C at the RIKEN Laboratory in Japan at beam energy of 95 A MeV. The data were taken simultaneously with n-n and n-p coincidences, that are still under analysis. Even if the results are preliminary, agreement is found with a three-body quantum model including the Coulomb interaction of the particles with their emitter, known from the literature. Differences in the shape of the correlation functions are observed between the Au and C targets.

Nuclear intensity interferometry is a powerful method to translate correlation functions for pairs of fermions (p-p, p-n, n-n) into space-time information concerning nuclear reactions [1]. Both quantum interference effects and final-state interactions must however be properly evaluated in order to obtain such space-time information. To get rid of mutual Coulomb interaction and of Pauli exclusion contribution, it is required simultaneous measurements of p-p, n-n, p-n correlations with precise measurements of two momentum vectors at small relative momentum.

A nuclear interferometry experiment has been carried out at the RIKEN cyclotron in Japan on simultaneous p-p, n-p and n-n correlation functions at 95A MeV in order to investigate the preequilibrium component in particle emission. The proton multidetector array (**EMRIC**) [3] has been upgraded by replacing its photomultipliers and implementing its full remote control from the mechanical and electronical point of view. Actually the data are analysed in collaboration with a Lund University group (B. Jakobsson et al), which had responsibility for the neutron detectors.

The experimental setup consisted of 13 CsI crystals, as proton detectors with face of 4x4cm^2 and 10 cm thickness, arranged in spherical symmetry at 60 cm from the target and put at 45 degrees with respect to the beam direction, together with 23 liquid scintillators, as neutron detectors. The relative angle between neighbouring CsI was 3 degrees. Twelve liquid scintillators were

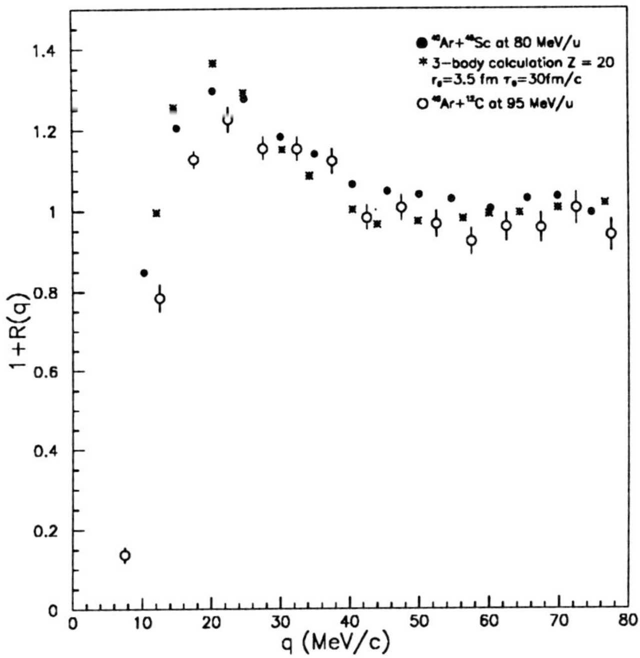

Figure 1: p-p correlation function for the reaction Ar ^{40}+C^{12}

placed behind holes in the CsI array in order to guarantee a minimum angular separation between the detected pairs. The remaining neutron detectors were put at different angles in order to get further n-n correlation information.

The analysis on p-p correlation has been accomplished by using the **GEANT** simulation code from CERN, with the aim to determine all the experimental corrections, as for instance the proton multiple scattering, the Si detector dead layers for the CsI calibration etc.; the cross-talk between two neighboring neutron detectors was measured through the use of brass shadow bars.

The calibration of CsI detectors was performed by putting a two-element Si telescope as close as possible to the target, and registering the punch-through points of p, d, t, ^{3}He and ^{4}He. Liquid scintillators were calibrated via the standard Time-of-Flight method and by determining the offset by means of the prompt gamma ray peak.

It has been shown that there are several advantages of using nucleons for intensity nuclear interferometry in heavy ion collisions [2]. Namely two-proton emission has much more phase space available than two-meson emission and the enhanced ^{2}He resonance leads to strong sensitivity of the correlation function

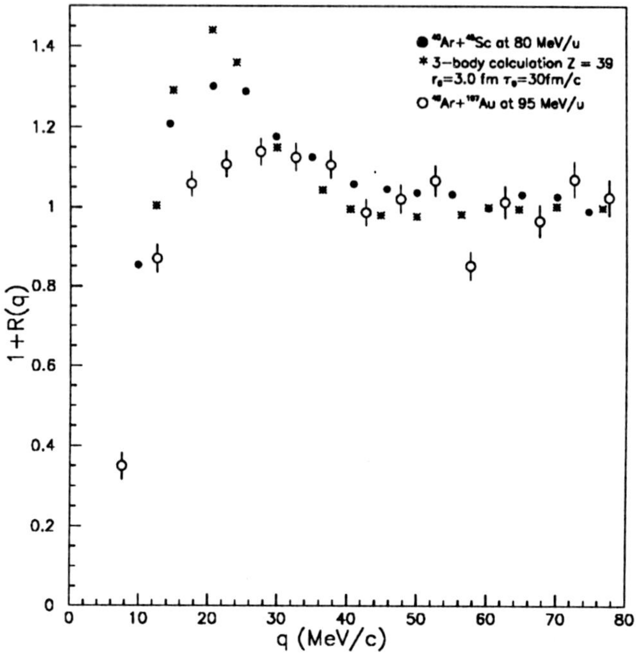

Figure 2: p-p correlation function for the reaction Ar ^{40}+Au^{197}

to the source size .

Here we report on the first preliminary data obtained for p-p coincidences. Work is in progress for the n-n and n-p correlation functions.

The p-p correlation function for C target, is shown in fig. 1, where our data are compared with theoretical calculations from Martin et al [4] and previous experimental data for the same projectile but a different target at slightly lower energy [5]. In fig. 2 we show the preliminary results for Au target, data available with lower statistics. At first glance the correlation function shape is qualitatively in agreement with the same theoretical calculations and with previous measurements at different beam energies [6].

The previous comparison is meaningful, as far as the p-p source parameters, radius r_0 and lifetime τ, are deemed insensitive to small differences in the energy available to the nuclear system. The preliminary conclusion from our data should be the confirmation of the influence of the emitter Coulomb field on two-proton correlation functions for both targets. However we did not calculate explicitely the importance of this effect for our target-projectile combination and for the actual 95 A MeV beam energy. A clear suppression of the low relative momentum peak is observed for heavier Au target data in comparison to the lighter C target ones. When the data for neutron-proton and two-neutron correlation functions will be available, we should draw more specific conclusions.

Actually our interferometry program focuses on one experiment with a combination of an effective fermion interferometer and an event characterizing device, in order to determine the impact parameter and reaction plane. To this aim a forward wall of 32 high-resolution phoswich detectors, made of a thin (up to 700 μm) plastic scintillator optically coupled to a 10 cm thick BaF2 [7], has been added to the interferometry setup and a new experiment started at Laboratorio Nazionale del Sud in Catania at 45 A MeV of ^{58}Ni incident beam.

ACKNOWLEDGEMENTS

The authors gratefully acknowledge very useful discussions with B. Jakobsson, A. Fokin and A. Kuznetsov. Thanks also to G. Poli for technical support. One of us (V. Bellini) wishes to thank the Organizing Committee, namely Prof. A. Insolia, for the very friendly welcome to the Conference.

1. S. E. Koonin *Phys. Lett.* **B70** (1977) 43
2. W. Bauer et al. *Annu. Rev. Nucl. Part. Sci.* **42** (1992) 42
3. F. Merchez et al. *Nucl. Instrum. Methods* **A275** (1989) 133
4. L. Martin et al. *Nucl. Phys.* **A604** (1996) 69
5. M. A. Lisa et al. *Phys. Rev. Lett.* **71**(1993) B 2863
6. M. Cronqvist et al *Phys. Lett.* **B317** (1993) 505
7. G. Lanzano et al. *Nucl. Instrum. Methods* **A323** (1992) 694

PARTICLE CORRELATIONS WITH HEAVY IONS AT LHC ENERGIES

B. ERAZMUS, L. MARTIN, C. ROY, K. WERNER
SUBATECH
(UMR, Université, Ecole des Mines, IN2P3/CNRS)
4, rue Alfred Kastler, La Chantrerie
44070 Nantes, France
E-mail: erazmus@subatech.in2p3.fr

R. LEDNICKY
Institute of Physics, Na Slovance 2
18040 Prague 8, Czech Republic

V. LYUBOSHITZ
JINR Dubna, P.O.Box 79,
Moscow, Russia

K. MIKHAILOV, A. STAVINSKY
ITEP, B.Cheremushkinskaya 25,
117259 Moscow, Russia

J. PLUTA
Institute of Physics, Warsaw University of Technology,
00-662 Warsaw, Poland

Yu. SINYUKOV
Institute for Theoretical Physics of National Acad. Sci.,
Kiev 252143, Ukraine

for the ALICE COLLABORATION

The ALICE detector will offer very good conditions to study the space-time characteristics of particle production in heavy-ion collisions at LHC from measurements of the correlation function of identical and non-identical particles at small relative velocities. The correlations - induced by Coulomb and nuclear final-state interactions - of non-identical particles appear to be directly sensitive to the space-time asymmetries of particle production allowing, in particular, a measurement of the mean relative delays in particle emission at time scales as small as few fm/c.The problem of Coulomb interaction of the correlated particles is particularly important in the case of the large effective volumes formed in ultra-relativistic heavy-ion reactions.

1 Introduction

Particle correlations at high energies usually measure only a part of the space-time emission region since the sources, despite their fast longitudinal motion, emit the correlated particles with small relative velocities mainly at close space-time points.

The features of emitting sources can be investigated in the frame of an approach which includes the dynamics of the emission process as well as the effects of quantum statistics and final state, Coulomb and strong interactions. It is important that particle correlations contain information about the dynamical evolution of the emission process, such as proper time of decoupling (freeze-out) duration of particle emission and the presence of collective flows. This information is, in principle, closely related to the QGP formation [1,2,3]

2 Simulations

Our correlation analysis were performed using the string Model for Ultra-relativistic Hadronic Interactions VENUS (version 5.14) [4,5]. Based on the string model, one first determines connected regions of high energy density. These regions are referred to as quark matter (QM) droplets. For such regions, the initially produced hadrons serve only as a mean to produce the proper fluctuations in the energy density. Presently the model is not well suited for the LHC energy region. However, it provides space-time and momentum space characteristics of the freeze-out points of the emitted particles, which can be used as an approximation accounting for the basic features of the multiparticle production, including the fast longitudinal motion of the source as well as the resonance production.

The events were generated without the phase of the particle rescattering and droplet formation which normally leads to the increase of the space-time characteristics of particle emission. Instead, we introduce the scale factor which allows to extend the production region or introduce the shifts in emission times of various particle species in a controlled way.

The generation of events using VENUS model is followed by the calculation of the correlation function for the selected pairs of identical and non-identical particles. In this way we can study the sensitivity of the ALICE detector to particular correlation effects [6].

3 Experimental resolution

In order to account for effects due to the experimental resolution we used a program based on parametrisation of the full detector simulation [7]. A realistic

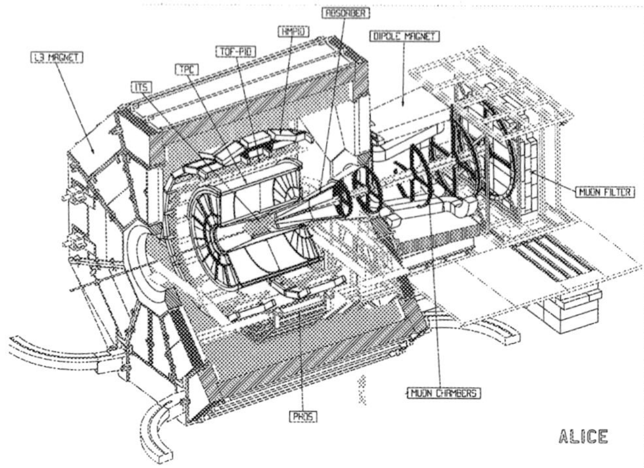

Figure 1: Alice Detector.

geometry of the ALICE central barrel has been taken into account (Fig. 1). This program does include contributions due to multiple scattering, measurement precision and detector alignment as well as fluctuations in energy loss. Effects due to the track finding algorithms were not included since the tracking efficiency in ALICE is reather good and starts only to decrease at very low transverse momentum.

The two-track resolution was not taken into account in the simulations. However, it is expected to have rather small effect on correlation functions of like particles and practically no effect on unlike particle correlations at small relative velocities. The effect of the experimental resolution on the relative momentum is summarized in Table 1.

Table 1: The ALICE resolution: $\sigma(Q_i - Q_i^{'})$, MeV/c. A cut $Q_{inv} < 0.05 GeV/c$ has been applied.

Particles	Q_{side}	Q_{out}	Q_{long}	Q_{inv}
$\pi^+\pi^+, \pi^+\pi^-$	0.3	3.8	0.7	1.2
π^+K^+, π^+K^-	0.4	4.0	1.1	1.8
π^+p, π^-p	0.4	3.9	1.3	2.0
K^+K^+, K^+K^-	0.5	6.5	2.6	3.6
K^+p, K^-p	0.6	8.2	3.2	4.8
pp	0.6	11.6	5.0	6.0

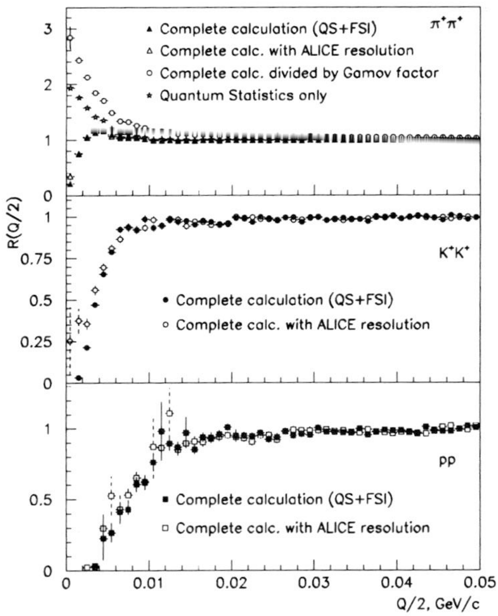

Figure 2: The simulated correlation functions for two identical charged pions, kaons and protons without and with the experimental resolution accounted for. The simulations were done using VENUS model with the extended space-time freeze-out coordinates to set $\langle r \rangle = 30$ fm and $\langle t \rangle = 20$ fm/c in the reaction c.m.s. (scale-factor = 5). For two pions the Gamow corrected correlation function is compared with that including the effect of quantum statistics only.

4 Correlation of identical particles

The correlation function of identical particles is sensitive to the relative space-time distances between the emission points due to the effects of Bose-Einstein or Fermi-Dirac statistics [8,9] and the strong and Coulomb final state interactions [10,11]. The relative importance of these three factors will be different according to the particle species.

In Fig. 2 we show the two-particle correlation functions for pions, kaons and protons calculated taking into account the effects of quantum statistics and both Coulomb and nuclear final state interactions. In the case of large effective sources, the correlations of two protons are, due to their relatively

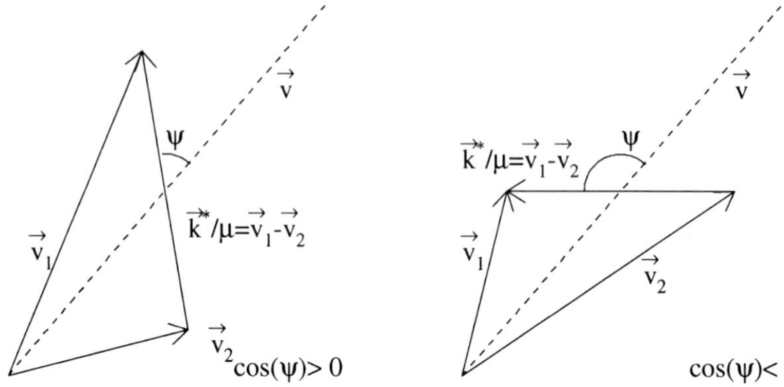

Figure 3: We can determine which sort of particles was produced earlier and which later by studying the correlation functions of two non-identical particles separately for the angles less and greater than 90° between the relative velocity $\mathbf{k}^*/\mu$ ($\mathbf{k}^* = \mathbf{p}_1^* = -\mathbf{p}_2^*$ and μ is the reduced mass of the two particles) and the total pair velocity $\mathbf{v}$.

small Bohr radius of 58 fm, stronger than those of pions and kaons. Thus, the correlations of protons appear to be a sensitive probe of the evolution of the collision in the case of large emitting region.

The correlation functions are only weakly affected by the experimental resolution even in the case of large sizes expected at LHC energies.

5 Correlations of non-identical particles

The non-identical particle correlations can be used for the determination of the space-time parameters of the emission process in a similar way as the identical ones using the sensitivity of the Coulomb and strong interaction effects to the space and time intervals of particle emissions. On the other hand, the correlations of non-identical particles appear to be directly sensitive to the sequence of particle emission [12]. To study this sensitivity one should construct the correlation functions $R_+(\mathbf{k}^*\mathbf{v} \geq 0)$ and $R_-(\mathbf{k}^*\mathbf{v} < 0)$ corresponding to the positive and negative values of $\mathbf{k}^*\mathbf{v}$, respectively (Fig. 3). On the absence of the space-time asymmetries, the ratio R_+/R_- is equal unity. In Fig. 4 we illustrate the effect of the asymmetry in the case of the emission of K^+K^- pairs. The structures (a deep or a peak) and their amplitude appearing in this ratio depend on the sign and the magnitude of the time delay introduced in

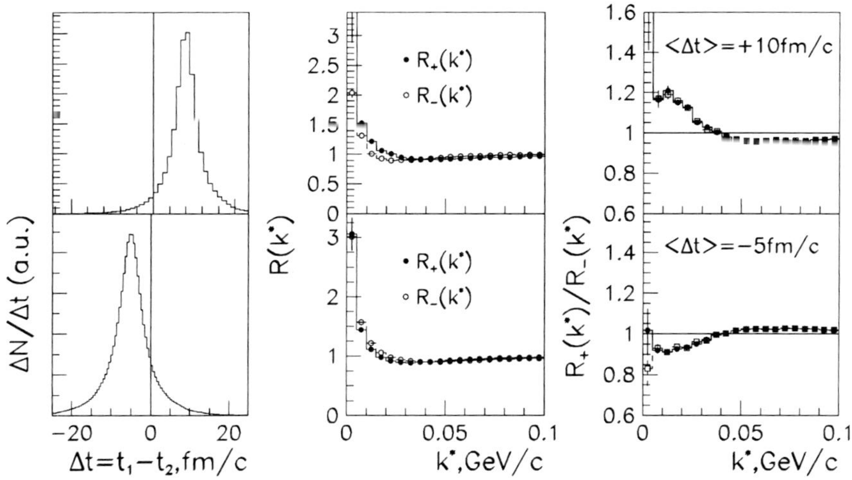

Figure 4: The distributions of the difference of the K^+ and K^- emission times $\Delta t = t_{K^+} - t_{K^-}$ simulated by VENUS with the shifts $\langle \Delta t \rangle$ =+10 and -5 fm/c introduced *ad hoc*, the corresponding correlation functions R_+ ($\mathbf{v} \cdot \mathbf{k^*} \geq 0$) and R_- ($\mathbf{vk^*} < 0$) and their ratios calculated for K^+K^- pairs. The ratios distorted by the effect of experimental resolution are represented by open symbols.

the calculations. The observation of a delay in the emission of K^+ and K^- can be considered, in particular, as a strong indication of strangeness distillation.

Fig. 5 demonstrates the scaling of the space-time asymmetry with the inverse Bohr radius for the like- and unlike-sign of πK, πp and Kp pairs with the corresponding Bohr radii of 248.5, 225.5 and 83.6 fm, respectively. The correlation functions for the πK, πp pairs were calculated using one simulated event only.

Due to its very good resolution, the ALICE experiment will be sensitive to the delays in the particle emission times of the order of a few fm/c even in the frame of event-by-event analysis in the case of pairs involving pions.

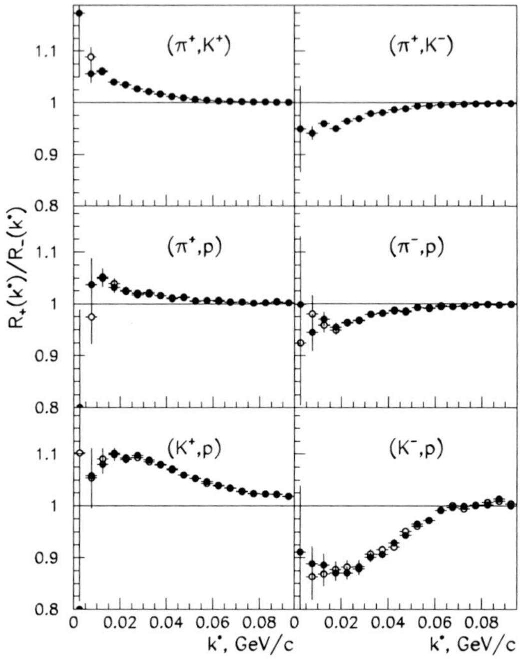

Figure 5: The same ratios R_+/R_- as in Fig. 4 calculated with $\langle \Delta t \rangle$ = -10 fm/c for different pairs of non-identical particles, taking into account (open symbols) and neglecting (full symbols) the effect of experimental resolution.

6 Coulomb corrections for interferometry analysis

For the large expanding hadron systems, expected at LHC energies, the usual Gamow correction substantially deviates from the true Coulomb effects (see Fig. 2). A simple modification of the standard zero-distance correction has been proposed [13]. This finite-size analytical Coulomb correction allows one to extract the pure quantum statistics correlation function (e.g. in a hydrodynamic model) of two identical charged pions or kaons in the case of relatively large effective emission volumes. This procedure can be useful for a fast interferometry analysis and reasonable preliminary estimate of the model parameters even at RHIC and LHC energies.

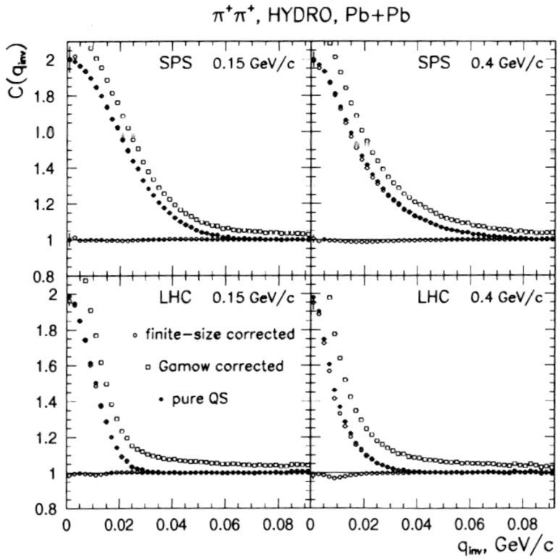

Figure 6: The comparison of the Coulomb corrected two-pion correlation functions at SPS and LHC energies for p_t=0.15 and 0.4 GeV/c, simulated with the hydrodynamical calculations. The points near the unity correspond to the ratio of the reconstructed pure quantum statistics correlation function to the finite-size corrected correlation function.

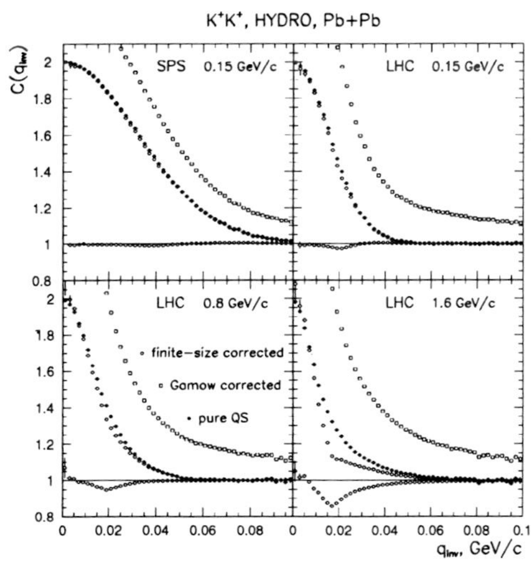

Figure 7: The same as Fig. 6 for the two-kaon correlation functions at SPS and LHC energies for p_t=0.15, 0.8 and 1.6 GeV/c.

In Figs. 6 and 7 we present the results of calculation of the pion and kaon correlation functions at SPS and LHC energies for various mean transverse momentum of the particles using the hydronynamic model[14]. The discrepancy between the pure quantum statistics calculations and the finite-size corrections appears to be significant only for kaons at high transverse momentum.

7 Conclusions

We analyzed the ability of the ALICE detector for determination of the space-time characteristics of particle production in heavy-ion collisions at LHC.

Correlation functions of various particle pairs can be measured and will be only slightly distorted by the detector resolution effects.

We demonstrated that non-identical particle correlations contain an information complementary to that provided by identical particles. In particular, the unlike particle correlations allow to determine the space-time asymmetries in the particle emission process.

The simulations of correlations of various particle pairs were performed using the event generator VENUS adapted arbitrarily for LHC conditions and including the expected resolution of the ALICE detector. The results show that the relative time delays of the order of a few fm/c can be measured.

The two-particle Coulomb interaction gives an important contribution to correlations in the case of large sizes and lifetimes. An analytical correction has been proposed allowing to extract the pure quantum statistics correlation function and to rapidly estimate the parameters describing the emitting source.

References

1. S. Pratt, *Phys. Rev. Lett.* **53**, 1219 (1984); *Phys. Rev.* D **33**, 1314 (1986); S. Pratt, T. Csorgo, J. Zimanyi, *Phys. Rev.* C **42**, 2646 (1990).
2. A.N. Makhlin, Yu.M. Sinyukov, *Yad. Fiz.* **46**, 637 (1987); *Z. Phys.* C **39**, 69 (1988); Yu.M. Sinyukov, *Nucl. Phys.* A **498**, 151c (1989).
3. G. Bertsch, M. Gong, M. Tohyama, *Phys. Rev.* C **37**, 1896 (1988).
4. K. Werner, *Phys. Rep.* **232**, 87 (1993).
5. K. Werner and J. Aichelin, *Phys. Rev.* C **52**, 1584 (1995).
6. B.Erazmus, R.Lednicky, V.Lyuboshitz, L.Martin, K.Mikhailov, J.Pluta, Yu.Sinyukov, A.Stavinsky, K.Werner, ALICE Note, INT-95-43.
7. ALICE Technical Proposal, CERN/LHCC 95-71, LHC/P3 (1995), Chapter 11.
8. G. Goldhaber, *Phys. Rev.* **120**, 300 (1960).
9. M.I. Podgoretsky, *Fiz. Elem. Chast. Atom. Yad.* **20**, 628 (1989) (*Sov. J. Part. Nucl.* **20**, 266 (1989)).

10. S.E. Koonin, *Phys. Lett.* B **70**, 43 (1977).
11. R. Lednicky, V.L. Lyuboshitz, *Yad. Fiz.* **35**, 1316 (1982); *Sov. J. Nucl. Phys.* **35**, 770 (1982); Proc. Int. Workshop on Particle Correlations and Interferometry in Nuclear Collisions, CORINNE 90, Nantes, France, 1990 (ed. D. Ardouin, World Scientific, 1990)p.42; JINR report P2 546-92 (1992) 1; *Heavy Ion Physics* **3**, 93 (1996).
12. R. Lednicky, V.L. Lyuboshitz, B. Erazmus, D. Nouais, *Phys. Lett.* B **373**, 30 (1996).
13. Yu.M. Sinyukov, R. Lednicky, S.V. Akkelin, J. Pluta, B. Erazmus, *Phys. Lett.* B **432**, 248 (1998).
14. Yu.M. Sinyukov, *Nucl. Phys.* A **498**, 151 (1989); S.V. Akkelin, Yu.M. Sinyukov, *Phys. Lett.* B **356**, 525 (1995); *Z. Phys.* C **72**, 501 (1996).

EXPERIMENTAL ISSUES FOR HBT STUDIES AT RHIC

THOMAS J. HUMANIC and HELEN L. CAINES
Department of Physics, The Ohio State University
Columbus, Ohio 43210
E-mail: humanic@mps.ohio-state.edu

The Relativistic Heavy Ion Collider at Brookhaven will begin producing colliding beams by the middle of next year. We will undoubtedly have our first preliminary HBT analysis from one or more of the RHIC detectors shortly thereafter. In this paper we will attempt to identify some experimental issues involved in carrying out HBT studies at RHIC in the context of the STAR detector and will try to address what realistic expectations we can have for these studies.

1 Introduction

RHIC should easily provide the most interesting physics yet to probe with Hanbury-Brown-Twiss (HBT) interferometry and the means to carry out HBT studies with a wide variety of particle types[1]. In addition, we have the advantage at RHIC that we can apply the lessons we have learned about extracting physics from HBT studies in extensive AGS and SPS experiments (e.g. Ref.[2,3]). All of the four main RHIC experiments (BRAHMS, PHENIX, PHOBOS, and STAR) will have the capability of carrying out HBT studies with charged pions, kaons and protons and some of the experiments will in addition be able to study more exotic probes such as K^0_S-K^0_S and Λ-Λ correlations.

The main question to be answered when the analysis of the data has been completed in an HBT experiment is "what have we measured?" This question applies to both the physical and the experimental interpretations of the data. This paper concentrates on the experimental interpretation of the data which is necessary to understand before one can move on to consider the physical implications of what was measured. The experimental issues involved in HBT at RHIC will be studied in the context of the STAR experiment[4] which has the capability of using a broad range of HBT probes. For the present work, only charged pion pairs and K^0_S pairs near mid-rapidity in STAR will be considered.

In STAR, charged hadrons are measured at mid-rapidity ($-1<\eta<1$) in cylindrical-geometry tracking detectors (whose axis is defined by the colliding beams) which are the Silicon Vertex Tracker (SVT) at smaller radii and the Time Projection Chamber (TPC) at larger radii. The SVT is composed of three barrel-geometry layers of silicon drift detectors (SDD's). The combination of the SVT and TPC form a global tracker. Matching tracks between the SVT and TPC to form global tracks allows better overall momentum resolution (very important for HBT interferometry studies) and better particle identification using dE/dx than either detector alone. Note that a fourth barrel of silicon strip detectors (SSD's) located at a radius between the SVT and TPC will also be present in STAR but at this time is not implemented in the STAR software and thus will not be included in the present study. K^0_S's are measured in the SVT+TPC detector by observing the decay channel $K^0_S \rightarrow \pi^+\pi^-$.

2 Calculational Method

The goal of the present work is to make "realistic" Monte Carlo calculations for STAR HBT capabilities for charged pion pairs and K^0_S pairs using the most recent (and available) versions of STAR software and to determine what distortions are introduced into the experimental correlation functions by the detector. The STAR software chain used is the following: a) detector Monte Carlo which takes a HIJING event and converts the output list of particles via GEANT into smeared hits in the detectors (no cluster-finding used), b) track reconstruction in the SVT and TPC from the detector hits, c) track matching in the SVT and TPC to make global tracks, d) V^0 reconstruction of oppositly charged pions to find K^0_S's and e) the extraction of the "experimental" momenta of the reconstructed particles. The HBT effect is then imposed pairwise on the reconstructed particles and correlation functions are generated (see below). Several "non-realistic" assumptions are also made in the present work which to some extent can be justified: 1) perfect charged π-K separation using dE/dx--this should be a good approximation for $p_T <$ 500 MeV/c and for stringent cuts being made on dE/dx, 2) 100% purity in reconstructed K^0_s's--this limit can be approached in STAR by making stringent cuts on the V^0 reconstruction parameters and the invariant mass and by using neural-network pattern recognition (this purity comes at the expense of reconstruction efficiency--in the present work we get two reconstructed pure K^0_s's per event), 3) cluster finding software is not used, only smeared space points as already mentioned above--however recent cluster-finding codes have shown that almost 100% efficiency for finding space points can be obtained, and 4) silicon strip detector (fourth) layer in STAR not included in software-- when included the SSD should improve dE/dx, tracking, and K^0_s reconstruction.

The method used to impose HBT correlations on particle pairs and to assess the distortions due to the detector on the correlation function and extracted boson source parameters is now discussed. After running the STAR Monte Carlo chain on a HIJING event (as described above) one has generated a "list" of reconstructed particles for that event. Each particle in the list has associated with it a HIJING (unprimed) four-momentum (E, **p**) and a reconstructed (primed) four-momentum (E',**p'**) which reflects the distortions due to detector effects and due to imperfect reconstruction software. From this list, an "experimental" two-boson correlation function is formed using the following four steps:

1) choose two like particles from the list and calculate their true (HIJING) and distorted momentum differences (in the LCMS frame, i.e. $p_{1z}+p_{2z}=p_{1z}'+p_{2z}'=0$)

$$\mathbf{Q}=\mathbf{p}_1-\mathbf{p}_2,\quad Q_0=E_1-E_2,\quad Q_{inv}^2=\mathbf{Q}^2-Q_0^2$$

$$\mathbf{Q}'=\mathbf{p}_1'-\mathbf{p}_2',\quad Q_0'=E_1'-E_2',\quad Q_{inv}'^2=\mathbf{Q}'^2-Q_0'^2\ ,$$

2) calculate the true HBT weight (i.e. using HIJING momenta), W(Q), for the two particles assuming a gaussian boson source and assuming values for the source parameters R_{To}, R_{Ts}, R_L, λ (Bertsch-Pratt parameterization [2]) and include the final-state Coulomb effect between

the two bosons using a Gamow factor, $G(Q_{inv})$

$$W(Q)=[1+\lambda\exp(-R_{Ts}^2Q_{Ts}^2-R_{To}^2Q_{To}^2-R_L^2Q_L^2)]G(Q_{inv}) \qquad (1)$$

where $G(Q_{inv})=2\pi\eta/[\exp(2\pi\eta)-1]$ and $\eta=m\alpha/Q_{inv}$ and where m is the boson mass and $\alpha=1/137$,

3) calculate the "experimental" correlation function, $C_{exp}(Q_{bin}')$, in terms of the distorted momenta by binning pairs of bosons (weighed by Eq.1) in terms of the distorted momentum differences and make a Coulomb correction (in the denominator) also using the distorted momentum differences

$$C_{exp}(Q_{bin}')=A[R(Q_{bin}')/B(Q_{bin}')] \qquad (2)$$

where $R(Q_{bin}')=\sum_i W(Q_i)$ and $B(Q_{bin}')=\sum_j G(Q_{inv\,j}')$ and the sums over i and j extend over boson pairs whose distorted momentum differences fall into a particular momentum-difference bin such that $Q_{bin}'=\sum_i Q_i'/N$ where N is the number of boson pairs that falls in the particular bin. The quantity A is a normalization constant determined by the condition that $C(Q_{bin}')=1$ for large Q_{bin}', and

4) fit a gaussian correlation function, $C(Q')$ to the "experimental" correlation function calculated from Eq.2 to extract the "distorted" source parameters, R_{To}', R_{Ts}', R_L', λ', and compare the results with the true source parameters input into the calculation via Eq.1,

$$C(Q')=A'[1+\lambda'\exp(-R_{Ts}'^2Q_{Ts}'^2-R_{To}'^2Q_{To}'^2-R_L'^2Q_L'^2)] \qquad (3)$$

where A' is a normalization parameter determined in the fit.

In order to carry out the proceedure outlined above, it is necessary to guess reasonable values to use for the input boson source parameters, R_{To}, R_{Ts}, R_L, λ, which are to be inserted into Eq.1. For the present study two sets of input source parameters were used: 1) $R_{To}= R_{Ts}= R_L=6$ fm,

λ=0.6, and 2) $R_{Ts}=R_L$=6 fm, $R_{To}=3R_{Ts}$=18 fm, λ=0.6. Set #1 represents a lower-limit or conventional physics case since it approximates measured values for CERN Pb+Pb collisions[2] and preliminary predictions for RHIC Au+Au collisions from a simple rescattering calculation[5]. Set #2 represents an upper-limit case where a phase transition to Quark Matter takes place resulting in a long-lived boson source as predicted by the hydrodynamics calculations of Rischke and Gyulassy[1]. In the present study $\pi\pi$ calculations were performed for both sets of parameters whereas for $K^0{}_sK^0{}_s$ calculations only set #1 was used.

3 Results

Figures 1-5 present the results of the present study. Figure 1 shows a sample m_T distribution for pion momenta reconstructed from HIJING in the STAR tracking detectors (SVT+TPC) as described above. An exponential slope parameter of 188 MeV is extracted by fitting the expression

$$(1/m_T)dN/dm_T = C \exp(-m_T/B)$$

where B is the slope parameter and C is a normalization constant.

Figure 2 shows the momentum resolution for rectangular components for reconstructed pions in STAR and projections of the three-dimensional two-pion correlation function produced from these reconstructed pions weighted by Eq.1 for the "conventional physics" case (input parameter set #1). The difference between the "real" (unprimed) and "recon" (primed) components are plotted and a gaussian is fitted to extract σ in each case. As is seen σ is slightly less than 5 MeV/c for each component. The distored source parameters obtained by fitting Eq.3 to the correlation function are also shown. The statistics for this correlation function corresponds to one hour of STAR running assuming a data acquisition rate of one event per second, yielding about 8 million pion pairs. The average k_T of these pions is 250 MeV/c. Error bars are statistical only.

The momentum binning used in this case is 10 MeV/c for all components.

The distorted source parameters from Figure 2 are compared graphically with the input source parameters from set #1 and two other calculations in Figure 3. The dashed line at 6 on the ordinate of the plot represents both the input radius parameters and the input λ*10. The three calculation types indicated on the abcissa of the plot are the following: a) "all effects"--pion pairs weighted by Eq.3 and Coulomb corrected as described above, i.e. the parameters shown in Figure 2, b) "no Coulomb effects"--pion pairs weighted by the symmetrization part of Eq.3 but not the Gamow part and with no Coulomb correction to the

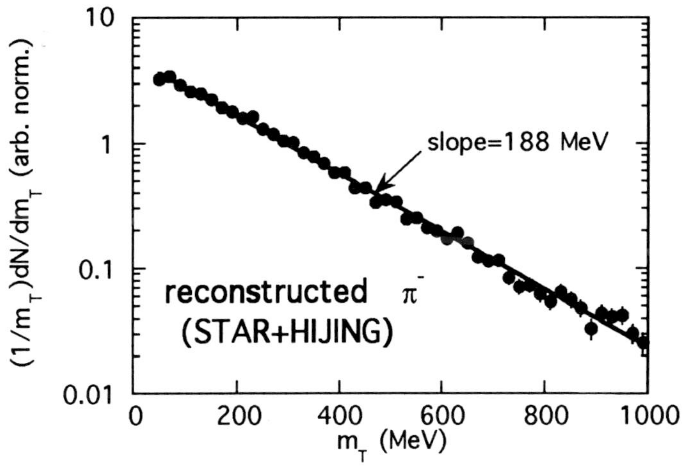

Figure 1: Sample m_T distribution for pion momenta reconstructed from HIJING events in the STAR detector (SVT+TPC).

correlation function, and c) "binning effects only"--take out all detector effects and study the smearing introduced into the correlation function by the 10 MeV/c binning only, i.e. weight pairs with Eq.3 but form the correlation function with HIJING (unprimed) momenta. As seen in Figure 3, the "binning effects only" calculation has a negligible effect on

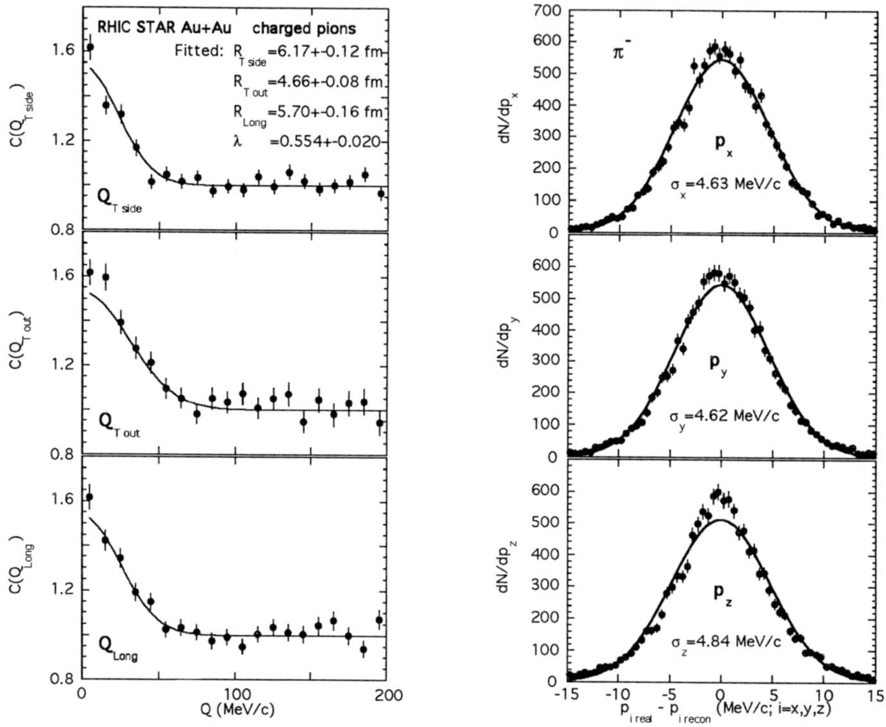

Figure 2: Momentum resolution for rectangular components for reconstructed pions in STAR and projections of two-pion correlation function from STAR for input parameter set #1 (conventional physics).

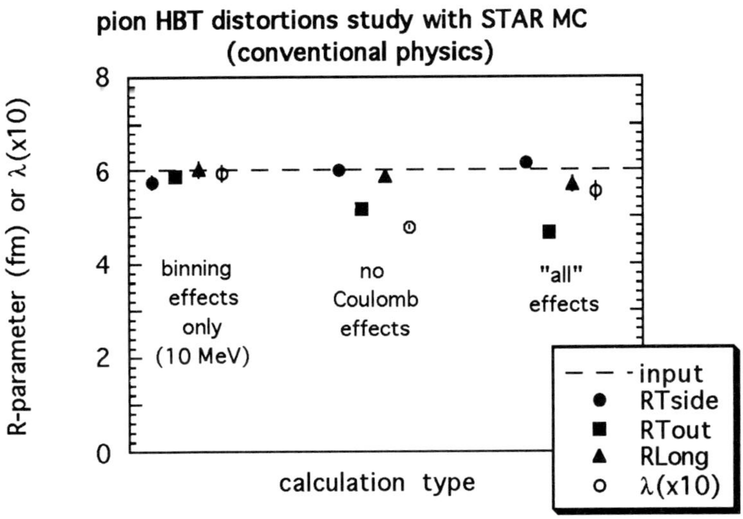

Figure 3: Distorted source parameters for conventional phyiscs input (set #1) compared for several calculations.

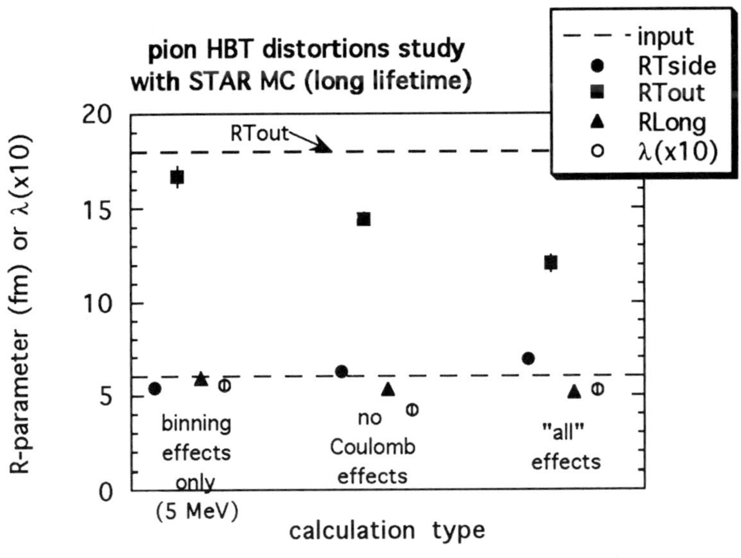

Figure 4: Same as Figure 3 except for long-lifetime input parameters (set #2).

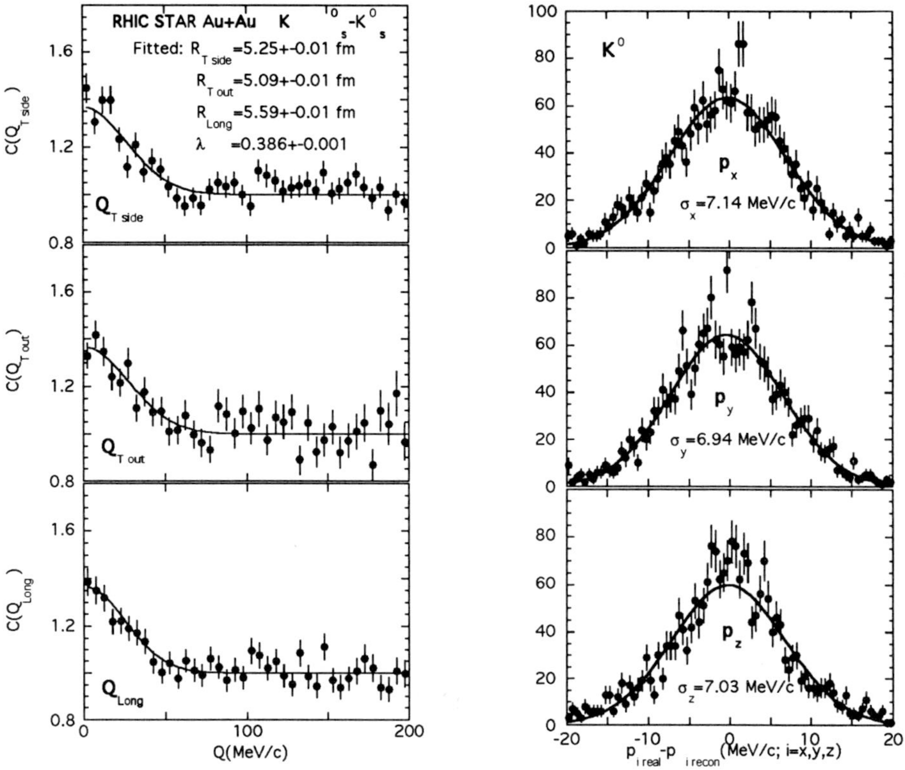

Figure 5: Same as Figure 2 except for K^0_s's.

the source parameters and thus acts as a check on the method being used. The "no Coulomb effects" case shows some distortions in the extracted R_{Tout} and λ parameters due to the detector momentum smearing. Including "all effects" increases the distortion in R_{Tout} but making a "wrong" Gamow correction via Eq.2 using the distorted momenta compensates for the momentum smearing effect in λ. Figure 4 is similar to Figure 3 except that the long-lifetime input parameters (set #2) were used. The distortions are qualitatively similar to those seen in Figure 3 but the effect on R_{Tout} is larger.

Figure 5 shows the results of the present study for $K^0{}_s$'s (Ref.[6] describes an earlier study for $K^0{}_s$'s.).As seen, the momentum resolution width is about 7 MeV/c which is approximately equal to $\sqrt{2}*\sigma_{pion}$ as expected since a $K^0{}_s$ is reconstructed from its decay into two pions. The correlation function with distorted source parameters using input parameters from set #1 is also shown. The statistics for this correlation function corresponds to about 100 days of STAR running and results in about 8 million pairs. The average k_T of these $K^0{}_s$'s is 525 MeV/c. Comparing this calculation with the similar calculation for pions seen in Figure 2, the distortions in the radius parameters are qualitatively the same whereas the distortion in λ is significantly greater for $K^0{}_s$'s. This is due to two effects: 1) the momentum resolution for $K^0{}_s$'s is larger than for pions and, 2) there is no "wrong" Gamow correction to apply to compensate for the momentum resolution effects.

Summarizing the results shown above, distortions in the extracted boson source parameters induced by detector effects can be significant at RHIC. Thus, in most cases it is necessary to use an accurate detector Monte Carlo to correct the experimental correlation function, as has already been applied successfully in SPS experiments[2].

Acknowledgment

This work was supported by the National Science Foundation.

References

1. D. Rischke and M. Gyulassy, *Nucl.Phys.*A **608**, 479 (1996).
2. I. G. Bearden et al. (NA44 Collab.), *Nucl.Phys.*A **610**, 240c (1996).
3. L. Ahle et al. (E866 Collab.), *Nucl.Phys.*A **610**, 213c (1996).
4. J. W. Harris et al. (STAR Collab.), *Nucl.Phys.*A **566**, 277c (1994).
5. T. J. Humanic, *Phys.Rev.*C **57**, 866 (1998).
6. D. Keane and K. Wilson, *STAR note* #0224 (1995)

ESA PROJECTS ABOUT INTERFEROMETRY FROM SPACE

JEAN-MARIE MARIOTTI
European Southern Observatory, Karl-Schwarzschildstr. 2, Garching bei Munchen, D-85748 (Germany)
E-mail: jmariott@eso.org

In parallel with the development of large ground-based interferometric facilities (Very Large Telescope Interferometer in Chile, Keck Interferometric Array in Hawaii,...), space agencies such as ESA and NASA have considered the advantage of space for astronomical observations at very high angular resolution. The rationale behind these efforts is that space is large, quiet and cold. This provides, on the long term, definite advantages for the development of astronomical interferometry over ground-based observatories which are limited in extension (current projects have baselines up to a few hundred meters), are severely penalized by the atmospherical turbulence at visible wavelengths and by the background thermal emission in the infrared. ESA has started several years ago to develop a strategy in order to take advantage of the characteristics of space environment. This has led to the selection of an "Interferometry Cornerstone" mission in the future scientific plan of ESA, called "Horizon 2000+". The final goal of this mission has not been selected yet among various alternatives. The possible options are GAIA, a mission dedicated to very high accuracy astrometry, and IRSI/DARWIN a mission dedicated to the detection of Earth-like planets around stars of the solar neighborhood. ESA has also conducted a study called MOFFIT, for a very long baseline imaging interferometer, either Moon-based or free-floating. Similar projects are also in study in the US. This paper will describe the main features of these candidate missions, with an emphasis on their scientific goals.

A VERY LOW FREQUENCY RADIO TELESCOPE ON THE FAR SIDE OF THE MOON

G. WOAN
Department of Physics and Astronomy
University of Glasgow
Glasgow G12 8QQ
United Kingdom
E-mail: graham@astro.gla.ac.uk

There is growing interest in the notion of constructing a space-based radio interferometer to carry out astronomy at low frequencies inaccessible from Earth.[1,2,3] One possible site for this instrument is the far side of the Moon[4] as this offers a stable, electrically quiet site well shielded from the Earth and, for about half the time, the Sun. The performance of such a telescope is limited by scattering in the solar wind and the interstellar medium, both of which severely restrict the maximum angular and temporal resolution achievable at frequencies below about 10 MHz. Furthermore, the Moon itself influences the observations, through its own electrical properties and exosphere. With these effects in mind, ESA has recently completed a study into the design of such a Moon-based radio interferometer. Here 1 review the scientific goals and engineering challenges of this design, and how modern digital signal processing techniques blur the distinction between phased arrays and interferometers.

1 Introduction

It is particularly pleasing to be presenting a paper on low frequency radio astronomy at a meeting on HBT interferometry. Hanbury Brown's original radio intensity interferometer operated at the relatively low frequency of 125 MHz, so the subject of the paper is closer to his pioneering work than you may realise. It's also interesting to note that, a little later, the Moon itself played an important role in our struggle to measure angular sizes of radio sources, via the edge diffraction pattern from its limb during occultations.

Surprisingly, the portion of the electromagnetic spectrum below about 30 MHz is one of the 'last frontiers' of astronomy. The desire over the past few decades to increase angular resolution and sensitivity in radio astronomy has generally been met by an emphasis on high observing frequencies, leaving the lower frequencies to languish in undeserved neglect. Inherent difficulties in calibration, together with the ever-increasing level of radio interference in the band have only added to the difficulties.

There are, of course, limitations to what can be achieved at these very low frequencies. The solar wind and interstellar medium impose a plasma frequency cut-off at a few tens of kilohertz, making astronomy below these frequencies impossible. Also, the Earth's ionosphere makes routine ground-based astronomy impossible below about 10 MHz, although work has been done down to about 2 MHz from selected sites, such as Tasmania. In addition to these propagation constraints there is also a large amount of strong, transient interference present in the band, both natural and man-made. The natural signals, such as the Earth's auroral kilometric emissions (AKR) and the transient bursts from the Sun, are subjects for investigation in themselves, but their power hampers more sensitive observations of interstellar and extragalactic phenomena. Of course, man-made interference is a particular problem in the band. Even at the Moon, shortwave signals from Earth dominate the spectrum.[5]

The are clearly many technical challenges to overcome before useful astronomy can be carried out at these frequencies, so what sort of science do we expect in return? Although not the main topic of this paper, here is a flavour of the science goals:

- We expect to detect bright new sources of coherent emission including cyclotron masers driven by accretion and outflow, and pulsars. Coherent emission is also expected from relativistic jets in extragalactic objects such as quasars. Plasmas containing charged dust grains are expected to scatter quasi-coherently when illuminated with radio waves longer that the Debye length. This should be particulary evident in interstellar dust clouds and in star and planet formation regions.

- Bremsstrahlung (free-free) absorption from the diffuse ionised hydrogen in the interstellar medium is particularly strong at these frequencies. The spatial distribution of this component within the Galaxy is not well known at present and a primary goal of any instrument would be to map absorption over the entire sky.

- Observations of optically thick clouds, at a known thermal temperature and distance, can also be used to determine the local galactic synchrotron emissivity along the line-of-sight to the cloud, and hence the 3-dimensional distribution of low energy cosmic ray electrons in the Galaxy.

In addition, there are a number of very promising observations of the Sun, planets and the interplanetary medium that can be carried out at very low frequencies, with the aim of furthering our understanding of solar variability and the transient phenomena that cause geomagnetic storms.

Assuming that we want to build a telescope to do astronomy well below 30 MHz, how should we proceed? There are three clear approaches. Firstly, we could build a telescope at an exceptionally good site on Earth and battle with the ionosphere and the terrestrial interference. Secondly, we could build a 'free-flying' space telescope well away from the the Earth, and use the inverse square law to attenuate the radio interference. Thirdly we could build a telescope relatively close to the Earth, though outside its ionosphere, and hide it behind a large screen. Here we will consider this third approach, the screen in question being, of course, the Moon.

2 Propagation effects

The performance of a Moon-based radio telescope is limited by the way in which very low frequency radio signals propagate through the interstellar and interplanetary media. Although we may be observing well above the plasma frequency, diffractive turbulence in these plasmas blurs our view of background sources by scattering (initially plane) waves over a characteristic scattering angle, θ_s. The effect has approximately the same wavelength dependence as the plasma refractive index (λ^2). At frequencies, ν, below about 30 MHz the scattering from the interplanetary medium (IPM) and interstellar medium (ISM) are observed to be approximately [6,7]

$$\text{IPM}: \quad \theta_s \sim 100/(P\nu_{\text{MHz}})^2 \quad \text{arcmin}, \tag{1}$$

$$\text{ISM}: \quad \theta_s \sim 22/\nu_{\text{MHz}}^2 \quad \text{arcmin}, \tag{2}$$

where P is the 'impact parameter' of the line of sight to the source relative to the Sun in AU, and the ISM value is for high galactic latitudes (see Figure 1). This has a major effect on the design of a radio interferometer. Clearly, resolutions of only $\sim 1°$ will be possible at frequencies ~ 1 MHz, so baselines longer than a few tens of kilometers will yield little at these frequencies. At higher frequencies however (~ 10 MHz) sub-arcminute resolutions should be possible, so some degree of baseline flexibility is necessary.

Scattering also brings temporal broadening, as rays from the source can propagate along a number of paths before they are received. Taking small angle approximations, the range in propagation times generated by a scattering medium a distance z from the observer is

$$\Delta\tau_b \simeq \frac{z\theta_s^2}{2c}. \tag{3}$$

Again, this expression can be given an approximate calibration using observations of pulsars and angular broadening at frequencies of a few tens of megahertz,[8] to give

$$\text{IPM}: \quad \Delta\tau_b \sim 0.1\,\nu_{\text{MHz}}^{-4} \quad \text{sec}, \tag{4}$$

$$\text{ISM}: \quad \Delta\tau_b \sim 10^8\,\nu_{\text{MHz}}^{-4} \quad \text{sec}. \tag{5}$$

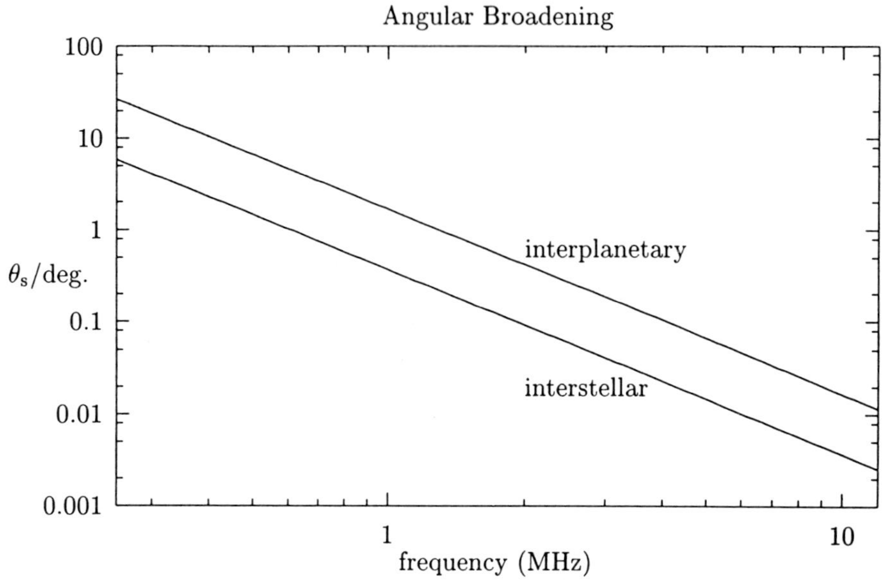

Figure 1: Interplanetary and interstellar angular broadening.[4]

We see immediately that the comparatively vast distances involved in interstellar propagation make the effects of interstellar temporal broadening severe. At 1 MHz the temporal broadening is approximately 5 years, so there is no possibility of observing interstellar transient phenomena, such as the pulses from pulsars, at these frequencies.

In addition to these scattering effects, the interstellar medium gradually becomes optically thick with decreasing frequency, due to free-free absorption. The turnover frequency at which a plasma of electron number density n_{e} and temperature T_{e} becomes optically thick is [9]

$$\nu_{\mathrm{T}} = 5.21 \times 10^{-7}\, T_{\mathrm{e}}^{-0.64} \left(\int n_{\mathrm{e}}^2 \, \mathrm{d}x \right)^{0.48} \quad \mathrm{GHz}, \tag{6}$$

where x is in parsecs along the line-of-sight. For the diffuse, warm, interstellar medium, $T_{\mathrm{e}} \simeq 10^4$ K and $n_{\mathrm{e}} \simeq 3 \times 10^4$ m^{-3}, so the distance one can see through the ISM is approximately

$$l \simeq (34 \nu_{\mathrm{MHz}})^{2.1} \quad \text{parsec}. \tag{7}$$

The galactic disc is $\sim$ 1 kpc thick, so the whole sky appears 'foggy' at frequencies below 1–2 MHz [10] although there will still be directional variations.

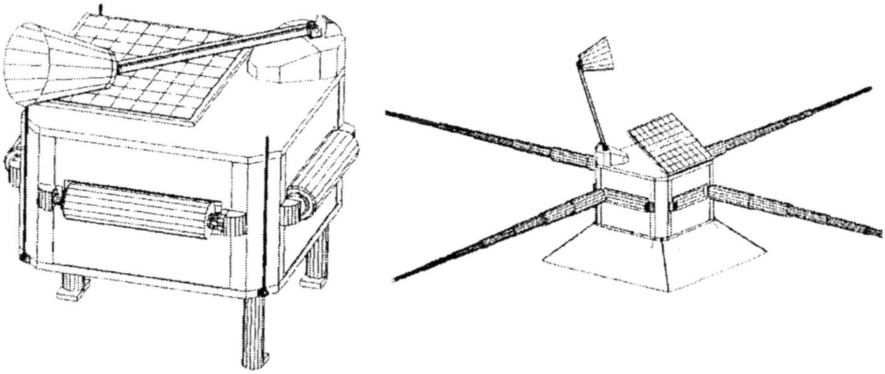

Figure 2: One receiving element: stowed for launch (left) and deployed (right).[4]

This absorption will limit our view of galactic and extragalactic objects at the lower extremes of the band, but will in turn reveal the spatial distribution of the warm component of the ISM.

These three propagation effects largely define the maximum useful physical size and data-rate of the antenna and correlator, and the scope of the astronomy we can perform. We should now consider where the telescope should be placed.

3 Far-side radio astronomy – The VLFA

The Moon is attractive as a base for radio astronomy not only because it offers a platform well away from the Earth's ionosphere, but also because its far side is the only place in the solar system permanently directed away from the Earth. This shielding effect is probably the Moon's most attractive feature. During night-time on the far side the telescope will also be shielded from the Sun, so giving a double advantage. However, these advantages come with significant technical challenges.

In 1995 the European Space Agency began a preliminary design study for the 'Very Low Frequency Array' (VLFA) [4] – a Moon-based radio telescope operating in the frequency band below 16 MHz. The straw-man design for this comprises a 300 element interferometer capable of operating at any spot frequency between 500 kHz and 16 MHz with a bandwidth of 100 kHz. Each element of this array is a $25 \times 25 \times 25$ cm autonomous unit that communicates with a central station via a microwave link (Figure 2). The central station

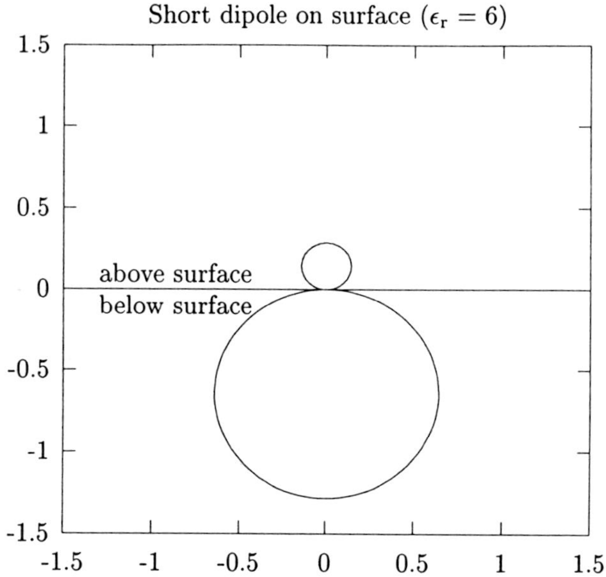

Figure 3: Relative gain of a short dipole placed horizontally on the surface of the Moon ($\epsilon_r = 6$).

itself communicates with Earth via a relay satellite in a halo orbit about the Earth-Moon L2 Lagrangian point. A pair of short (4 m) crossed dipoles extend horizontally from each unit to receive the astronomical signal. The design of these battery operated units is critical, as they have to operate in the dark, at surface temperatures as low as $-200\,^\circ$C for about half a month at a time.

3.1 The lunar regolith and exosphere

We know from the Apollo missions that the dusty lunar surface (or regolith) has a very low d.c. electrical conductivity – about $10^{-14}\,\Omega^{-1}\,\mathrm{m}^{-1}$. This allows us to place the antennas horizontally on the regolith without a strong surface reflection nulling their power patterns. However, the regolith does have a relative permittivity of up to $\epsilon_r \simeq 6$, so that there is an 'image' dipole beneath the surface $-(\epsilon_r - 1)/(\epsilon_r + 1)$ times the real dipole in the vacuum. The effect of this is to reduce the gain of the dipole in the vacuum direction to about 2/7 its free-space value (Figure 3) and to increase its gain in the regolith direction by 9/7. This is not quite as serious as it sounds, as the *directivity* above the surface is still that of a dipole, and the signals are still strong enough to feed a good, actively matched, preamplifier. It does however focus attention on the lobe of the power pattern pointing towards the surface, and force us to consider the possibility of stray signals being reflected from structures beneath the surface fo the Moon. Careful site tests would clearly have to be made.

Contrary to popular belief, the Moon does have an atmosphere, or more properly an exosphere, with a total mass of about 10^4 kg. The free electron content of this exosphere will determine the local plasma frequency on the lunar surface and the minimum frequency at which radio observations can be made. Dual frequency measurements made by the Luna 19 and 20 spacecraft have been interpreted by Bauer [11] as a photoelectron layer with a number density on the sunlit side of up to 2×10^3 cm^{-3} and scale-height of several kilometers. This gives the illuminated side of the Moon a plasma frequency ~ 0.5 MHz, which is sufficient to severely affect the performance of the VLFA. This, together with need to shield the telescope from solar radio bursts, ensures that observations must be carried out during the night.

3.2 Interferometer design

Despite the theme of this meeting, the VLFA is not designed to be an intensity interferometer. The microwave links to the 300 elements ensure phase stability over the array. Indeed, the straw-man design goes further, and intentionally blurs the definition of an interferometer by working with many very narrow bandwidth ($\sim$ 1 kHz) channels and carrying out all the signal processing in DSP chips. The choice between a phased array and an interferometer can now be made in software, depending on whether the DSP block is programmed to synthesise 'point-like' or 'wave-like' beams on the sky. For most of the time though, the array will be operated as a 300-element interferometer, observing the sky in 'Moon-rotation aperture synthesis' mode. The sensitivity of the instrument is fundamentally determined by the system noise, all of which comes from the sky at these frequencies. When mapping extended emission, the filling factor of the array, i.e., the fraction of its area covered by antennas, is also very important. The effective area of a hertzian dipole scales as λ^2, so the filling factor, and therefore sensitivity, of the array drops with increasing observing frequency. Furthermore, the dipoles have to be positioned carefully so that their effective areas do not overlap at the lowest frequencies. One possible array configuration that addresses these constraints is shown in Figure 4. Each 86 km spiral arm contains 100 dipole antennas, spaced with a power law. The entire array is contained within a circle of 40 km radius, giving a maximum baseline of 68.75 km. A spiral 'Y' shaped design is used to deliver long baselines, prevent overlap at the lower frequencies, and to recover the filling factor in the centre

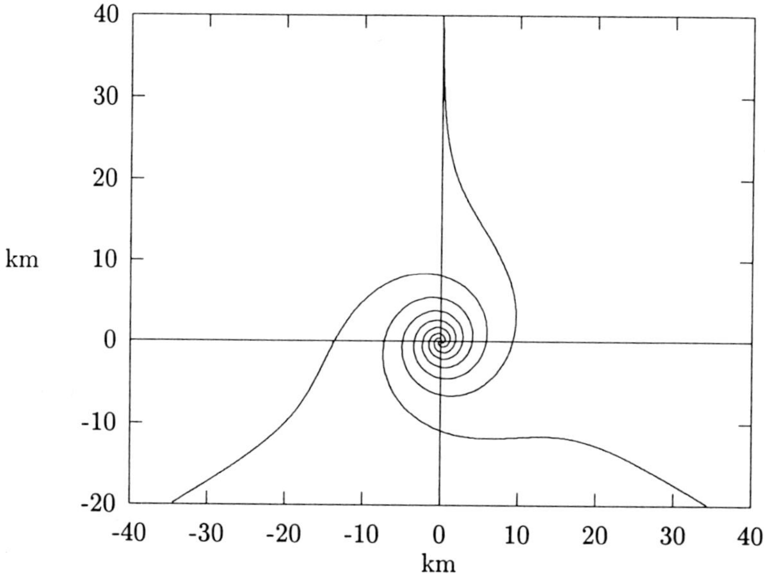

Figure 4: Layout of the spiral array design.[4]

of the array at higher frequencies.

The array needs to be positioned on a relatively flat region of the lunar far side where the central relay station has a clear view of all the elements. A possible site is the 100 m diameter Tsiolkovsky crater (20°S, 129°E) where the 300 receiver elements can be deployed robotically. The whole package has an estimated mass of 2720 kg, low enough to be transported to the Moon by a single Ariane 5E launch.

4 Conclusions

In terms of hardware and scope, radio interferometry has come a long way in the past 50 years, but the basic ideas have changed very little. When these ideas are applied to space-based radio astronomy at the lowest frequencies, we see that the Moon offers a well-shielded platform on which a multi-element connected interferometer could be built with the intention of performing all sky surveys of high resolution and sensitivity, well away from the influences of the Earth. At frequencies below a few megahertz it is the interplanetary medium that ultimately limits the angular resolution of the instrument, and the interstellar medium that restricts how far out we can see.

The VLFA Moon design study is one of a number of investigations of how to deploy a low frequency radio telescope in space. Another very promising idea is for a free-flying constellation of antennas much further from the Earth. Underpinning all these designs is the belief that a thorough understanding of the principles of interferometry will be central to any exploration of the low frequency universe.

Acknowledgments

I thank the other ESA team members involved in the VLFA design study on which this paper is based: Pierre-Yves Bèly, Robin Laurance, Sergio Volonte, Roberto Ambrosini, Arnold van Ardenne, Colin Barrow, Jean-Loius Bougeret and Juan-Maria Marcaide. I also thank Enrique Rey and the CASA team, and Siegfried Bauer for some illuminating conversations on the lunar exosphere.

References

1. J.P. Basart, J.O. Burns, B.K. Dennison, K.W. Weiler, N.E. Kassim, S.P. Castillo, B.M. McCune, *Radio Science* **32**, 251 (1997).
2. J.P. Basart, J.O. Burns, B.K. Dennison, K.W. Weiler, N.E. Kassim, S.P. Castillo, B.M. McCune, *Radio Science* **32**, 265 (1997).

3. K.W. Weiler, in *Low Frequency Astrophysics from Space*, eds. N.E. Kassim & K.W. Weiler, Springer–Verlag Lecture Notes in Physics **362**, 8 (1990).
4. P.Y. Bely, R.J. Laurance, S. Volonte, R.R. Ambrosini, A. Ardenne,C.H. Barrow, J.L. Bougeret, J.M. Marcaide, G. Woan, *ESA report SCI(97)2*, European Space Agency (1997).
5. J.K. Alexander, J.C. Novako, F.R. Grena, R.R. Weber, *Astron. Astrophys.* **40**, 365 (1975).
6. W.C. Erickson, *Ap.J.* **139**, 1290 (1964).
7. P.J. Duffett-Smith, A.C.S. Readhead, *Mon. Not. R. Astr. Soc.* **174**, 7 (1976).
8. J.M. Cordes, in *Low Frequency Astrophysics from Space*, eds. N.E Kassim & K.W. Weiler, Springer–Verlag Lecture Notes in Physics **362**, 165 (1990).
9. M.A. Gordon in *Galactic and Extragalactic Radio Astronomy* (Second Edition), eds. G.L. Verschuur & K.I. Kellerman, Springer-Verlag (1988).
10. R.J. Reynolds, in *Low Frequency Astrophysics from Space*, eds. N.E Kassim & K.W. Weiler, Springer-Verlag Lecture Notes in Physics **362**, 121 (1990).
11. S.J. Bauer, *Anzeiger Abt II* **133**, 17 (1996).

An HBT Excitation Function at the AGS

M.A. Lisa[j], for the E895 Collaboration
N.N. Ajitanand[m], J. Alexander[m], D. Best[a], P. Brady[e], T. Case[a], B. Caskey[e], D. Cebra[e], J. Chance[e], I. Chemakin[d], P. Chung[m], V. Cianciolo[i], B. Cole[d], K. Crowe[a], A.C. Das[j], J. Draper[e], S. Gushue[b], M. Gilkes[l], M. Heffner[e], H. Hiejima[d], A. Hirsch[l], E. Hjort[l], L. Huo[g], M. Justice[h], M. Kaplan[c], J. Klay[e], D. Keane[h], J. Kintner[f], D. Krofcheck[k], R. Lacey[m], J. Lauret[m], E. LeBras[m], H. Liu[h], Y. Liu[g], R. McGrath[m], Z. Milosevich[c], D. Olson[a], S. Panitkin[h], C. Pinkenburg[m], N. Porile[l], G. Rai[a], H.-G. Ritter[a], J. Romero[e], R. Scharenburg[l], L. Schroeder[a], R. Soltz[i], B. Srivastava[l], N.T.B. Stone[b], T.J. Symons[a], S. Wang[h], R. Wells[j], J. Whitfield[c], T. Wienold[a], R. Witt[h], L. Wood[e], X. Yang[d], W. Zhang[g], Y. Zhang[d]
[a]Lawrence Berkeley Lab, [b]Brookhaven National Lab, [c]Carnegie Mellon University, [d]Columbia University, [e]U.C. Davis, [f]St. Mary's College, [g]Harbin Institute, China, [h]Kent State University, [i]Lawrence Livermore National Lab, [j]The Ohio State University, [k]University of Auckland, NZ, [l]Purdue University, [m]SUNY at Stony Brook

The E895 Collaboration has measured Au+Au reactions at 2, 4, 6, and 8 AGeV with a very large acceptance Time Projection Chamber, at the Brookhaven AGS. Studies of reactions at these energies are sensitive to nuclear matter at the highest baryon density. It has been suggested that such conditions may trigger the transition to the Quark Gluon Plasma (QGP). Signatures of such a transition may include large lifetimes as measured by HBT techniques. Here, we present the first excitation function of pion inteferometry at intermediate AGS energies. Analyses of the correlation function in the Bertsch-Pratt and Yano-Koonin parametrizations provide insight into the π^--emitting source size, lifetime, and longitudinal flow. No large jumps in the HBT signals are observed as the beam energy is increased. Preliminary comparisons with RQMD calculations are given.

1 Introduction

1.1 E895 Motivation

A primary goal of high-energy heavy ion physics is to create and study the quark-gluon plasma (QGP), a phase of matter in which partonic– instead of hadronic– degrees of freedom describe the system. Transport, hydrodynamic, and nucleation theories[1,2,3,4] suggest that energy densities achieved in central heavy ion collisions at AGS energies may be sufficient to create the QGP.

If the QGP is created in a heavy ion collision, the timescale for particle emission is expected to be longer than a scenario in which only ordinary hadronic degrees of freedom play a role[5,6], due to the extra time of hadronization. Thus, one proposed signature for QGP formation has been a large apparent source lifetime as measured by pion HBT measurements. However, HBT analyses of very heavy ion collisions at the maximum AGS energy [7,8]

(10.6 AGeV) and at CERN SPS [9,10] (158 AGeV) do not indicate emission timescales longer than about 4 fm/c, which is consistent with the time distribution from normal hadronic physics models. Thus, it may seem pointless to look for long lifetimes at energies *below* maximum AGS energy.

However, recent hydrodynamical calculations by Rischke and collaborators [6] suggest that some signatures of QGP creation– including flow and large source lifetimes from HBT– may best be studied with excitations functions, which would scan the energy density of the system through the QGP threshold energy, or "softest point" in the Equation of State [6,11,12]. For a source created at this energy, the lifetime is longer because the system does not expand and cool as rapidly as it would if there were no phase transition. For collisions at energies very much above this threshold energy, the system (which is in the QGP phase) expands and cools rapidly, and the lifetime effect is diminished.

Finally, any excitation function study serves as a stringent test of transport models, which represent our understanding of the dynamics of heavy ion collisions. As these models prepare to deal with data from ever higher bombarding energies, we would do well to constrain– and understand the relative importance of– physical assumptions in them, such as effects due to the nuclear meanfield.

1.2 Measurement and Experimental Issues for HBT

Using the Time Projection Chamber [13,14] used in the EOS experiments at the Bevalac, the E895 collaboration has measured about 0.5-1 million collisions at 2, 4, 6, and 8 AGeV at the Brookhaven AGS. The results presented here represent roughly 50% (25%) of the total statistics at 4 and 8 (2 and 6) AGeV.

The TPC was located in the MPS magnet operated with a field of 0.75 or 1 T. The active volume of the TPC is a rectangular region 154 cm x 96 cm x 75 cm in the beam, bend, and drift directions, respectively. Electrons liberated by charged particles passing through the TPC drift to 15360 pads arranged in 128 rows at the bottom of the TPC. The signal on each pad is sampled and digitized every 100 ns (140 time buckets), providing about 2 million 3-dimensional pixels in which the ionization is measured. "Hits" are reconstructed from the pixels, corresponding to a track crossing a padrow. Tracks are formed from the hits, giving continuous tracking with nearly 4π acceptance in the c.m.

Event reconstruction results are processed by the HBT analysis after passing a series of quality cuts. The most important for this analysis are an event-wise primary vertex cut, to ensure that the interaction comes from the target, and a track-wise cut requiring that the track project back to the primary interaction; this latter removes many π^- from weak decays (e.g. Λs) as well as ghost tracks from unused hit combinatorics.

Particle identification in the TPC is performed by correlating the magnetic rigidity of a track with its specific ionization in the gas. In this space, the electron band overlaps somewhat the π^- band; with cuts, we estimate the level of e^- contamination at about 5%. We find that relaxing the PID cuts to purposely allow more electrons into our π^- sample affects only the HBT λ, not the radii. Negative kaon contamination is insignificant at these energies.

The momentum resolution from the tracking software itself is about 1%, but more important is the contribution from multiple coulomb scattering in the target (a 3% interaction target was used), which worsens the overall resolution to 1.5-3%. The resolution in terms of relative momentum is as poor as 12 MeV/c in 8 AGeV reactions. Prescriptions exist for correcting the correlation signal for resolution effects (e.g. that developed by the NA44 collaboration [15]). While we have attempted such corrections previously with some success[16], subtle imperfections in the E895 simulator made our implementation unstable. In the present analysis, we do not correct the data for resolution, but instead smear the RQMD predictions by our resolution in order to compare correlation functions directly.

Tracks which are very close in the TPC can merge and appear as a single track, leading to a large suppression in the correlation function at low relative momentum. To correct for this, we require (in both the numerator and denominator of the correlation function) that the two tracks have well-seperated hit positions on at least 15 padrows. (In applying this cut to the mixed events, we account for event-to-event differences in primary vertex position.) Further increasing the required hit seperation distance or number of padrows does not affect the measured correlations. Simulations show that this also obviates the need for a further track-splitting cut; false "pairs" from split tracks are eliminated by this cut.

1.3 Coulomb correction

Throughout this work, the Coulomb correction applied to the correlation functions is obtained by averaging the square of the Coulomb wave function [17,18] over a spherical Gaussian source of 5 fm radius. This radius is consistent with those extracted from 3-dimensional fits to the correlation function (see below); changing the radius used in the Coulomb correction by ± 1 fm has negligible effect on the fit parameters. Our correction is the same as used by the E877 collaboration [7], and equivalent to that used by NA49 [10]. Identical Coulomb corrections are applied to the data and to RQMD model predictions.

2 1-D Correlation Functions at Midrapidity

For a first analysis of the correlation function, we concentrate on π^- emitted at midrapidity ($y = y_{c.m.} \pm 0.25$) from central collisions ($\sim 10\%\sigma_T$). In order to compare our results with those of E866/917[8,19], we require p_T¿100 MeV/c.

The quality of the data is apparent in Figure 1, which shows correlation functions at the four bombarding energies, as a function of $Q_{inv} \equiv (\vec{Q}^2 - (Q^0)^2)^{1/2}$. While interpretation of the parameters from a Gaussian fit to the 1-dimensional correlation function is not straightforward, and the χ^2 values for the lower energies are significantly larger than unity (see Fig. 1), the bombarding energy systematics of the parameters can be investigated and the values compared to those extracted by others. Therefore, also shown are fits to the conventional Gaussian form $C(Q_{inv}) = 1 + \lambda e^{-Q^2_{inv} R^2_{inv}}$, where the lowest Q-bin is excluded from the fit). The bombarding energy dependence of the R_{inv} and λ is shown in Figure 2. The data show no obvious jumps flagging a phase transition, and lead smoothly to the result of E866[8].

Also shown for the energies measured by E895 are the results of fits to the correlation function using the RQMD model[20] (v2.3) and the correlator code (CRAB) of Pratt[21]. Sets of events with b=0-5 fm, geometrically weighted, were generated with RQMD run in the cascade mode, and also with the mean field on. As compared to the results for the cascade events, the calculations with the mean field show a stronger E_{beam} dependence of the fit parameters, and, especially for λ, appear to reproduce the data a bit better.

The fitted 1-D radius parameter appears relatively constant at roughly 5.5 fm for $E_{beam} = 4-11$ AGeV, but is considerably larger at 2 AGeV (however we note again that the χ^2 value is larger as well). This trait is not predicted by the RQMD model. Further analysis of the 2 AGeV data with more statistics will be required to confirm this possible increase at low energy, but it is interesting to note that the radius extracted (through somewhat different analysis) by the FOPI collaboration for Au+Au collisions at 1.06 AGeV is similarly large ($R_{inv} \approx$7.5 fm)[23].

3 Excitation Function of Multi-dimensional HBT

To explore the correlations in more detail, we have performed multi-dimensional fits to the correlations. In the system center of mass frame, we decomposed the relative momentum vector in the Bertsch-Pratt parametrization[22], and fit the 3-dimensional correlation function with

$$C(q_{out}, q_{side}, q_{long}) = 1 + \lambda e^{(-R^2_{out} q^2_{out} - R^2_{side} q^2_{side} - R^2_{long} q^2_{long} - 2R^2_{ol} q_{out} q_{long})} \quad (1)$$

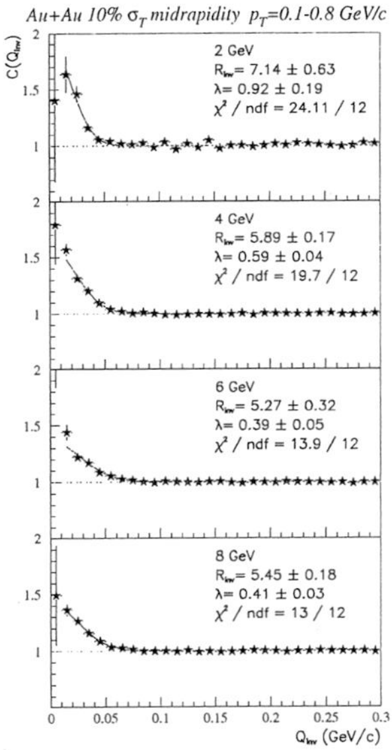

Figure 1: Correlation functions in Q_{inv} for midrapidity π^- with p_T >0.1 GeV/c from central collisions at the four bombarding energies measured. Data is shown by stars, while Gaussian fits are indicated by the lines.

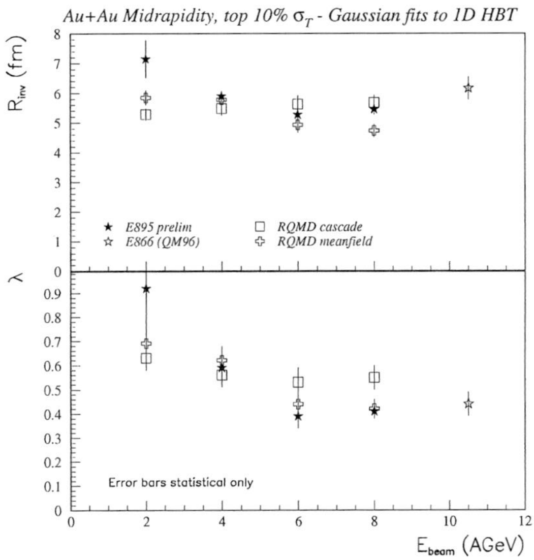

Figure 2: The excitation function of Gaussian fit parameters for 1-dimensional correlation functions measured at midrapidity by E895 (filled stars) and E866[8] (empty star). Also shown are results to fits of correlations obtained from the output of the RQMD model with (crosses) and without (squares) the meanfield. Note that the E895 data is not corrected for momentum resolution, and the RQMD calculations have been smeared by the E895 resolution.

The particles were ordered such that $q_{long} > 0$. Figure 3 shows projections of the correlation functions and the fits for correlations measured at midrapidity and $p_T = 0.1 - 0.8$ GeV/c from central collisions. The projections in a given component are integrated over the other two components in the range ±30 MeV/c, and the curves in the figure are projections of the fits weighted by number of event-mixed pairs in each bin. Statistics were insufficient to perform the multidimensional analysis for the 2 AGeV dataset and required a relaxed rapidity window for the 6 AGeV set.

The cross-term [24] R_{ol}^2 (which may be negative) was included in all fits, but consistently comes out with very large uncertainty. At midrapidity and for low p_T, this term is expected to vanish. Fixing $R_{ol} = 0$ typically reduces the other radii by an amount which is within their uncertainties. Figure 4 shows the evolution of the 3-d fit parameters with bombarding energy, along

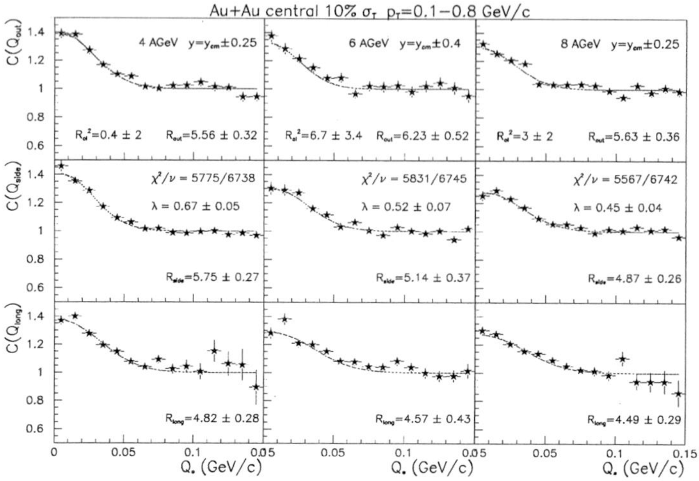

Figure 3: Projections of the measured correlation function decomposed in the Bertsch-Pratt convention for pions measured near mid-rapidity from central collisions at 4, 6, and 8 AGeV bombarding energy (columns left to right). Curves are projections of 3-dimensional fits. Projections in a q-component for both the data and the fits are integrated over the range $\pm 30 MeV/c$ in the two other components.

with the results of fits for RQMD events. The effect of momentum resolution (Section 1.2) on the correlation functions from RQMD is a reduction of $\sim 15\%$ in the extracted radii and 0.25 in λ.

Overall, the RQMD model reproduces the data fairly well, with or without meanfield effects. Some sensitivity to the meanfield may be present in the R_{long} parameter, which is smaller at the lower energies (and in better agreement with the data) when the meanfield is on. This may be due to enhanced sidewards expansion as compared to longitudinal flow, when the meanfield acts, which would dilute the system earlier and lead to smaller freezeout extension along the beam axis [25]; this possibility requires further study with the RQMD.

For almost all phasespace cuts studied at all beam energies, $R_{out} > R_{side} > R_{long}$, although this first inequality is violated for the 4 AGeV midrapidity cut shown. The decreasing difference between R_{side} and R_{long} observed in our data appears consistent with an evolutionanly trend with increasing energy; at the maximum AGS energy $R_{side} \approx R_{long}$ [7], and for SPS energies $R_{side} < R_{long}$ [9].

The decrease in λ is understood by enhanced production of long-lived resonances at higher energy. Although many π^- from decays of, e.g. Λ's, may be eliminated from the HBT analysis with a cut on the decay vertex, detailed

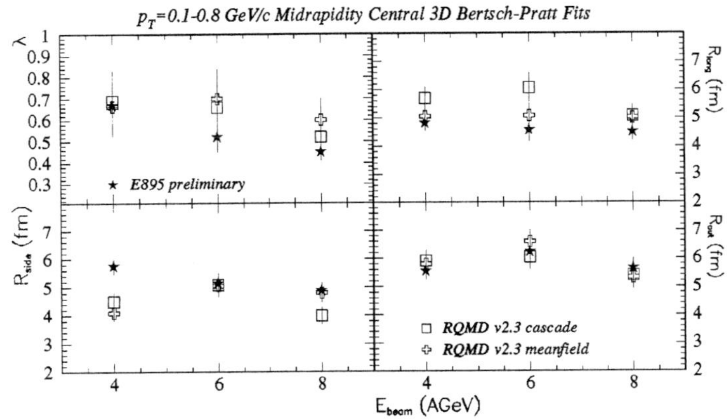

Figure 4: Bertsch-Pratt fit parameters to the 3-d correlations as a function of bombarding energy. Open symbols are the results of fits to RQMD-generated correlations.

simulations indicate that 5-10% of the π^- that pass our cuts come from weak decays. Especially at the lower energies, resonance production in the RQMD is suppressed when the meanfield is on; however this effect is not large enough to make a substantial difference to the correlation fits.

4 Multi-dimensional HBT at low p_T for 4 AGeV Central Collisions

A more detailed probe of the collision may come from studying the correlation function for small slices in phasespace. Here, we discuss only the 4 AGeV data at very low p_T (0-0.15 GeV/c) in narrow slices of rapidity; these correlations are not affected by the strong $\vec{x}-\vec{p}$ correlations in the transverse direction, and are likely to probe the longest-lived source.

Bertsch-Pratt (BP) fit parameters are shown in the left panel of Figure 5, for data and the two RQMD event sets, as a function of rapidity. At all rapidities, $R_{out} > R_{side}$, suggesting π^- emission over a finite timescale. To better characterize the timescale, we have also decomposed and fit the correlation function using the Yano-Koonin-Podgoretskiĭ (YK) parameterization [27]. There are several advantages to this parameterization: 1) the numerical values of the fit parameters are frame-independent, 2) the timescale R_0 is fit directly as opposed to being related to the difference of two large numbers $R_{out}^2 - R_{side}^2$, and 3) the velocity (rapidity) of the emitting source is also directly fit, thus providing direct information on the dynamical nature of the source.

On the right panels of Figure 5 are shown the YK timescale parameter R_0, the longitudinal radius $R_{||}$, and the source rapidity, as a function of y_{π^-}. The fits indicate an emission timescale on the order of 6.5 fm/c, in good agreement

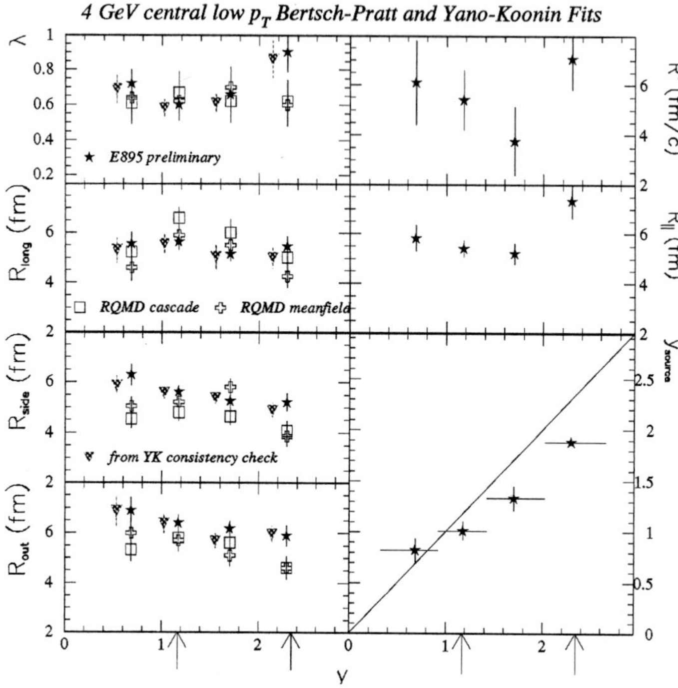

Figure 5: Bertsch-Pratt (left) and Yano-Koonin (right) fit parameters to measured π^- correlation functions for $p_T = 0 - 0.15$ GeV/c, as a function of lab rapidity. Bertsch-Pratt fits to RQMD-generated correlations are also shown. *Derived* Bertsch-Pratt parameters from the Yano-Koonin consistency relations are shown by inverted triangles, and are offset in rapidity for clarity; see text for details. Arrows indicate the beam and c.m. rapidity.

with the π^- time spectrum coming directly from RQMD, which has a Gaussian width of 7.5 fm/c. The longitudinal radius (in the local source frame) shows a large rise at beam rapidity, suggesting that π^- found at y_{beam} come from the middle of the fireball as well as the forward regions. This qualitative picture is also consistent with the lower right-hand panel, which plots the source rapidity as a function of y_{π^-}. A source characterized by boost-invariant longitudinal flow would show $y_{source} = y_{\pi^-}$ (solid line in figure), as is seen at SPS energies[28]. A static source (no longitudinal flow) would show a horizontal line $y_{source} = y_{c.m.}$, while the measured correlations at 4 AGeV correlations show evidence of non-boost invariant longitudinal flow.

It is clear that the correlation function contains the same information in both the YK or BP momentum decompositions, and a set of consistency relations has been worked out by Heinz[27], that relate the sets of parameters. In order to check for consistency of the analysis and stability in the fits, we have used these relations to derive the *expected* BP parameters from our measured YK parameters. The results are shown by the inverted triangles in the figure. (We note that the consistency relations give $R_{side}^{expected} = R_{\perp}$, so there is no need to plot the YK parameter $R_{\perp}$ seperately.) Within uncertainties, we see good consistency between the two (independent) decompositions.

Finally, also shown in Figure 5 are BP fit results to correlation functions from the RQMD. R_{side} and R_{out} are generally underpredicted at 4 AGeV for all rapidity. An identical analysis at 8 AGeV (not shown) shows that these parameters are well reproduced, with or without the meanfield on. As seen in Figure 4, R_{long} is better reproduced by RQMD with meanfield on at midrapidity, but the model shows a decrease in R_{long} away from $y_{c.m.}$ (similar to that seen at SPS energies), which is not observed in the data; data and model correlations disagree in a similar way at 8 AGeV.

5 Summary

In summary, we have presented the first intensity interferometry excitation function at AGS energies. For central collisions, no sharp jump in HBT fit parameters is seen as the bombarding energy is increased. Due to both increased resonance production and resolution effects, the intercept parameter λ falls with energy, in agreement with RQMD predictions. The ordering $R_{out} \gtrsim R_{side} > R_{long}$ likely indicates a source of finite lifetime (i.e. duration of emission). The relative size of R_{side} and R_{long} seems to evolve from a small R_{long} below maximum AGS energy to a large R_{long} at SPS. This may be due to the increased longitudinal motion coupled to a finite "relaxation time" before which particles are emitted.

Yano-Koonin fit parameters are seen to be consistent with the BP parameters. The source of very low p_T π^- from central collisions at 4 AGeV shows strong, but not boost-invariant, longitudinal flow, and an emission timescale of about 6 fm/c, in qualitative agreement with RQMD predictions.

Althought discrepancies between the data and model predictions exist and must be investigated, it appears that "standard" hadronic physics, as represented by RQMD, describes the π^- correlations well. Possible sensitivity in R_{long} to meanfield effects in the model will be investigated with more statistics.

Acknowledgments

This work supported by the U.S. Department of Energy under contract DE-AC03-76SF00098 and grants DE-FG02-89ER40531, DE-FG02-88ER40408, DE-FG02-87ER40324, by the U.S. National Science Foundation under grants PHY-

9722653, PHY-9601271, PHY-9225096, and by the University of Auckland Research Committee, NZ/USA Cooperative Science Programme CSP 95/33.

References

1. B.A. Li and C.M. Ko, Nucl. Phys. **A601**, 457 (1996).
2. J.I. Kapusta, A.P. Vischer, R. Venugopalan, Phys. Rev. C**51**, 901 (1995).
3. D. Rischke *et al.*, J. Phys. G14, 191, (1988); Phys. Rev. D**41**, 111 (1990)
4. N.K. Glendenning, Nucl. Phys. **A512**, 737 (1990).
5. S. Pratt, Phys. Rev. **C49** 2772 (1994); Phys. Rev. **D33**, 1314 (1986).
6. D.H. Rischke, nucl-th/96039; Nucl. Phys. **A610**, 88c (1996).
7. J. Barrette, *et al.* (E877), Nucl. Phys **A610**, 227c (1996); J. Barrette, *et al.*, Phys. Rev. Lett **78**, 2916 (1997).
8. M.D. Baker, *et al.* (E802), Nucl. Phys. **A610**, 213c (1996); B.A. Cole, *et al.* (E802), Nucl. Phys. **A590**, 179c (1995)
9. I.G. Bearden, *et al.* (NA44), Nucl. Phys. **A610**, 240c (1996)
10. K. Kadija, *et al.* (NA49), Nucl. Phys. **A610**, 248c (1996).
11. C.M. Hung and E.V. Shuryak, Phys. Rev. Lett. **75**, 4003 (1995).
12. D.H. Rischke, S. Bernard, J.A. Maruhn, Nucl. Phys. **A595**, 346 (1995).
13. H.G. Pugh, G. Odyniec, G. Rai, and P. Seidl, report LBL-22314 (1986).
14. G. Rai, *et al.*, IEEE Trans. Nucl. Sci. **37**, 56 (1990).
15. H. Bøggild, *et al.* (NA44), Phys. Lett. **B302**, 510, (1993); D. Hardtke, Ph.D. dissertation, The Ohio State University (1997).
16. M.A. Lisa *et al.* (E895), in Proceedings of Winter Workshop, Snowbird 1998. Plenum Press (in press).
17. S. Pratt, T. Csörgő, and J. Zimányi, Phys. Rev. **C42**, 2646 (1990); S. Pratt, Phys. Rev. **D33**, 72 (1986).
18. A. Messiah, **Quantum Mechanics**, North-Holland Publishing (1961).
19. J.H. Lee, contribution to these proceedings.
20. H. Sorge, Phys. Rev. **C52**, 3291 (1995).
21. S. Pratt, Quark Matter '93, *Nucl. Phys.* A **566**, 103c (1993)
22. G. Bertsch, M. Gong, and M. Tohyama, Phys. Rev. **C37** 1896 (1988); S. Pratt, Phys. Rev. **D33**, 1213 (1986).
23. D. Pelte, *et al.* (FOPI), Proceedings of the International Research Workshop, Poiana Brasov, Romania, World Scientific (1997), and contribution to these proceedings.
24. S. Chapman, P. Scotto, and U. Heinz, Phys. Rev. Lett. **22**, 4400 (1995).
25. H. Sorge, private communication.
26. S. Chapman, J.R. Nix, and U. Heinz, Phys. Rev. **C52**, 2694 (1995).
27. U. Heinz, Nucl. Phys. **A610**, 264c (1996).
28. R. Ganz, contribution to these proceesings.

Concluding Remarks

Walter Greiner
Institut für Theoretische Physik,
Johann Wolfgang Goethe–Universität,
Postfach 111932, D–60054 Frankfurt, Germany

A week of hard work lies behind us. We had an interesting conference with an appealing program. Many short contributions of 15 to 20 minutes, sometimes on minor points, supported the messages of principal speakers and demonstrated the research activity of the field. Quite a number of speakers just came for their own talk. I wonder what they did before and afterwards. Some wanted to be at two conferences at the same time. Cloning may here have its chances in the future.

The conference was opened with a historical lecture by Professor HANBURY BROWN, who afterwards attended all the talks. It must be particularily rewarding for him to see how his idea of nearly 50 years ago has grown and stimulated many young physicists in areas other than astronomy and astrophysics. GORDON BAYM, a major player in the field of HBT–correlation physics in heavy ion collisions, gave an excellent overview of this interesting research sector. I won't say much more and refer to his contribution.

The hard work was delightfully interupted by astrophysical lectures (N. PANAGIA, G. PALUMBO, J.–M. MARIOTTI, G. WOAN) which embedded the physics we are doing into astronomical science. The excursion up to the crater of Etna on Wednesday introduced us into the landmark of Catania and the extraordinary nature around it.

Now let me comment on a few items which I found interesting. I begin with

Astrophysical News

GIORGIO PALUMBO reported about *Active Galactic Nuclei* (AGNs). Figure 1 represents the unified model of AGN. At the center of the galaxy there is a Black Hole (BH, $10^8\ M_\odot$) which is powered by matter falling on it distributed in an *accretion disc*, around the disc a *thick molecular torus* absorbs radiation depending on the inclination of the observers line of sight. Small gas concentrations, ionized by the high energy photons generated by the BH are the source of optical Broad lines (BLR) and Narrow lines (NLR). Rotating clockwise one can explain various AGN types: QSOs, BL Lac–BLAZARS or Seyfert 2, Liner, Sey 1. The BH can be probed by x–ray lines emitted by the accretion disc and

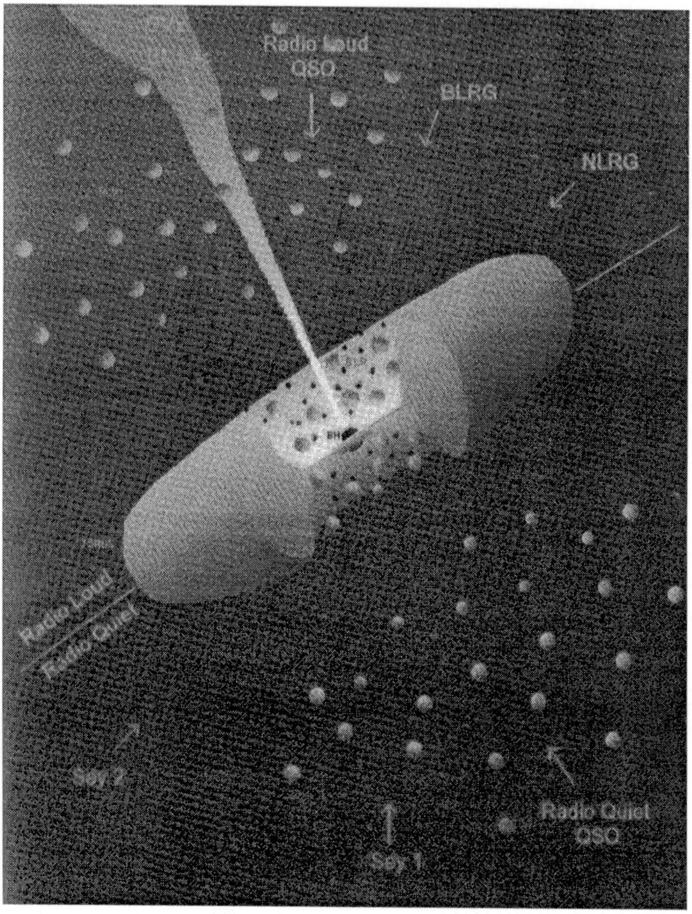

Figure 1: Vision of Active Galactic Nuclei (AGNs). Top part of the diagram: radio loud AGNs, bottom part: radio quiet AGNs.

Figure 2: Supernova 1987A: fluorescence echo in the inner circumstellar ring.

γ–rays from the jet when pointed towards the observer.

N. PANAGIA spoke about the important results from supernova 1987A in the Large Magellanic Cloud (LMC). They are

- The study of this supernova has confirmed (*first* experimental confirmation!) that massive stars explode as a result of the collapse of its core.
- The fluorescence echo in the inner circumstellar ring (Fig. 2) provides an accurate and *direct* (i.e. purely geometrical) determination of the distance to the LMC. This is a *fundamental* step for the distance scale of the Universe. In particular, all accurate distance measurements of galaxies rely on Cepheids as distance indicators; their calibration as such is crucially dependent on the LMC distance.
- The interaction between SN ejecta and inner circumstellar ring provides essential information about the SN explosion mechanism (distribution of chemical abundances, distribution of momentum in the ejecta, detection of

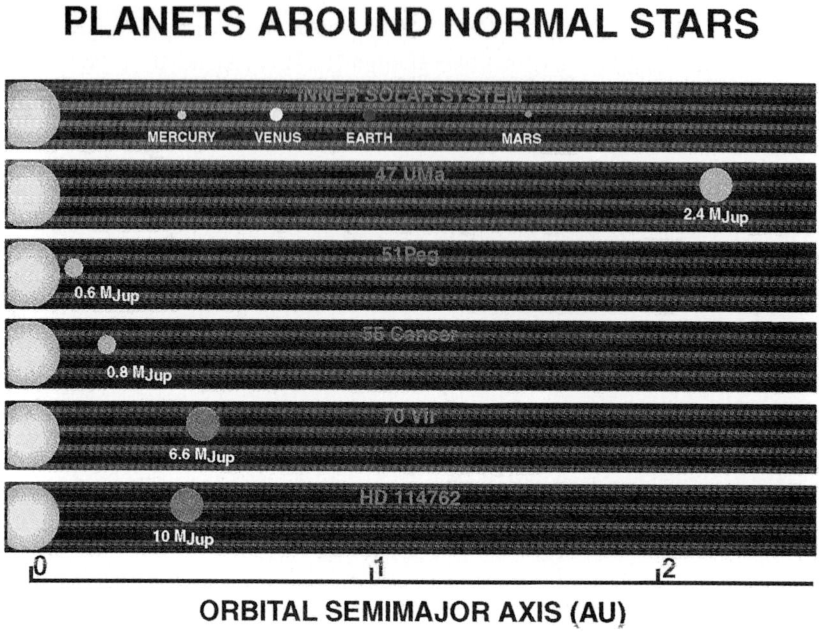

Figure 3: Planets discovered around solar–type stars ($T \approx 6000$ K) in the solar neighborhood. The typical distances of these stars from our sun is ≈ 20–60 light years.

unseen matter surrounding the system SN+rings) and, equally important, will provide a marvelous "experiment" of collision of two media at relative velocity up to 30 000 km s^{-1} (fundamental to understand hydrodynamics of relatively high energy gases).

- The study of the stellar population around SN 1987A is providing a unique opportunity to understand the processes of star formation (the SN progenitor was only ≈ 11 M years old when it exploded) and its dependence on the environment (≈ 20 000 stars with ages ranging from ≈ 1 Myr up to 6 Gyrs are revealed within a radius of ≈ 30 pc from SN 1987A). For example, clear evidence is found for segregation of massive stars from lower mass stars, which indicate the presence of mutually exclusive processes of star formation for stars of different masses. These results are going to drastically revise the current view of star formation theories and our understanding of star formation both at low and high redshifts.

The advantage of the *Hubble Space Telescope*, particularly its resolution power as compared to ground based telescopes has also been demonstrated by N. Panagia. In this connection the Galaxy Centaurus A has received his attention, but also the detection of type Ia supernovae up to redshifts close to unity, which

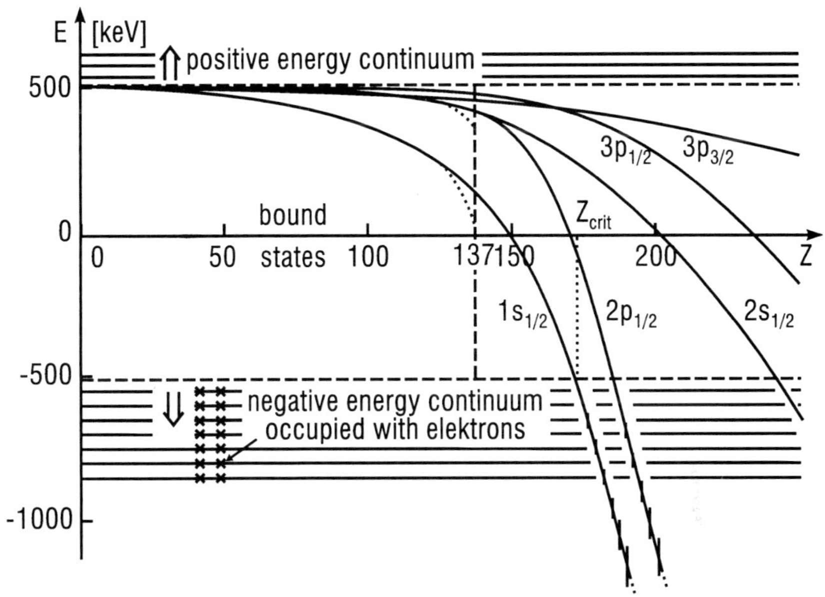

Figure 4: In strong Coulomb fields with central charge $Z > Z_{\text{crit}} \approx 164$ the $1s$ and the $p_{1/2}$–levels are "diving" into the Dirac sea and spontaneous e^+e^- pair creation sets in. The e^- is captured into a bound orbit and the e^+ is emitted.

allow the measurement of the expansion and deceleration of the universe.

ESA projects about interferometry from space versus the Very Large Teloscope Interferometers in Chile and Hawaii were discussed by J. M. MARIOTTI. He captured my special attention when showing the planets discovered around normal stars (fig. 3).

G. WOAN described the growing interest in the notion of constructing a space-based radio interferometer to carry out astronomy at low frequencies inaccessible from Earth. One possible site for such an instrument is the far side of the Moon, as this offers a stable, electrically quiet site well shielded from the Earth and, for about half the time, also from the Sun.

Because we have heard from G. PALUMBO about the discovery of black holes, let me comment in this connection on the status of the supercritical field idea in QED (fig. 4). A similar effect happens if a star shrinks to become a black hole (fig. 5). It leads to the frequently discussed Hawking–radiation, which can possibly be never observed. Therefore it is of utmost importance to verify eventually the analogous effect of vacuum decay in strong Coulomb fields (see Greiner, Müller, Rafelski, QED of Strong Fields, Springer, Berlin, Heidelberg, 1985).

At GSI various experimental groups lead by famous experimenters claimed for 15 years that narrow e^+e^- resonances were discovered, but when the APEX–

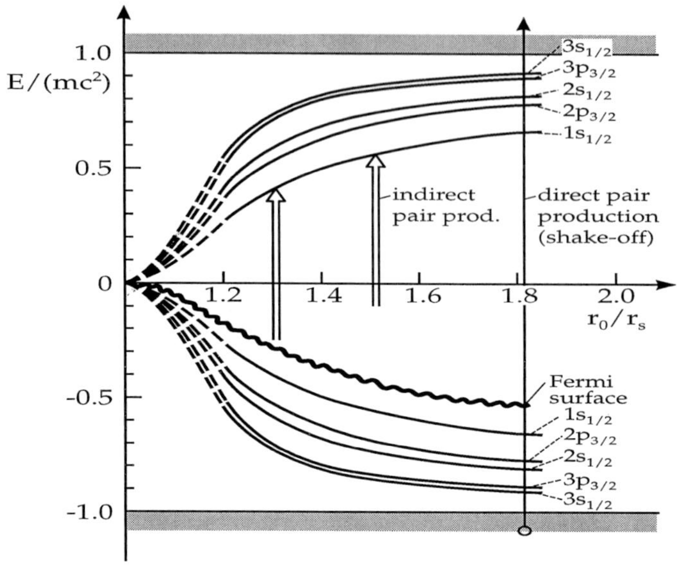

Figure 5: When a star of radius r_0 shrinks towards the Schwarzschild radius the energy gap between upper and lower continuum of fermions vanishes. Then vacuum fluctuations lead to pair creation. This is the origin of Hawking-radiation.

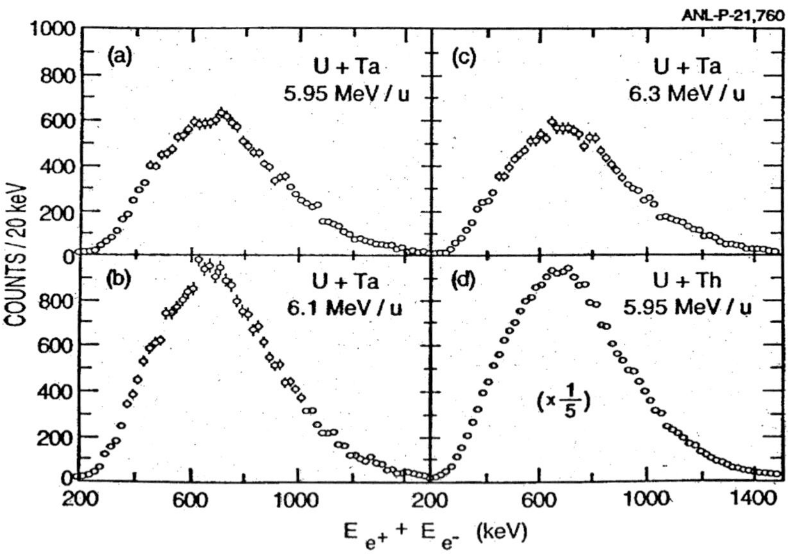

Figure 6: The e^+e^-–coincidence spectra of the APEX–group show "No evidence for sharp sum–energy lines".

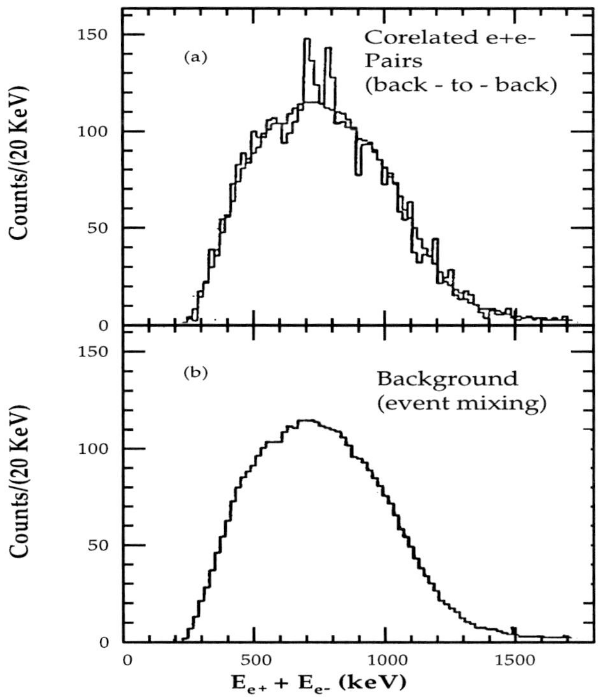

Figure 7: APEX–data analysis by J. S. Greenberg and G. Xu shows narrow line structure.

group (Argonne NL) measured with thick targets the e^+e^- sum–energy spectrum, no such line–structure was reported (fig. 6).

In contrast, Greenberg's (Yale) analysis of the same experiment showed again the narrow line structure (fig. 7). Nevertheless, the EPOS–group of GSI claimed after the APEX–data became known that their line structures, seen for 15 years, were statistical fluctuations. There are difficulties in believing this because the same group reported experiments carried out at different times and analysis done by different people, but showing the same narrow line structure (fig. 8). Recently, the Orange collaboration lead by P. Kienle (Sophie Heinz et al., Eur.Phys.J.A1, 27 (1997)) showed that indeed some of the formerly observed narrow e^+e^-–lines are due to pair conversion (fig. 9). This seems to indicate that my (nearly 15 years old) suggestion that the narrow e^+e^-–line structures are due to pair conversion is correct. It has nothing to do with the overcritical vacuum decay, which is thus still waiting for its experimental verification.

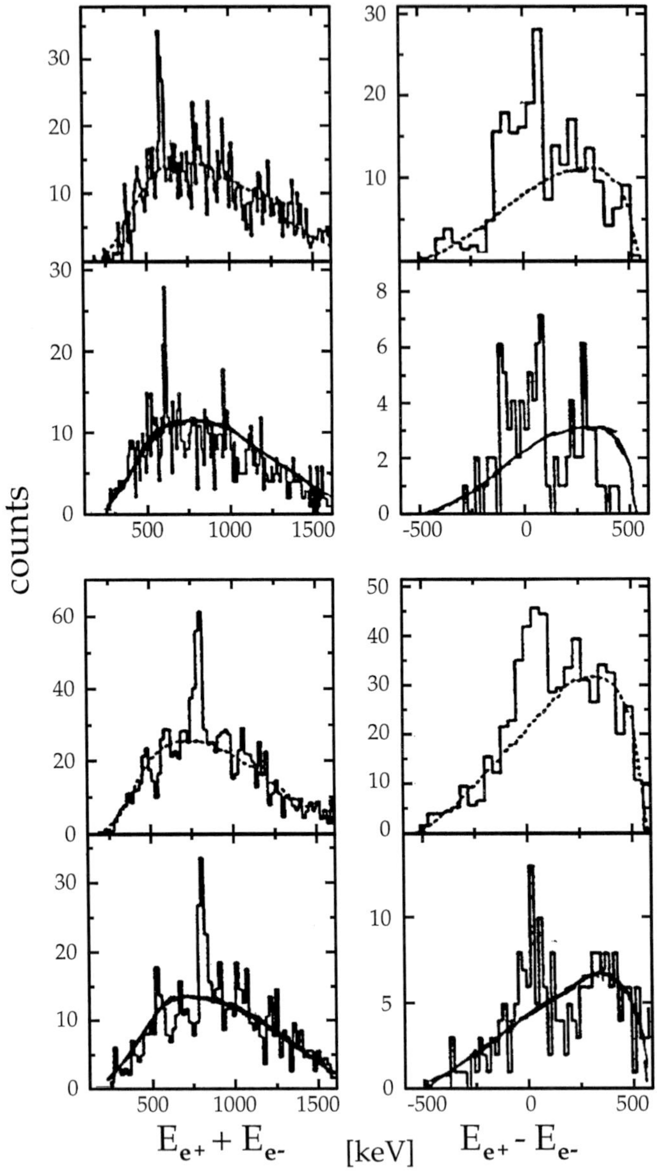

Figure 8: EPOS experiments carried out for the same U+Th collision window at different times by different people: the same narrow line structure is seen (H. Bokemeyer, Habilitationsschrift Frankfurt am Main, 1990; GSI report GSI-90-11)

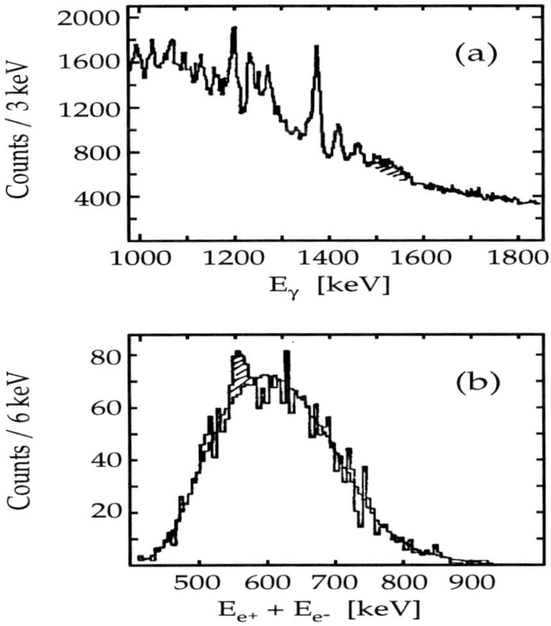

Figure 9: The shaded narrow e^+e^- peak in the lower spectrum corresponds to the Dopplershifted γ–transition.

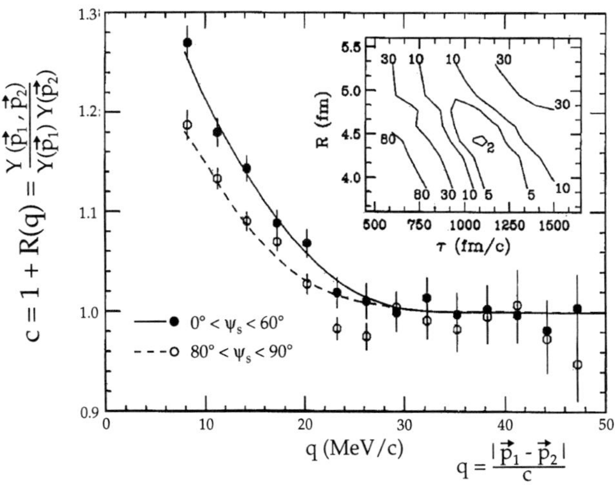

Figure 10: Angle–gated nn–correlation functions for $^{18}O+^{24}Mg$ according to N. Colonna.

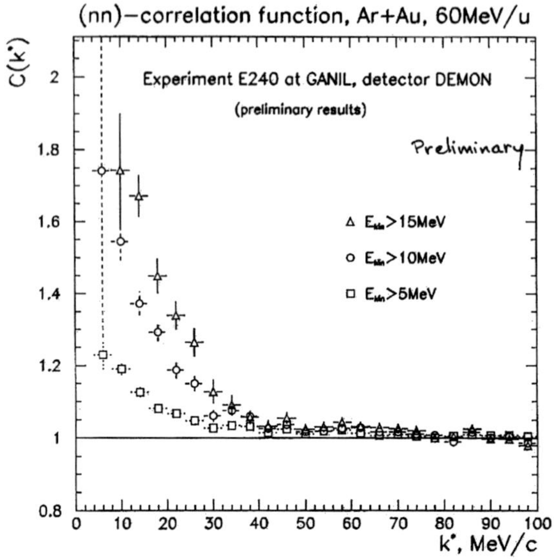

Figure 11: nn-correlation function for Ar+Au at 6 AMeV by J. Pluta et al. and reported by D. Ardouin.

Neutron–Neutron Correlations

We had reports on this topic by N. COLONNA, S. GAFF and D. ARDOUIN. Figure 10 shows angle-gated nn correlation functions measured in a Tandem experiment at Legnaro, Italy, for the reactions $^{18}O+^{24}Mg$ at 130 MeV total energy ($\approx$ 7.2MeV/nucleon). In this reaction a ^{42}Ca compound nucleus is formed with $E^* \approx$ 100MeV. The *lifetime of the compound nucleus* (and therefore its total decay width) can be measured with two-neutron interferometry. To resolve the space–time ambiguity, angle–gated correlation functions were used, with the angle ψ defined as $\cos\psi = (\vec{q}\cdot\vec{P}_{\rm tot})/(q\,P_{\rm tot})$, where $q = (\vec{p}_1 - \vec{p}_2)/2$. The solid and open symbols represent respectively longitudinal ($\psi < 60^o$) and transverse ($80 < \psi < 90$) correlation functions measured in the experiment, and corrected for cross–talk. The solid and dashed line represent calculations with the Koonin-Pratt formalism, assuming emission from a hard sphere of radius R and lifetime τ. The values of R and τ that minimize the χ^2 are $R = 4.4$ fm and $\tau = 1100$ fm/c. The contour plot of χ^2 is shown in the panel as a function of R and τ.

As a test for the time scale of neutron emission, two neutron HBT interferometry was applied to the reaction $^{40}Ar+^{165}Ho$ at $E/A = 25$ MeV by S. GAFF. After appropriate corrections for neutron scattering, the correlation function indicated a lifetime that was about a factor two shorter than the lifetime predicted for the statistical decay of an equilibrated nucleus. Preequilibrium emission may

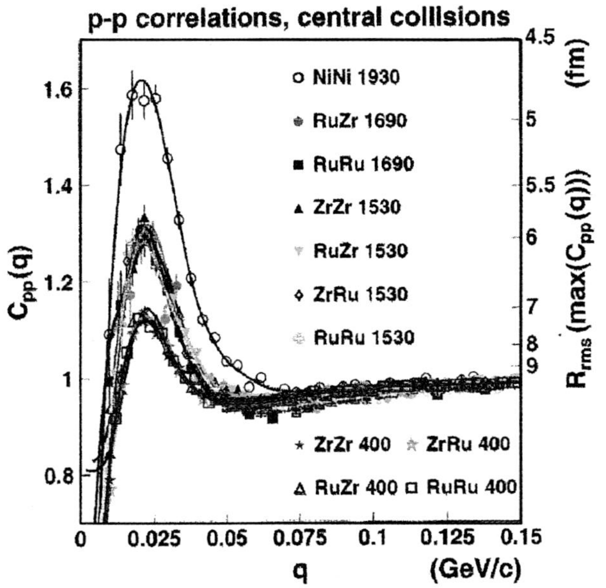

Figure 12: pp-correlation functions for various systems at various energies according to R. Kotte.

explain some of the difference.

D. ARDOUIN showed data of Pluta et al. for nn–correlation functions for Ar+Au at $E/A = 6\,\mathrm{MeV}$ (fig. 11). Most of the earlier correlations measured with neutrons contain mistakes due to incorrect corrections for cross-talks between neighbouring detectors (n–scattering). Here, using two parallel neutron walls, one is able to apply kinematical checks and corrections. The same method has been used at MSU (see S. Gaff's talk).

The measured correlations and their total momentum dependence, although preliminary, agree with standard calculations in a better way than earlier measurements. At this energy ($E = 60\,A\mathrm{MeV}$), neutrons are not originating from compound nuclei (see works from Dünnweber in 1990) where quantum statistics seemed to play the dominant role (too large time separations to allow for strong interaction). However neutron correlations are a unique tool to study the behaviour of a simple fermionic system.

Proton–Proton Correlations

We heard a fine talk by R. KOTTE on pp–correlations from central collisions for various systems (fig. 12). Three groups of correlation functions can be seen:

a) The small system ^{58}Ni+^{58}Ni at 1930 AMeV

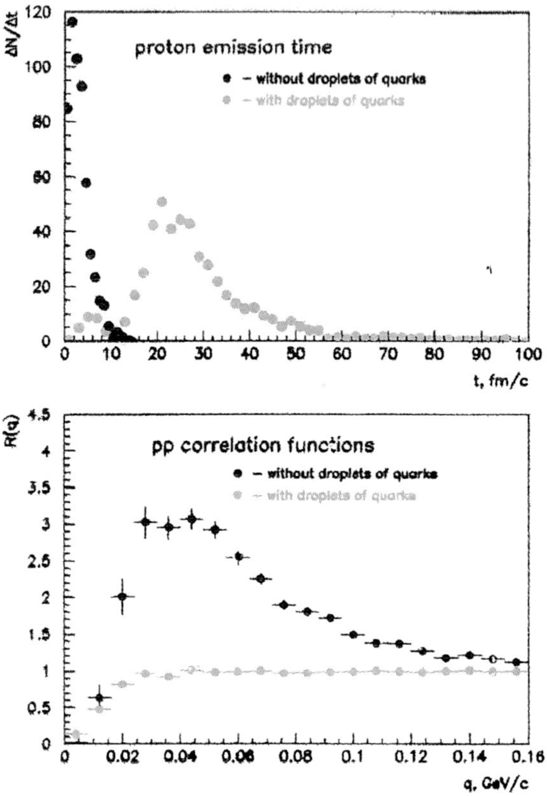

Figure 13: pp-correlations according to the Venus+Nantes correlation model calculations for S+Pb at 200 GeV/c, reported by B. Erazmus.

b) Larger systems ^{96}Ru+^{96}Ru, ^{96}Ru+^{96}Zr, ^{96}Zr+^{96}Ru, and ^{96}Zr+^{96}Zr at the same energy ($1.5 - 1.7\,A$GeV)

c) The same larger systems at smaller energy (400 AMeV). He concludes that the difference between a) and b) is due to the size of the source alone (participant number), but the difference between b) and c) can only be attributed to different time scales (e.g. different radial expansion times) and/or stronger momentum–coordinate space correlations which increase the correlation peak. The space–time extension for different beam energy seems to play a role. I recommend to perform and compare with UrQMD–calculations in order to confirm/check these suggestions.

D. Ardouin also communicated pp-correlations for 50 AMeV Xe+Sn collisions by D. Gourio et al. (Ganil) utilizing for the first time a 4π detector. It allows to measure independently size and time correlations.

An interesting investigation was presented by B. Erazmus. The Nantes–

group investigated correlation functions under the presence/absence of Quark–Gluon–Plasma droplets (fig. 13).

In the case of the QGP–droplet formation (high energy density regions), the distribution of the time emission of protons is larger leading to larger distances between particles. The two proton correlation function presents different shapes when using versions of VENUS with and without droplets. In the former case the correlation is weak due to the large distances between protons, induced by the broad time distribution. In the case without droplets, the sharp distribution of emission times gives short distances between protons increasing the two–particle correlations. Obviously, two–proton correlation functions can be considered as a possible probe of the collision dynamics.

Coulomb Effects

Various speakers addressed this issue: D. Pelte, H. W. Barz, S. Pratt. If, as D. Pelte did it, the correlation function of $\pi\pi$–pairs is described with wave packets of the type $\psi(r,t) = C_1 \exp[\frac{i}{\hbar}(pr - Et) - \frac{(r-R(t))^2}{4s\sigma_0}]$ where $s = \sigma_0(1 + i\frac{\hbar}{2m\sigma_0^2}t)$, a new parameter σ_0 (the width of the wave packet at $t = 0$) enters the theoretical description. It is related to the coherence length Λ of the pion by $\sigma_0 = \Lambda/(4\pi)$. The pions obtain an additonal energy, the zero point energy $E_0 = 3/8\,\hbar^2/(m\sigma_0^2)$ which determines the decay of the wave packet in time ($t = 0$ characterizes the time of creation). The dynamics of the wave packets is influenced at later times by Coulomb forces of charged pions only if the Coulomb energy V_{Coul} is of the same order of magnitude as the zero point energy, i.e. $\xi = V_{\mathrm{Coul}}/E_0 = 8e^2/(3\sqrt{\pi}\hbar^2)m\sigma_0 \approx 1$. D. Pelte showed that Coulomb effects on the pion-pion correlation function are only visible if the pion coherence length is large. In other words, in this case, Coulomb effects can only be seen if the pion pairs are created through the decay of long-living resonances. I know that some workers in the field, particularily from Regensburg, claim that the introduction of wave packets leads "only" to a rescaling of the parameters deduced by plane wave analysis. D. Pelte gives convincing evidence that this – in case of pion interferometry – is not correct.

H. W. Barz included both the final state interaction and the mean field effects in his calculations and compared his calculations in particular with the data from the FOPI–collaboration.

Multiple Source Structure

Sven Soff, utilizing UrQMD–calculations, showed that in general the pion source consists of several subsources. These subsources have different freeze-out times and freeze-out radii, e.g. $\bar{t}_\pi(string) < \bar{t}_\pi(\rho) < \bar{t}_\pi(\Delta) < \bar{t}_\pi(\omega)$ and also $\bar{R}_\pi(string) < \bar{R}_\pi(\rho) < \bar{R}_\pi(\Delta) < \bar{R}_\pi(\omega)$.

An impression of the various freeze-out time distributions is obtained from fig. 14.

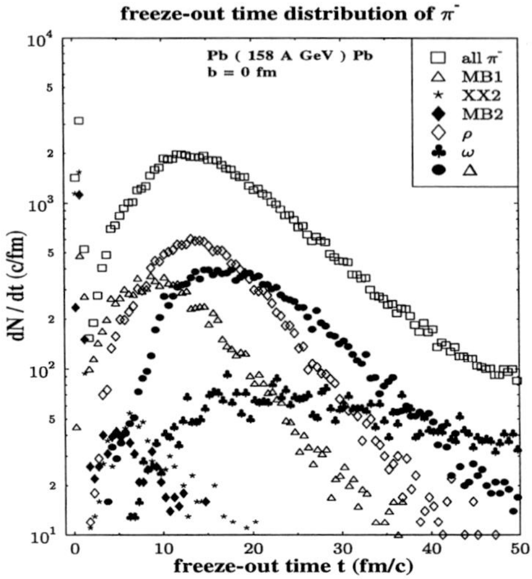

Figure 14: $dN/d\tau_f$ of π^- from different subsources as a function of the freeze-out time (according to UrQMD–calculations by S. Soff).

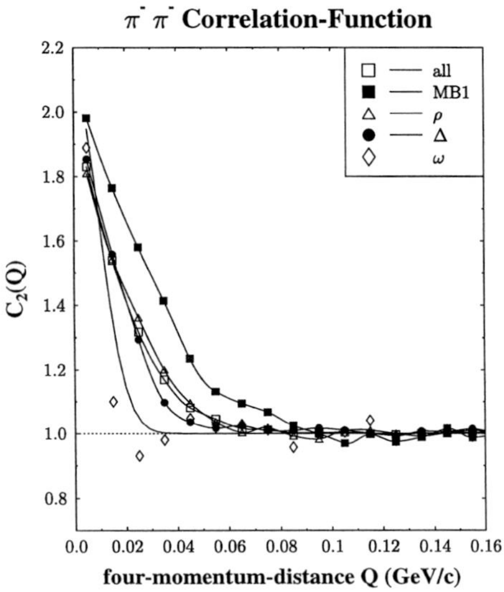

Figure 15: $\pi^-\pi^-$ correlation functions for various subsources according to S. Soff.

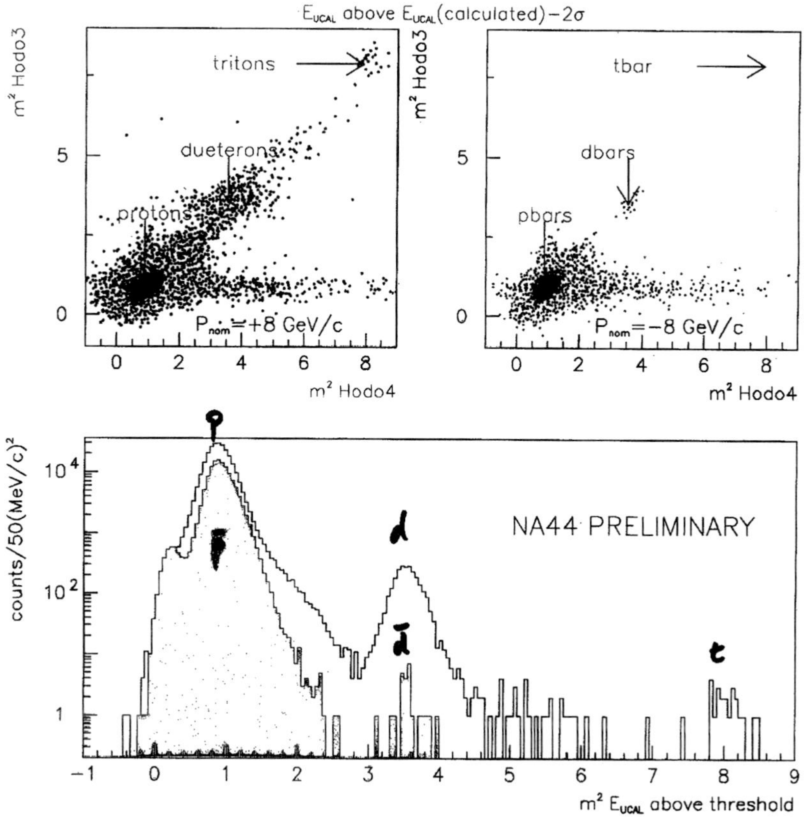

Figure 16: Clusters and Antimatter clusters as reported by Gaardhoje.

If the correlation functions could be experimentally constructed from pairs of individual subsources (e.g. only $\pi(\rho)$, $\pi(\omega)$, etc.) one could identify different widths of the correlation functions, depending on the subsource (fig. 15), thus providing a new kind of spectroscopy.

Coalescence of Clusters and Anticlusters

I was particularly fascinated by J. J. GAARDHOJE's talk about coalescense of clusters like deuterons, tritons, ...and antimatter clusters like $\bar{p}$, antideuterons, antitritons,... . An impression of his measurement gives fig. 16.

I mention that there are rumours that Klaus Pretzl (Bern) has seen antihelium. Gaardhoje compares the measurements of the NA44–group with Mishustin-Bonndorf coalescence calculations, including flow in a very primitive

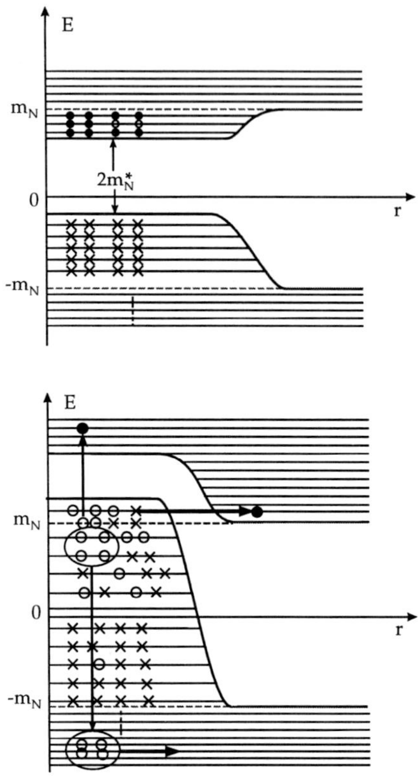

Figure 17: Level structure for the positive and negative energy states within a meson field theory.

form. He finds good agreement.

What fascinates me is that this might be the beginning of research of bulk antimatter (antiperiodic system) reaching far into the future. Indeed, already during 1990-1992 Mishustin, Satarov, Stöcker and myself have put forward the idea that antimatter clusters might be directly produced out of the vacuum and not, as is standardly done - see the Mishustin–Bondorf paper quoted by Gaardhoje - by coalescense of antiprotons and antineutrons in phase space. The idea is the following: In meson field theory where baryons interact via scalar and vector mesons a level structure as shown in fig. 17 results. The levels in the lower potential, diving "upward" from the Dirac sea are also filled with baryons.

Holes therein are antibaryons in bound states. We are thus dealing with a highly correlated vacuum, which becomes even more correlated under compression (lower part of fig. 17).

Due to heat and nuclear dynamcis many holes are created in the lower potential. They correspond to antibaryons sitting in bound states. One can directly

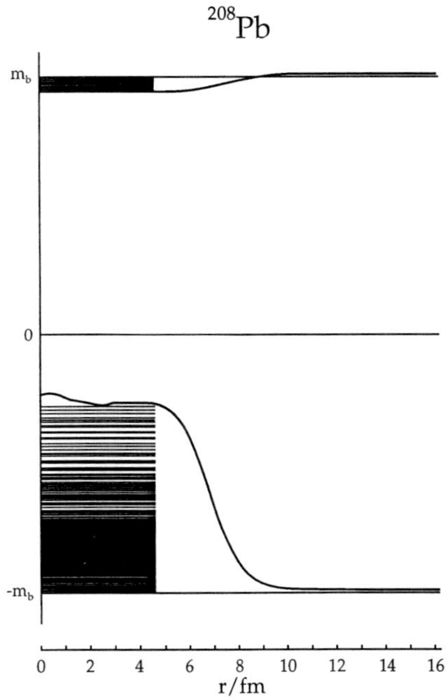

Figure 18: There are between 20000 and 40000 bound states of negative energy in the lower potential. Typical shell structure can be seen [P. K. Panda, S. K. Patra, J. Reinhardt, J. Maruhn, H. Stöcker, W. Greiner, Int. J. Mod. Phys. E, Vol. 6, No.2, (1997), 307]. Chirally symmetric models lead to similar, but not identical results.

see how the antimatter cluster structure is preformed in that way. During the expansion phase of a high energy nucleus–nucleus collision the antimatter clusters are partly annihilated by the matter around them, but partly they may escape. One is thus lead to antimatter cluster production directly out of the highly correlated vacuum. These ideas need to be quantified, and to be studied in detail. But no doubt, here opens eventually a fascinating future for nuclear physics! In passing, I mention that there are about 20000 – 40000 such bound states of negative energy within the lower potential well (fig. 18) and that there are differences wether one works within a Walecka–type meson field theory or e.g. within a chirally symmetrical theory.

Photon–Photon–Coincidences

The issue of measuring the space–time extent of the source emitting high energy photons in heavy ion collisions at intermediate energies was addressed by

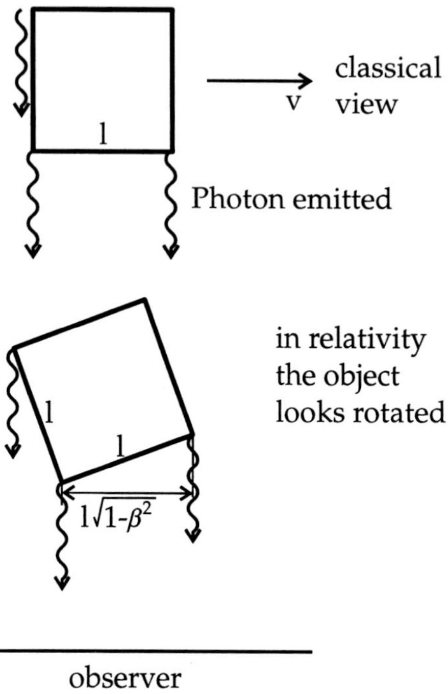

Figure 19: The principle of Terrell's correct interpretation of the observation of a moving body by light.

R. BARBERA (Catania). He focused in particular on the experimental problems in conjunction with the analysis techniques.

Also D. ARDOUIN communicated $\gamma\gamma$–correlation work by Y. Schultz et al. Since this is published already four years ago, I won't comment on it any further.

Instead I draw your attention to the Terrell–effect, who corrected in 1959 a misinterpretation of an effect of special relativity (see fig. 19): A moving body, if observed by light (photons) will not be seen as shrunk, but as rotated body (for a detailed discussion see the chapter on Special Relativity in W. Greiner, Theoretische Physik, Mechanik I, Verlag Harri Deutsch, Thun 1993, published in English by Springer (New York)).

Be aware of this effect when you interprete your correlation data, especially by the photon–photon correlation data, but also those of pions (in that case this effect is somewhat modified).

CERES–Data

and that again and again talked about "hole". It is much advertised. Everybody knows what I mean when I say that the "cocktail" of processes cannot explain all

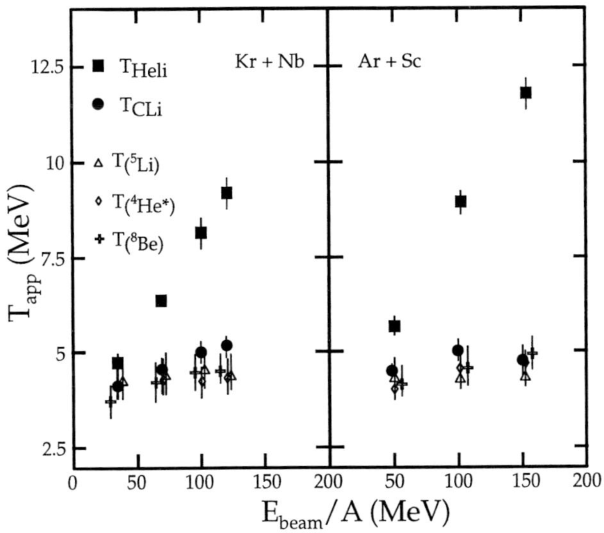

Figure 20: W. Lynch's measured temperatures according to the various methods used.

the cross section measured for e^+e^- pair production and leaves a hole around $0.3 \leq m \leq 0.8\,\mathrm{GeV/c}^2$ when compared to the data. Since wishful thinkers advertise that one is seeing a density–dependence of the ρ–meson mass, this is advertised as "chirality restoration".

R. ALZETTA claims that coherent fluctuations of the electromagnetic field induced by the collective quantum fluctuations of the nuclear matter field (giant resonance type) gives rise to a significant e^+e^- production in high energy heavy ion collisions. He says that his calculations are parameter free and fully explain the mentioned "hole" in the data. I need to see a detailed paper about these calculations before I will trust it, but let me say that such conventional explanation (and better data) are to be explored before I start believing into the possible observation of "chirality restoration"

Multifragmentation

W. LYNCH gave a very clear and interesting report on the determination of break–up conditions for nuclear multifragmentation. Correlation measurements provide opportunities to examine the break–up conditions relevant for various modes of particle production. In the case of nuclear multifragmentation it is hoped that one may thereby deduce wether the conditions attained may be relevant to low density phase transitions. Correlation functions, however, may also indicate that cooling processes and other time dependent phenomena are

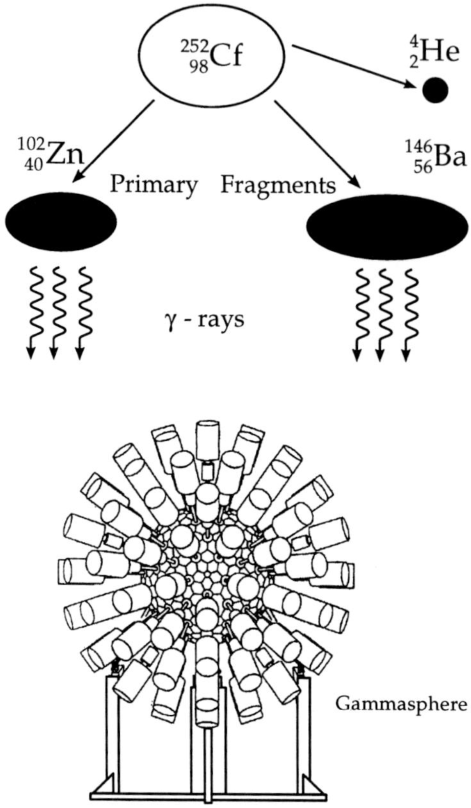

Figure 21: α–Ternary fission of ^{252}Cf and a sketch of the multi–photon–detector Gamma–sphere.

important. Thus one can, via such measurements, select reactions which either minimize or maximize the role played by such time dependent phenomena.

Of particular interest, in connection with the liquid-vapor phase transition claimed to be observed at GSI by J. Pochodzalla, U. Lynen and their collaborators, was W. Lynch's discussion of the temperature measurements: Two methods exist: One by measuring the excited states population of various nuclear isotopes, the other one by measuring ^{3}He–Li or C–Li production ratios. He showed that these various measurements of temperatures agree, except that based on the ^{3}He-Li ratio (fig. 20).

While Carbon-Lithium temperatures follow the trend of the excited state thermometers, ^{3}He–Li–temperatures do not. Moreover, comparison of the ^{3}He multiplicities to statistical multifragmentation calculation indicate that the experimental ^{3}He multiplicities exceed the predictions of those calculations, which are believed to be correct because they reproduce the multiplicities of the in-

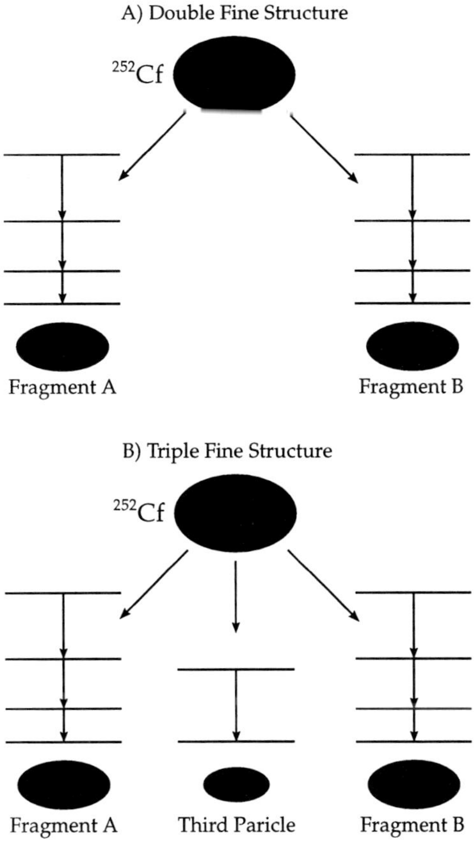

Figure 22: Binary and Ternary fission and a schematic view of the identification principle of the fragments through the observation of the emitted γ–rays.

termediate mass fragments. This suggests *another source for the ^{3}He emission,* i.e. eventually dynamical or non-equilibrium ^{3}He emission.

If true, this would shed new light on the liquid–vapor phase transition measurements mentioned earlier (see in this context also the talk by G. IMME).

Let me finally draw your attention to very interesting research utilizing the Gamma–sphere, which may culminate in *spontaneous cold multifragmentation* of heavy nuclei.

J. Hamilton, V. Ramaya and their many international collaborators observe the γ–rays of the fission fragments of ^{252}Cf (figs. 21,22). These γ–rays are the "footprints" of the fission products and serve for their identification.

In this way cold ternary fission out of the ground state of ^{252}Cf was discovered as indicated in fig. 23, i.e. the break up of ^{252}Cf into ^{146}Ba+^{10}Be+^{96}Sr.

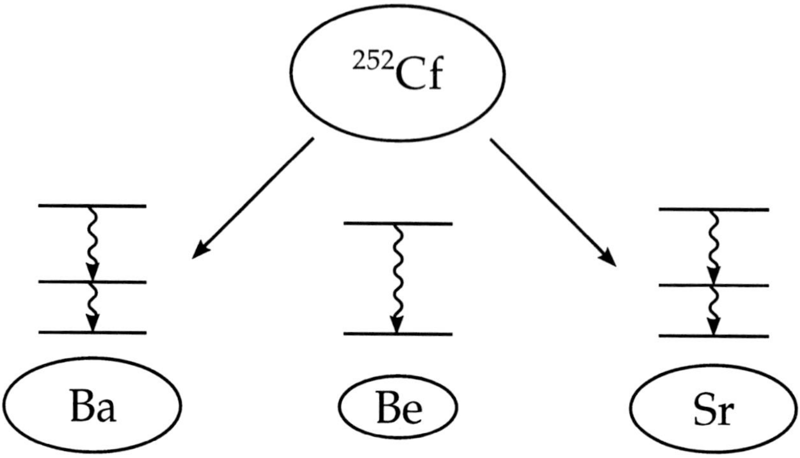

Figure 23: Ternary fission of ^{252}Cf.

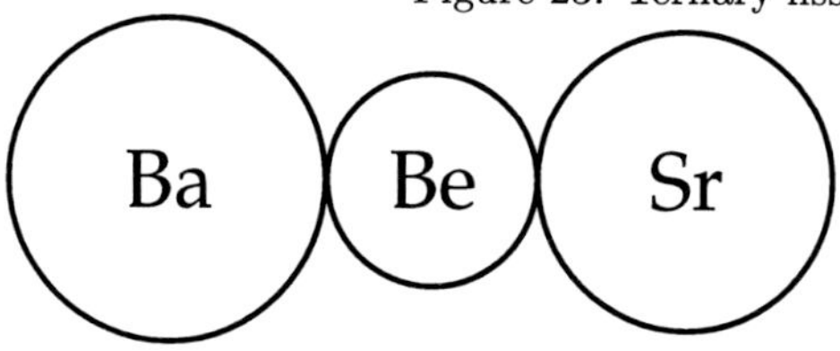

Figure 24: The existence of a long-living nuclear molecule. The life–time is $\tau \geq 10^{-12}$ s!

The observed ^{10}Be–γ–line shows no Doppler–shift, which one would expect because of the large kinetic energy gained by ^{10}Be when running down the Coulomb barrier. Also, the observed ^{10}Be–γ–line is by about 6 keV smaller than the one of free ^{10}Be. What does this tell us? Obviously, the ^{10}Be–line must be emitted while sticking together with the Ba and Sr fragments in form of a molecule (see fig. 24). This complex nuclear molecule must live longer than about 10^{-12} s, because this is the typical γ–emission time of this ^{10}Be–2^+–state.

This is an exciting discovery, which should also hold for the U+Cm–system and other "giant" nuclear systems near or below the Coulomb barrier. The large extension of such molecules due to the cluster structure lowers the strong Coulomb force and gives nuclear shell structure a chance. This is what is needed for studying the fundamental spontaneous vacuum decay I mentioned earlier. Therefore I plead for exploring further this new spontaneous multifragmentation of heavy nuclear systems, also in soft heavy ion encounters as e.g. U+Cm!

Where to go? What is the future of HBT?

Let me close with the following remarks:

1. Just showing measured correlation functions is not enough!
2. They should be compared systematically with dynamical theory like UrQMD, RQMD, VENUS, ...
3. Decomposition of the space–time and momentum structure of sources by setting appropriate windows is a formidable task.
4. Is there any substructure in the correlation function ?
5. Can the sources for Δ's, resonances, strings be depicted by invariant mass identification ?
6. ...

Finally let me express our warmest appreciation of a wonderful, very interesting conference and a thank you to the organizers: Prof. A. INSOLIA, Prof. S. ALBERGO, Prof. S. COSTA, Prof. C. TUVE.

You were great hosts! The program, the housing, the food, the wine, the excursions and the weather were great! Let me also express special thanks to the conference secretaries Ms. ANGELA TORRISI, Ms. AMALINDA MAGRI, Ms. ANGELA DISILVESTRO, Ms. ELISABETH DESILVESTRO and Mr.TONY VENASCO. Your help was efficient and highly appreciated. To you all I wish a pleasant trip home!

CRIS '98

2nd Catania Relativistic Ion Studies

Measuring the Size of Things in the Universe:
HBT Interferometry and Heavy Ion Physics

Acicastello, Italy, June 8-12, 1998

Program:

Monday, June 8, 1998

09:00-09:10 **Salvatore Costa** (Conference Scientific Secretary)
Welcome address

09:10-09:20 **Paolo Castorina** (Mayor of Acicastello)
Welcome address

09:20-09:35 **Emilio Migneco** (Vicepresident of INFN)
Opening Remarks

Chairperson: **E. Migneco** (Catania)

09:40-10:25 **R. Hanbury Brown** (Professor Emeritus)
Photons, Waves and Stars

10:35-11:00 **COFFEE BREAK**

11:00-11:45 **G. Baym** (Urbana-Champaign)
Hanbury Brown and Twiss Intensity Interferometry: from Stars to Nuclear Collisions

12:00-17:00 **LUNCH and RECREATIONAL ACTIVITIES**

Chairperson: **R. Potenza** (Catania)

17:00-17:25 **N. Panagia** (Baltimore)
Probing the Universe with the Hubble Space Telescope
17:30-17:55 **N. Colonna** (Bari)
Two-Neutron Interferometry In Low- and Intermediate-Energy Heavy Ion Collisions
18:00-18:25 **S. Gaff** (Duke)
Two-Neutron HBT Interferometry to Measure Time-Scales in the Reaction 40Ar + 165Ho at E/A=25 MeV
18:30-18:50 **J. Schmidt-Sorensen** (Lund)
Recent Progress on Three-particle Interferometry Study

20:00-21:00 **WELCOME PARTY**

Tuesday, June 9, 1998

Chairperson: **V. Pirronello** (Catania)

09:00-09:45 **U. Heinz** (Regensburg)
Hanbury Brown-Twiss Interferometry in High Energy Nuclear and Particle Physics
09:55-10:20 **G. Palumbo** (Bologna)
Measuring Black Holes

10:25-10:50 **COFFEE BREAK**

10:50-11:15 **R. Ganz** (MPI-München)
HBT Correlation in 158 AGeV Pb+Pb Collisions
11:20-11:45 **J. Sullivan** (LANL)
Two Particle Correlations from NA44
11:50-12:10 **B. Lasiuk** (Yale)
Particle Correlation Studies at NA49
12:15-12:35 **U. Wiedemann** (Regensburg)
Bose-Einstein Weights for Event Generators of Heavy Ion Collision

12:40-17:00 **LUNCH and RECREATIONAL ACTIVITIES**

Chairperson: **G. Palumbo** (Bologna)

17:00-17:25 **D. Pelte** (Heidelberg)
Pion Interferometry Using Wavepackets
17:30-17:50 **H.-W. Barz** (Rossendorf)
Combined Effect of Nuclear Coulomb Field and Radial Flow on Two-meson Correlations
17:55-18:15 **D. Hardtke** (LBNL)
Central Coulomb Effects on Interferometry Measurements
18:20-18:45 **J.J. Gaardhoje** (NBI)
Probing Sizes with Coulomb Effects and Coalescence

21:30-23:00 **CONCERT**

Wednesday, June 10, 1998

Chairperson: **J. Schukraft** (CERN)

09:00-09:45 **R. Weiner** (Marburg & Orsay)
Hadron versus Photon Interferometry - Attempt of an "Underview"
09:55-10:20 **D. Ferenc** (Zagreb)
How To Measure Direct Photon Yields Using Photon Correlations?

10:25-10:50 **COFFEE BREAK**

10:50-11:15 **R. Barbera** (Catania)
Photon-photon Intensity Interferometry in Heavy-ion Collisions at Intermediate Energies
11:20-11:40 **V. Lyuboshitz** (Dubna)
Connection between the Interference Correlations of Identical Particles in Heavy Ion Physics and Astrophysics: Analogy and Difference
11:45-12:05 **S. Soff** (Frankfurt)
Multiple Sources for Meson- and Lambda-Radiation in High Energy Nucleus-Nucleus Collisions
12:10-12:30 **R. Alzetta** (Cosenza)
Giant Resonances and Low-Mass (e+,e-) Pair Production in High-Energy Nucleus-Nucleus Collisions

12:35-on **LUNCH and FREE FOR EXCURSIONS**

Thursday, June 11, 1998

Chairperson: **R.Pucci** (Catania)

09:00-09:45 **S. Pratt** (MSU)
Fifty Ways to Measure Source Size
09:55-10:20 **D. Ardouin** (Nantes)
Determination of Space and Time Extents of Particle Sources in Heavy-Ion Collisions Using Particle Correlations

10:25-10:50 **COFFEE BREAK**

10:50-11:15 **W. Lynch** (MSU)
Determination of Breakup Conditions for Nuclear Multifragmentation
11:20-11:40 **K. Morawetz** (Rostock)
Formation of Correlations at Short Time Scales and Consequences on Interferometry Methods
11:45-12:10 **D. Miskowiec** (GSI)
Separation between Sources of Pions and Protons in Central Au+Au Collisions at the AGS (E877)

12:15-17:00 **LUNCH and RECREATIONAL ACTIVITIES**

Chairperson: **D. Vinciguerra** (LNS)

17:00-17:20 **G. Immè** (Catania)
Temperature and Size of Expanding Nuclei in Au+Au Collisions
17:25-17:45 **R. Kotte** (Rossendorf)
Proton-Proton Correlations in Central Collisions of Ni+Ni at 1.93 A GeV and the Space-Time Extent of the Emission Source
17:50-18:10 **J.H. Lee** (BNL)
Transverse Momentum Dependent Two-pion Correlations in Au+Au Collisions at 11.6 A GeV/c
18:15-18:35 **V. Bellini** (Catania)
Simultaneous nn,np and pp Interferometry in Ar+Au, C Collisions at 95 A MeV

20:30-22:30 **CONFERENCE DINNER**

Friday, June 12, 1998

Chairperson: **A. Agodi** (Catania)

09:00-09:25 **B. Erazmus** (Nantes)
Particle Correlations with Heavy-ions at LHC Energies
09:30-09:55 **T. Humanic** (OSU)
Experimental Issues for HBT Studies at RHIC

10:00-10:30 **COFFEE BREAK**

10:30-10:55 **J.-M. Mariotti** (ESO-Garching)
ESA Projects about Interferometry from Space
11:00-11:25 **G. Woan** (Glasgow)
A Very Low Frequency Radio Telescope on the Lunar Far Side

11:30-17:00 **LUNCH and RECREATIONAL ACTIVITIES**

Chairperson: **G. Belvedere** (Catania)

17:00-17:20 **M. Lisa** (OSU)
An HBT Excitation Function at the AGS -- Experiment E895
17:25-18:10 **W. Greiner** (Frankfurt)
Conference Summary Talk
18:20-18:30 **Antonio Insolia** (Conference Chairperson)
Closing Remarks

20:00-22:00 **FAREWELL PARTY**

Attilio AGODI

Dipartimento di Fisica
Università di Catania

Corso Italia, 57
Catania I-95129

Italy

phone + 39 95 7195 289

fax + 39 95 383023

attilio.agodi@ct.infn.it

Sebastiano ALBERGO

Dipartimento di Fisica
Università di Catania

Corso Italia, 57
Catania I-95129

Italy

phone +39 95 7195 258

fax + 39 95 383023

albergo@ct.infn.it

Renzo ALZETTA

I.N.F.N. Sezione Cosenza
Universita` della Calabria

dpt. Fisica Universita` della Calabria
I-87030 Arcavacata di Rende (Cosenza)

Italy

phone 39-984-493168

fax 39-984-493187

alzetta@fermi.fis.unical.it

Daniel ARDOUIN

Laboratoire SUBATECH/ UMR 6457
Universite/E. Mines/CNRS-IN2P3

Cabinet du Doyen 2 rue de la Houssiniere
BP 92208, 44322 Nantes Cedex3;France

France

phone +33-2-51125205

fax +33-2-51125252

ardouin@sciences.univ-nantes.fr

Angela BADALA'

I.N.F. N.
Sezione di Catania

Corso Italia, 57
Catania I-95129

Italy

phone +39 95 7195 231

fax +39 95 371600

angela.badala@ct.infn.it

Roberto BARBERA

Dipartimento di Fisica
Università di Catania

Corso Italia, 57
Catania I-95129

Italy

phone +39 95 7195 232

fax +39 95 383023

barbera@ct.infn.it

Hanns-Werner BARZ

Forschungszentrum Rossendorf
Institut fur Kern- und Hadronenphysik,

FZ Rossendorf Pf 510119
Dresden D-01314

Germany

phone 49 351 2603263

fax 49 351 2603700

barz@fz-rossendorf.de

Gordon BAYM

Nuclear Physics Laboratory, University of Illinois
Urbana-Champaign, Department of Physics

1110 W. Green St.
Urbana, IL 61801, USA

USA

phone 1-217-333-4363

fax 1-217-244-7559

gbaym@uiuc.edu

Vincenzo BELLINI

I.N.F.N.
Laboratorio Nazionale del Sud

Via Santa Sofia
Catania I-95127

Italy

phone +39 95 542111

fax +39 95 7141815

bellini@lns.infn.it

Robert Hanbury BROWN

White Cottage
Penton Mewsey

Andover, Hampshire SP11 ORQ
England

Greit Britain

phone

fax

Zelinda CACCIA

Dipartimento di Fisica
Università di Catania

Corso Italia, 57
Catania I-95129

Italy

phone +39 95 7195259

fax +39 95 383023

zelinda.caccia@ct.infn.it

Giuseppe CARDELLA

I.N.F.N.
Sezione di Catania

Corso Italia, 57
Catania I-95129

Italy

phone +39 95 7195 253

fax +39 95 371600

cardella@ct.infn.it

Nicola COLONNA

I.N.F.N. -
Sezione di Bari

Via Amendola, 173
Bari, Italy 70126

Italy

phone 080-5443186

fax 080-5442470

colonna@ba.infn.it

Maurizio CONSOLI

I.N.F.N.
Sezione di Catania

Corso Italia, 57
Catania I-95129

Italy

phone +39 95 7195 483

fax +39 95 371600

Salvatore COSTA

Dipartimento di Fisica
Università di Catania

Corso Italia, 57
Catania I-95129

Italy

phone +39 95 7195 260

fax +39 95 383023

costa@ct.infn.it95

Angela DI SILVESTRO

I.N.F.N. Presidenza

P.zza dei Caprettari, 70
00100 Roma

Conference Secretariat

phone +39-6-6840031

fax +39-6-68307924

segreteria@presid.infn.it

Elisabetta DI SILVESTRO

I.N.F.N.
Sezione di Roma II

Via E. Carnevale
00173 Roma

Conference Secretariat

phone +39-6-72594570
fax +39-6-2025364

Barbara ERAZMUS

SUBATECH
Ecoles des Mines de Nantes

4 Rue Alfred Kastler
Nantes F-44307

France

phone 33 2 51 85 84 29
fax 33 2 51 85 84 79
erazmus@subatech.in2p3.fr

Jens Jorgen GAARDHOJE

Niels Bohr Institute

Blegdamsvej 17
Copenhagen 2100 (Denmark)

Denmark

phone (+45) 35 32 53 09
fax (+45) 31 42 10 16
gardhoje @ nbi.dk

Sally J. GAFF

TUNL
Duke University

Box 90308
Durham, NC 27708

USA

phone (919)-660-2631
fax (919)-660-2634
gaff@tunl.duke.edu

Rudolf GANZ

Max Planck Institut Munich

Foehringer Ring 6
Munich D-80805

Germany

phone +49 (0)89 32354 276
fax +49 (0)89 32354 526
ganz@mppmu.mpg.de

Roberta GHETTI

University of Lund, Dep of Physics
Div. of Cosmic & Subatomic Physics

Solvegatan 14
Lund S - 223 62 (Sweden)

Sweden

phone +46-46-222 7388
fax +46-46-222 4015
roberta.ghetti@kosufy.lu.se

Roberto GIORDANO

Dipartimento di Fisica
Università di Catania

Corso Italia, 57
Catania I-95129

Italy

phone +39-95-7195411
fax +39-95-383023

Walter GREINER

Institute fur Theoretische Physics
J.W.Goethe Universitat

Robert Mayer Strasse 8-10 Postfach 11 19 32
Frankfurt D-60054

Germany

phone +49 69 - 798 22332
fax +49 69 - 798 28350
greiner@th.physik.uni-frankfurt.de

Marisa GULINO

Dipartimento di Fisica
Università di Catania

Corso Italia, 57
Catania I-95129

Italy

phone +39 95 7195 231

fax +39 95 383023

David HARDTKE

Lawrence Berkeley
National Laboratory

MS 50A-1148, One Cyclotron Road
Berkeley, CA 94720 (USA)

USA

phone 510-486-7043

fax 510-486-4818

dhhardtke@lbl.gov

Ulrich HEINZ

Institut fuer Theoretische Physik
Universitaet Regensburg

D-93040 Regensburg

Germany

phone +49 - 941 - 943 2004(8)

fax +49 - 941 - 943 3887

ulrich.heinz@physik.uni-regensburg.de

Thomas HUMANIC

Department of Physics, Ohio State University
Smith Laboratory

Columbus, Ohio 43210 (USA)

USA

phone 614-292-4775

fax 614-292-4833

humanic@mps.ohio-state.edu

Josette IMME'

Dipartimento di Fisica
Università di Catania

Corso Italia, 57
Catania I-95129

Italy

phone +39 95 7195255

fax +39 95 383023

imme@alp1ct.ct.infn.it

Antonio INSOLIA

Dipartimento di Fisica
Università di Catania

Corso Italia, 57
Catania I-95129

Italy

phone +39 95 7195442

fax +39 95 383023

insolia@ct.infn.it

Krzysztof KARPIO

Soltan Institute for Nuclear Studies

Hoza 69 Warsaw,
Warsaw PL-00-681

Poland

phone +48-22 6283031 274

fax +48-22 6212804

karpio@fuw.edu.pl

Roland KOTTE

Research Center Rossendorf
Forschungszentrum Rossendorf

Institut f. Kern- und Hadronenphysik
Postfach 51 01 19 D-01314 Dresden

Germany

phone +49 351 260 2323

fax +49 351 260 3700

kotte@fz-rossendorf.de

Brian LASIUK

Yale University

272 Whitney Ave
New Haven, CT 06511 (USA)

USA

phone 203 432 5829

fax 203 432 8926

brian.lasiuk@yale.edu

Raffaele LE PERA

I.N.F.N. di Cosenza
Università della Calabria

dpt. Fisica Universita` della Calabria
Arcavacata di Rende (Cosenza) I-87030

Italy

phone 0984 493168

fax 0984 493187

lepera@fermi.fis.unical.it

J.H. LEE

Brookhaven National Laboratory

Bldg. 510D
Upton, NY 11973 (U.S.A.)

USA

phone (516)344 3889

fax (516)344 1334

jhlee@bnl.gov

Michael LISA

The Ohio State University
LBL

174 West 18th Ave
Columbus, Ohio 43210 (U.S.A.)

USA

phone 1-614 292 4775

fax 1-614 292 4833

lisa@mps.ohio-state.edu

Umberto LOMBARDO

Dipartimento di fisica
Università di Catania

Corso Italia, 57
Catania I-95129

Italy

phone +39 95 7195443

fax +39 95 383023

umbert.lombardo@ct.infn.it

Bill LYNCH

NSCL and Department of Physics and Astronomy
Michigan State University

East Lansing, MI 48824-1321 (USA)

USA

phone (517)333-6319

fax (517)353-5967

lynch@nscl.msu.edu

Vladimir LYUBOSHITZ

Joint Institute for Nuclear Research
Laboratory of High Energies

141980, Dubna, Moscow Region, Russia

Russia

phone 7 (09621) 62064

fax 7 (09621) 65891

lyubosh@sunhe.jinr.ru

Anna Linda MAGRI'

I.N.F.N
Sezione di Catania

Corso Italia, 57
Catania I-95129

Conference Secretariat

phone +39 95 7195485

fax +39 95 371600

annalinda.magri@ct.infn.it

Cettina MAIOLINO

I.N.F.N.
Laboratorio Nazionale del Sud

Via santa Sofia, 6
Catania I-95129

Italy

phone +39 95 542111

fax +39 95 7141815

maiolino@ct.infn.it

Jean-Marie MARIOTTI

European Southern Observatory

Karl-Schwarzschildstr. 2
Garching bei Munchen D-85748

Germany

phone +49 89 320 06 204

fax +49 89 320 06 23 62

jmariott@eso.org

Francesco MATERA

Dipartimento di Fisica
Universita` di Firenze

Largo E. Fermi 2
Firenze I-50125

Italy

phone 55 2307730

fax 55 229330

matera@fi.infn.it

Emilio MIGNECO

I.N.F.N.
Laboratorio Nazionale del Sud

Via Santa Sofia, 6
Catania I-95100

Italy

phone +39 95 542111

fax +39 95 7141815

migneco@ct.infn.it

Dariusz MISKOWIEC

GSI

Gesellschaft fuer Schwerionenforschung mbH
Planckstr. 1 Darmstadt

Germany

phone 0049 6159 712133

fax 0049 6159 712785

d.miskowiec@gsi.de

Klaus MORAWETZ

FB Physik, Universitat Rostock

Universitatumltsplatz 1
Rostock D-18051

Germany

phone +49 381 498 1611

fax +49 381 498 2857

klaus@darss.mpg.uni.rostock.de

Chiara NOCIFORO

I.N.F.N.
Sezione di Catania

Corso Italia, 57
Catania I-95129

Italy

phone +39 95 7195236

fax +39 95 371600

nociforo@ct.infn.it

Angelo PAGANO

I.N.F.N.
Sezione di Catania

Corso Italia, 57
Catania I-95129

Italy

phone +39 95 7195203

fax +39 95 371600

angelo.pagano@ct.infn.it

Armando PALMERI

I.N.F.N.
Sezione di Catania

Corso Italia, 57
Catania I-95129

Italy

phone +39 95 7195234

fax +39 95 371600

armando.palmeri@ct.in.f.it

Giorgio PALUMBO

Università di Bologna
Dipartimento di Astronomia

Via Zamboni 33
Bologna I-40124

Italy

phone 51 259424

fax 51 259407

ggcpalumbo@astbo3.bo.astro.it

Nino PANAGIA

Space Telescope Science
Institute/ESA

3700 San Martin Drive
Baltimore, MD 21218 USA

USA

phone 410 338 4916

fax 410 338 5090

panagia@stsci.edu

Dietrich PELTE

Physikalisches
Institut der Universitaet Heidelberg

Philosophenweg 12
Heidelberg D-69120

Germany

phone + 6221 516 414

fax + 6221 516 540

pelte@pel5.mpi-hd.mpg.de

Sara PIRRONE

I.N.F.N.
Sezione di Catania

Corso Italia, 57
Catania I-95129

Italy

phone +39 95 7195245

fax +39 95 371600

sara.pirrone@ct.infn.it

Valerio PIRRONELLO

Istituto di Fisica
Facoltà d'Ingegneria

Viale A. Doria, 6
Catania 95125 (Italy)

Italy

phone +39 95 7372434

fax +39 95 7141815

vpirrone@cdc.unict.it

Renato POTENZA

I.N.F.N.
Sezione di Catania

Corso Italia, 57
Catania I-95129

Italy

phone +39 95 7195241

fax +39 95 371600

renato.potenza@ct.infn.it

Scott PRATT

Cyclotron Laboratory
Michigan State University

Michigan State University
East Lansing, MI 48823

USA

phone (517) 333-6438

fax (517) 353-5967

pratt@nscl.msu.edu

Andrea RAPISARDA
I.N.F.N.
Sezione di Catania

Corso Italia, 57
Catania I-95129

Italy

phone +39 95 7195214
fax +39 95 371600
rapisarda@ct.infn.it

Laurent ROSSELET
University of Geneva

DPNC, 24 quai Ernest-Ansermet
Geneve 4, Switzerland CH-1211 (Switzerland)

Switzerland

phone ++41 22 767 23 30
fax ++41 22 782 48 97
aurent.rosselet@cern.ch

Giuseppe RUSSO
I.N.F.N.
Sezione di Catania

Corso Italia, 57
Catania I-95129

Italy

phone +39 95 7195271
fax +39 95 383923
giuseppe.russo@ct.infn.it

Andrea SAIJA
I.N.F.N.
Sezione di Catania

Corso Italia, 57
Catania I-95129

Italy

phone +39 95 7195111
fax +39 95 371600
andrea.saija@ct.infn.it

Salvatore SAMBATARO
Dipartimento di Fisica
Università di Catania

Corso Italia, 57
Catania I-95129

Italy

phone +39 95 7195244
fax +39 95 383023
salvatore.sambataro@ct.infn.it

Janus SCHMIDT-SORENSEN
Lund University

box 118
Lund, - 22100

Sweden

phone +46 46 222 7695
fax +46 46 222 4709
janus@quark.lu.se

Jurgen SCHUKRAFT
CERN

Switzerland

phone
fax
sks@cernvm.cern.ch

Sven SOFF
GSI Darmstadt
JWG-Universitaet Frankfurt

Postfach 11 05 52
Darmstadt D-64220

Germany

phone +49 69 798 22628
fax +49 69 798 28350
ssoff@th.physik.uni-frankfurt.de

Maria Leda SPERDUTO
I.N.F.N.
Laboratorio Nazionale del Sud

Via Santa Sofia, 6
Catania I-95100

Italy

phone +39 95 542111
fax +39 95 7141815
sperduto@ct.infn.it

John SULLIVAN
Los Alamos National Laboratory

MS H846
Los Alamos, NM 87545 (USA)

USA

phone +1 505 665 5963
fax +1 505 665 7920
sullivan@lanl.gov

Maria Concetta SUTERA
I.N.F.N.
Sezione di Catania

Corso Italia, 57
Catania I-95129

Italy

phone +39 95-7195262
fax +39 95-371600
sutera@ct.infn.it

Damiano TARANTINO
I.N.F.N.
Universita` della Calabria

Arcavacata di Rende
Cosenza I-87030

Italy

phone 0984 493168
fax 0984 493187 493014
alzetta3@fermi.fis.unical.it

Angela TORRISI
I.N.F.N.
Sezione di Catania

Corso Italia, 57
Catania I-95129

Conference Secretariat

phone +39 95 388450
fax +39 95 371600
angela.torrisi@ct.infn.it

Alessia TRICOMI
I.N.F.N.
Sezione di Catania

Corso Italia, 57
Catania I-95129

Italy

phone +39 95 7195111
fax +39 95 371600
alessia.tricomi@ct.infn.it

Cristina TUVE'
Dipartimento di Fisica
Università di Catania

Corso Italia, 57
Catania I-95129

Italy

phone +39 95 7195236
fax +39 95 383023
cristina.tuve@ct.infn.it

Tony VENASCO
I.N.F.N.
Sezione di Catania

Corso Italia, 57
Catania I-95129

Conference Secretariat

phone +39 95 7195487
fax +39 95 371600
tony.venasco@ct.infn.it

Domenico VINCIGUERRA
I.N.F.N.
Laboratorio Nazionale del Sud

Via Santa Sofia, 6
Catania I-95100

Italy

phone +39 95 516140
fax +39 95 7141815
vinciguerra@ct.infn.it

Sandor VOROS
University of Geneva

DPNC, 24 Quai Ernest-Ansermet
Geneve 4, Switzerland CH-1211

Switzerland

phone ++41 22 767 23 29
fax ++41 22 782 48 97
sandor.voros@cern.ch

Zuo WEI
I.N.F.N.
Laboratorio Nazionale del Sud

Via Santa Sofia, 6
Catania I-95100

Italy

phone +39 95 542111
fax +39 95 7141815
wei@lns.infn.it

Richard WEINER
University of Marburg

Wieselacker 8,
D-35041 Marburg

Germany

phone 49 6421 31269
fax
weiner@mailer.uni-marburg.de

Urs Achim WIEDEMANN
Institut fur Theoretische Physik
Universitaet Regensburg

Regensburg D-93040

Germany

phone +49 941 943 2018
fax +49 941 943 3887
urs.wiedemann@physik.uni-regensburg.de

Graham WOAN
Department of Physics and Astronomy
University of Glasgow

Glasgow G12 8QQ

Greit Britain

phone +44 141 330 5897
fax +44 141 330 5183
graham@astro.gla.ac.uk

SPONSORS

CRIS'98 has been supported by the following Institutions, Agencies and Companies:

- Istituto Nazionale di Fisica Nucleare
- Universitá di Catania
- Centro Siciliano di Fisica Nucleare e di Struttura della Materia
- Comune di Acicastello
- Comune di Catania
- Azienda Autonoma per il Soggiorno e il Turismo della Provincia Regionale di Catania
- Presidenza della Assemblea Regionale Siciliana
- Governo Regionale della Regione Sicilia
- Gruppo Averna, Caltanissetta